An INTRODUCTION to

CRYPTOGRAPHY

An INTRODUCTION to

CRYPTOGRAPHY

RICHARD A. MOLLIN

CHAPMAN & HALL/CRC

Boca Raton London New York Washington, D.C.

Library of Congress Cataloging-in-Publication Data

Mollin, Richard A., 1947–
    An introduction to cryptography / Richard A. Mollin.
        p. cm.
    Includes bibliographical references and index.
    ISBN 1-58488-127-5
    1. Coding theory. I. Title.
QA268 .M65 2000
003'.54—dc21   00-055482
                                                      CIP

Visit the CRC Press Web site at www.crcpress.com

© 2001 by Chapman & Hall/CRC

No claim to original U.S. Government works
International Standard Book Number 1-58488-127-5
Library of Congress Card Number 00-055482
Printed in the United States of America   4   5   6   7   8   9   0
Printed on acid-free paper

To Bridget — my muse

Contents

Preface

This book is intended for a one-semester introductory undergraduate course in cryptography. The text has been designed in such a way that the reader with little mathematical background can work through the text, and the reader with a firm mathematical background will encounter sufficient challenging material to sustain interest. Any mathematics required is presented herein in advance of the cryptographic material requiring it. The impetus for the writing of this text arose from this author's involvement in designing an undergraduate course in introductory cryptography for the Mathematics Department at the University of Calgary in 1998. No suitable text for that course was on the market, nor is there one at the time of this writing, hence the incarnation of this one. Essentially, the text is meant for *any* reader who wants an introduction to the area of cryptography. The core material is self-contained so that the reader may learn the basics of cryptography without having to go to another source. However, in the optional material sections, such as Section Two of Chapter Five, where the number field sieve is studied, the reader will require some knowledge of algebraic number theory such as that given in this author's previous book [145]. Also, for the advanced topics in optional Chapter Six, such as Section One on elliptic curve methods in primality testing, factoring, and cryptosystems, some algebraic background is required, but the basics of elliptic curves are developed in that section for the benefit of the reader.

For the instructor, a course outline is, simply put, the first four chapters (excluding the material in Section Three of Chapter Two on cryptanalysis of DES and the details of its AES successor — Rijndael) as the core material for a basic introduction to the area. Optional material (determined by the pointing hand symbol ☞) may be added at the discretion of the instructor, depending upon the needs and background of the students involved. The material in Section One of Chapter One is a motivator for the material in the text by giving an overview of the history of the subject, beginning with the first rumblings of cryptography in ancient Egypt four millennia ago and ending with our modern day needs and challenges. There is sufficient historical data on various ciphers, such as the Caesar Cipher, the Playfair Cipher, the German World War I ADFGVX Field Cipher, and one of Edgar Allan Poe's famous cryptograms, to provide a novel set of challenging exercises at the end of this first section to motivate the reader further. Section Two of Chapter One is similarly a history, this time of factoring and primality testing. This may be seen as a precursor to the core material in Chapter Four and the optional material in Chapter Five. We begin with the notion of a prime defined in Euclid's Elements and proceed through several primality testing and factoring techniques introduced by Fibonacci and developed by Fermat, Euler, and numerous others up to Lucas at the end of the nineteenth century. We continue with Kraitchik, Lehmer, and others whose work ushered in the twentieth century, setting the groundwork for algorithms to be developed in the computer age. Section Three of Chapter One develops the basics of computer arithmetic, and sets the stage for Section Four which explores complexity issues.

Section One of Chapter Two is an introduction to modular arithmetic. The language of congruences is developed from the basic definition through Wilson's Theorem, Fermat's Little Theorem, Euler's generalization, the Arithmetic of the Totient, the Chinese Remainder Theorem and its generalization together with numerous illustrations such as the Coconut Problem, the Egg Basket Problem, and the Units of Work Problem, culminating in a tool to be used in many of the cryptographic techniques developed in the text — the Repeated Squaring Method for Modular Exponentiation. Section Two of Chapter Two is devoted to the first of the two symmetric-key cryptosystems to be studied, namely block ciphers. From requisite definitions of the basic concepts such as enciphering and deciphering transformations, we provide detailed discussions, with examples and diagrams, of numerous block ciphers including: affine, substitution, running-key, as well as transposition and permutation ciphers; concluding with a complete, detailed description of the Data Encryption Standard, DES, together with many illustrative figures. Section Three of Chapter Two is optional, containing a discussion of modes of operation and cryptanalysis of block ciphers such as DES, and its AES successor — *Rijndael*. We give a complete, detailed, illustrated description of the Rijndael Cipher, which is preceded by a discussion of Feistel Ciphers of which DES is an example, and the reign of which Rijndael brought to an end. Section Four of Chapter Two deals with the other class of symmetric-key ciphers — stream ciphers. The reader is taken on a journey from the basic definition of a stream cipher through the notions of keystreams, seeds, generators, randomness, one-time pads, synchronous and self-synchronizing ciphers, linear feedback shift registers, and nonlinear combination generators, plus several illustrations and examples.

Chapter Three addresses public-key cryptosystems, beginning in Section One with exponentiation, discrete logarithms, and protocols. In particular, we present the Pohlig-Hellman Symmetric Key Exponentiation Cipher, one-way functions, coin flipping via both exponentiation and one-way functions, bit commitment protocols, the Pohlig-Hellman Algorithm for computing discrete logarithms, hash functions, message authentication codes, and the Diffie-Hellman Key-exchange Protocol. The latter motivates the discussion in Section Two where we look at public-key cryptosystems, beginning with the definition of trapdoor one-way functions, which leads to a discussion of the RSA Public-Key Cryptosystem. Then a definition of modular roots and power residues fleshes out our knowledge of congruences sufficiently to describe the Rabin and ElGamal Public-key Ciphers. Section Three deals with issues surrounding authentication, the need for which is motivated by an illustrated discussion of impersonation attacks on public-key cryptosystems. This leads into a definition of the notions surrounding digital signatures. As illustrations, we present the RSA, Rabin, and ElGamal Public-key Signature schemes. Section Four studies the Knapsack Problem. Once the basics are set up, we provide a definition of superincreasing sequences, illustrated by the Merkle-Hellman and Chor-Rivest Knapsack Cryptosystems. Chapter Three concludes with a comparison and a contrasting of symmetric-key and public-key ciphers, with a description of the modern approach using combinations of both types of ciphers for a more secure cryptographic envelope.

Chapter Four deals with primality testing, starting in Section One with an introduction to primitive roots, moving from the definition to a discussion of Gauss's Algorithm for computing primitive roots, Artin's Conjecture, and the fundamental primitive root theorem. We then engage in a development of the index calculus, leading to Euler's Criterion for power residue congruences. Our knowledge of quadratic residues is then advanced by the introduction of the Legendre Symbol and its properties, which allows us to prove Gauss's Quadratic Reciprocity Law. Another step up is taken with a definition of the Jacobi Symbol and its properties, which we need later in the chapter. Section Two inspects true primality tests including the Lucas-Lehmer Test, the Pocklington Theorem, Proth's Theorem, and Pepin's Test. The section concludes with a discussion of complexity of primality testing, including the introduction of the notion of a certificate. Probabilistic primality tests are the topic of Section Three, starting with the definitions of Euler Liars, pseudoprimes, and witnesses. Then we work through and illustrate the Solovay-Strassen Probabilistic Primality Test. Once the concept of strong pseudoprimes, liars, and witnesses are introduced, we are in a position to present the Miller-Rabin-Selfridge (strong pseudoprime) Test. The section concludes with a general discussion of Monte Carlo Algorithms, of which the latter two algorithms are examples.

Chapter Five on factoring is the first of two optional chapters. Section One involves an illustrated description of three factoring algorithms: Pollard's $p-1$ Method, the Brillhart-Morrison Continued Fraction Algorithm, and the quadratic sieve. A brief history of how the work of Legendre, Euler, Kraitchik, and Lehmer led to the development of these algorithms is also provided, as is a discussion of how the notions can be generalized. This motivates the topic of Section Two, which begins with an illustration of how Pollard's original idea for factoring with cubic integers led to the development of the number field sieve. Then a detailed description of the special number field sieve is given (with the factorization of the ninth Fermat Number as an illustration) along with a discussion of its complexity in relation to the general number field sieve.

Chapter Six contains advanced topics. The first section introduces elliptic curves from the basic definition and leads the reader through the development of the elliptic curve group structure with several figures to illustrate the geometry in relation to the algebraic development. Once the basics are established, we state (without proof) some deep results needed for the cryptography, including the Nagell-Lutz Theorem, Mazur's Theorem, Mordell's Theorem, Siegel's Theorem, and Hasse's Theorem. We then present, and illustrate with worked examples, Lenstra's Elliptic Curve Factoring Method as well as the Elliptic Curve Primality Test and some generalizations thereof. To prepare for the cryptosystems, we generalize the notion of discrete logs to elliptic curves, and describe both the ElGamal and Menezes-Vanstone Public-key Elliptic Curve Cryptosystems. The section concludes with a discussion of the security of elliptic curve ciphers. The next section takes a look at the concept of Zero-knowledge Proofs in its various formats. We illustrate with the Feige-Fiat-Shamir Identification

Protocol, the cut and choose protocol, and Hamiltonian Circuits. The section is concluded with a study of noninteractive zero-knowledge protocols and zero-knowledge proofs of discrete log. Section Three, the last section of the main text of the book, takes us into the realm of quantum cryptography. The basis upon which the latter rests is the Heisenberg Uncertainty Principle that we discuss at length. We demonstrate how this principle can be used to generate a secret key for a quantum cryptosystem, called quantum key generation. The amazing properties of both quantum computers and quantum cryptography are considered, including the proposals for nuclear magnetic resonance-based quantum computers. The latter are closer to the proposed DNA-based computers, which would be several orders of magnitude faster than the fastest supercomputers known today. To take the reader to the edge of the fantastic, we also look at quantum teleportation and all that implies. The building of a quantum computer (a prototype of which already exists) would have dramatic consequences, not the least of which would be the breaking of public-key cryptography, such as RSA cryptosystems.

The following highlights other aspects and special features of the text.

◆ Features of This Text

• The book is ideal for the student since it offers a wealth of exercises with nearly 300 problems. The more challenging exercises are marked with a ☆. Also, complete and detailed solutions to all of the *odd-numbered exercises* are provided at the back of the text. Complete and detailed solutions of the *even-numbered exercises* are included in a *Solutions Manual*, which is available from the publisher for the instructor who adopts the text for a course. The exercises marked with two stars are to be considered only by the reader who requires an *exceptional challenge*, or for the instructor to extract from the solutions manual to present to students as an additional feature.

• The text is *accessible* to anyone from the beginning undergraduate to the research scientist. Appendix A, described below, contains a review of all of the requisite background material. Essentially, the reader can work through the book without any serious impediment or need to seek another source in order to learn the core material.

• There are *more than 50 biographies* of the individuals who helped develop cryptographic concepts. These are given in the footnotes woven throughout the text, to give a human face to the cryptography being presented. A knowledge of the lives of these individuals can only deepen our appreciation of the material at hand. The footnote presentation of their lives allows the reader to have immediate information at will, or to treat them as digressions, and access them later without significantly interfering with the main mathematical text at hand. The *footnotes* contain not only the bibliographical information cited above, but also historical data of interest, as well as other information which the discerning reader may want to explore at leisure.

• There are *optional topics*, denoted by ☞, which add additional material for the more advanced reader or the reader requiring more challenging material which goes beyond the basics presented in the core data.

- There are more than *80 examples* throughout the text to illustrate the concepts presented, as well as in excess of 60 diagrams, figures, and tables.

- *Appendix A* contains fundamental facts for the uninitiated reader or the reader requiring a quick finger-tip reference for a reminder of the underlying background used in the text. We begin with the basic notions surrounding set theory, binary relations and operations, functions, the basic laws of arithmetic, as well as notions surrounding divisibility including the Euclidean Algorithm and its generalization, properties of the gcd and lcm, the fundamental theorem of arithmetic, the principle of induction, and properties of the binomial coefficient including the binomial theorem. Then we turn to some basic concepts in matrix theory, the fundamentals surrounding polynomials and polynomial rings (having already introduced the basic notions of groups, rings, and fields in Section One of Chapter Two), morphisms of rings, vector spaces, and sequences. We close with fundamental concepts needed in the text concerning continued fractions.

- For ease of search, the reader will find consecutive numbering, namely object $N.m$ is the mth object in Chapter N (or Appendix N), exclusive of footnotes and exercises, which are numbered separately and consecutively unto themselves. Thus, for instance, Diagram 2.76 is the 76th numbered object in Chapter Two; exclusive of footnotes and exercises; Exercise 3.36 is the 36th exercise in Chapter Three; and Footnote 4.9 is the ninth footnote in Chapter Four.

- The *bibliography* contains more than 200 references for further reading.

- The *list of symbols* is designed so that the reader may determine, at a glance, on which page the first defining occurrence of a desired notation exists, and the symbols are all contained on page 346.

- The *index* has more than 2,400 entries, and has been devised in such a way to ensure that there is maximum ease in getting information from the text.

- The webpage cited in the penultimate line of page xiv will contain a file for comments, and any typos/errors that are found. Furthermore, comments via the e-mail address on the bottom line of page xiv are also welcome.

◆ **Acknowledgments** The author is grateful for the proofreading done by the following people, each of whom lent their own (largely non-intersecting) expertise and valuable time: John Brillhart (U.S.A.), John Burke (U.S.A.), Kell Cheng (my graduate student), Jacek Fabrykowski (U.S.A.), Bart Goddard (U.S.A.), Franz Lemmermeyer (Germany), Renate Scheidler (U.S.A.), Peter Shiue (U.S.A.), Michel Spira (Brazil), Nikos Tzanakis (Greece), and Thomas Zaplachinski (Canada), a former student, now cryptographer.

Richard Mollin, Calgary, July 11, 2000

Preface for the Fourth Printing

The changes in this printing involve the corrections of a few typos, and refinement of some descriptions of topics for clarification, such as in Appendix A. Also, new developments are included such as the recent cryptanalysis of the Chor-Rivest Cryptosystem, the last remaining knapsack-based cipher. Thanks go to numerous readers for useful and instructive comments that help to polish the product. Readers are encouraged to visit the website given below for updates, and to email any comments to the address given on the last line.

December 12, 2001

website: http://www.math.ucalgary.ca/~ramollin/

e-mail: ramollin@math.ucalgary.ca

Chapter 1

Origins, Computer Arithmetic, & Complexity

1.1 What is Cryptography & Why Study it? — A History

Every area of study has its own language, which includes specific terms that facilitate an understanding of the objects being investigated. The science[1.1] of *cryptography* refers to the study of methods for sending messages in *secret* (namely, in *enciphered* or *disguised* form) so that only the intended recipient can remove the disguise and read the message (or *decipher* it). The original message is called the *plaintext*, and the disguised message is called the *ciphertext*. The final message, encapsulated and sent, is called a *cryptogram*. The process of transforming plaintext into ciphertext is called *encryption* or *enciphering*. The reverse process of turning ciphertext into plaintext, which is accomplished by the recipient who has the knowledge to remove the disguise, is called *decryption*[1.2] or *deciphering*. Anyone who engages in cryptography is called a *cryptographer*. On the other hand, the study of mathematical techniques for attempting to defeat

---

[1.1]Some would call it an *art*, or an art *and* a science (for instance see [173, p. 1]). However, herein we are concerned with the mathematical techniques that are embraced by this study, and the computational tools used to implement them. Hence, the term "science" is apt. Moreover, if we (broadly) define art as *creative skill or its application*, and science as *a systematized branch of knowledge operating upon objective principles*, then we will see in the following historical perspective that cryptography may be viewed as having begun as an *art* in the course of human development, but is very seriously a *science* in our modern world. Cryptography has, as its etymology, *kryptos* from the Greek, meaning *hidden*, and *graphein*, meaning *to write*. The (English) term *cryptography* was coined in 1658 by Thomas Browne, a British physician and writer.

[1.2]Some people (if not entire cultures) find the terms *encrypt* and *decrypt* to be repugnant since the terms may be interpreted as referring to dead bodies. Thus, *encipher* and *decipher* are becoming the standard in usage. Also, some sources call the process of turning ciphertext into plaintext *exploitation* (see [64, p. 92]).

cryptographic methods is called *cryptanalysis*. Those practicing cryptanalysis (usually termed the "enemy") are called *cryptanalysts*. The term *cryptology* is used to embody the study of both cryptography and cryptanalysis, and the practitioners of cryptology are *cryptologists*.[1.3] We will be primarily concerned in this text with cryptography, and not cryptanalysis.[1.4]

Cryptography may be viewed as *overt secret writing* in the sense that the writing is clearly seen to be disguised. This is different from *steganography*,[1.5] which conceals the very *existence* of the message, namely *covert secret writing*. For instance, *invisible ink* would be called a *technical* steganographic method as would suitcases with false bottoms containing a secret message. One of the most famous such methods, employed by the Germans during World War II, was the use of the microdot (invented by Emanuel Goldberg in the 1920's) as a period in typewritten documents. An example of *linguistic* steganography, the other branch of steganography, is the use of two typefaces to convey a secret message, a technique used by Francis Bacon,[1.6] which is described in his publication of 1623: *De Augmentis Scientarium*. Today, hiding (subliminal)[1.7] messages in television commercials would also qualify. Generally speaking, steganography involves the hiding of secret messages in other messages or devices. The modern convention is to break cryptography into two parts: *cryptography proper* or *overt secret writing*, and *steganography* or *covert secret writing*. The term *steganography* first appeared in the work *Steganographia*, by Johannes Trithemius (1462–1516) which he began writing in 1499. Trithemius's manuscript was circulated for over a century, and not published until 1606. In 1609, the Roman Catholic Church

---

[1.3]The term *cryptology* was coined by James Howell in 1645. However, John Wilkins (1614–72) in his book *Mercury, or the Secret and Swift Messenger*, introduced into the English language the terms *cryptologia* or *secrecy in speech*, and *cryptographia* or *secrecy in writing*. He also used *cryptomeneses* as a general term for any secret communications. Wilkins, who was a cofounder of the Royal Society along with John Wallis (1616–1703) later married Oliver Cromwell's sister, and became Bishop of Chester. The modern incarnation of the use of the word *cryptology* is probably due to the advent of Kahn's encyclopedic book [100], *The Codebreakers* published in 1967, after which the term became accepted and established as that area of study embracing both cryptography and cryptanalysis. The etymology of cryptology is the greek *kryptos* meaning *hidden* and *logos* meaning *word*.

[1.4]Examples of contemporary (research) cryptanalysis may be found in research publications such as the *Journal of Cryptology*, as well as in the proceedings of various regularly held conferences such as the annual American conference, *Crypto*, held in late August in Santa Barbara, California. Historical data on cryptology may be found in the journal *Cryptologia*.

[1.5]The etymology is *steganos* from the Greek meaning *impenetrable*.

[1.6]Francis Bacon (1561–1626) was born on January 22, 1561 in London, England. He was educated at Trinity College in Cambridge, and ultimately became a lawyer in 1582. After a brief unsuccessful stint in politics, he became Queen Elizabeth's counsel. In that capacity, he helped to convict Robert Devereux (1566–1601) the second earl of Essex, for treason, after which Essex was executed. After James I assumed the throne, Bacon again went the political route. Then after a sequence of subsequent legal positions, he was appointed lord chancellor and Baron Verulam in 1618, and by 1621, he was made Viscount St. Albans. In the intervening years 1608–1620, he wrote numerous philosophical works, and wrote several versions of his best-known scientific work, *Novum Organum*. In 1621, Bacon was accused of bribery and fell from power. He spent his final years writing his most valuable and respected works. He died on April 9, 1626 in London.

[1.7]...if indeed such messages exist. For instance, see http://www.reall.org/letters/1997-05-16-usnwr-subliminal-messages.html.

placed it on its index of Prohibited Books, where it remained for over two centuries. Nevertheless, it was reprinted numerous times, including as late as 1721. During Trithemius's lifetime, his *Steganographia* caused him to be known as a sorcerer, which did not sit well since he was an abbot at the abbey of Saint Martin at Spanheim, Germany. In fact, his fellow monks were so incensed that Trithemius was transferred to the monastery of Saint Jacob in Wurzberg, where he remained (writing and studying) until his death on December 15, 1516. Beyond these comments, we will not be concerned in this text with steganography, but rather with cryptography proper. (See [12] for more details on Steganography.)

The first recorded instance of a cryptographic technique was literally written in stone almost four millennia ago. This was done by an Egyptian scribe who used hieroglyphic symbol substitution[1.8] (albeit at the time not well-developed) in his writing on a rock wall in the tomb of a nobleman of the time, *Khnumhotep*. It is unlikely that the scribe was actually trying to disguise the inscription, but rather was trying to impart some increased prestige (cache) to his inscription of the nobleman's deeds, which included the erection of several monuments for the reigning Pharaoh *Amenemhet* II. In other words, the scribe was attempting to impress the reader, and perhaps impart some authority to his writing, somewhat in a fashion similar to the use of flowery or legalistic language in a modern-day formal document. Although the scribe's intent was not secrecy (the primary goal of modern cryptography) his method of *symbol substitution* was one of the elements of cryptography that we recognize today.[1.9] Subsequent scribes actually added the essential element of secrecy to their hieroglyphic substitutions (on various tombs), but the end-goal here seems to have been to provide a *riddle* or *puzzle* (and therefore an enticement to read the epitaph) which most readers could relatively easily unravel. Therefore, although the cryptanalysis required was trivial, and the cryptography of hieroglyphic symbol substitution not fully developed, one may reasonably say that the seeds of cryptology were planted in ancient Egypt. Given that cryptology was born quite early, it did not mature rapidly or continuously. It had several incarnations in various cultures, with probably fewer methods extant than the number lost in antiquity.

The oldest extant cryptography from ancient Mesopotamia is an enciphered cuneiform tablet, which has a formula for making pottery glazes, and dates from around 1500 B.C., found on the site of Seleucia on the banks of the Tigris river. Also, the Babylonian and Asyrian scribes occasionally used exceptional or unusual cuneiform symbols on their clay tablets to "sign-off" the message with a date and signature, called *colophons*. However, again, these substitution techniques were not intended to disguise, but rather to display the knowledge of cuneiform held by the individual scribe for later generations to see and admire.

In the Hebrew literature, there is also evidence of letter substitution. The most common is a technique called *atbash*, in which the last and the first letters of the Hebrew alphabet are interchanged, and the remaining letters similarly

---

[1.8]By a *substitution*, at this early juncture, we mean a permutation of the plaintext letters. We will give a rigorous mathematical definition later.

[1.9]The use of substitutions *without* the element of secrecy is called *protocryptography*.

permuted, namely the penultimate letter and the second are interchanged, and so on. This early form of protocryptography, which was used in the Bible, had a profound influence upon monks of the Middle Ages, and contributed to the evolution of the modern use of *ciphers*, which we may regard as a given "method" for transforming plaintext into ciphertext.[1.10] In the Bible, there is a well-known "cryptogram." It occurs in the Old Testament in the *Book of Daniel*, which was originally written in Aramaic, a language related to Hebrew. The scene is the great banquet given by King Balshazzar for a thousand of his lords. As it says in *Daniel 5:5*, "Suddenly, opposite the lampstand, the fingers of a human hand appeared, writing on the plaster of the wall in the king's palace." Needless to say, this distressed the king in a major way, and he sought his wise men to "decipher" the message. Either they could not or would not do so, since the message was bad news for the king, who was slain that very night. In any case, Daniel was brought before the king and easily interpreted the words for him. Daniel's accomplishment made him the first cryptanalyst, for which he became "third in the government of the kingdom." (*Daniel 5:29*)

There is also reference in the classical literature to secret writing. In Homer's *Iliad*, Queen Anteia, the wife of King Proteus of Argos, had failed to seduce the handsome Bellerophon (also known as *Bellerophontes*, one of the heros of Greek literature). Not taking rejection well, she lied to the king, telling him that he had tried to "ravish" her. The enraged king, not willing to go so far as to put him to death directly, instead sent him to his father-in-law, the Lycian king, with an enciphered message in a folded tablet that was to ensure Bellerophon's death. The Lycian king deciphered the tablet sent to him by his son-in-law Proteus, and sent Bellerophon on several dangerous tasks intended to result in his death. However, Bellerophon prevailed in each of the tasks, which ranged from slaying monsters to defeating Lycia's greatest warriors. Proteus's father-in-law concluded that he must indeed be under the protection of the gods, so the Lycian king gave him not only his daughter but half of his kingdom. This story contains the only reference in the *Iliad* to secret writing.

The first known establishment of *military cryptography* was given to us by the Spartans, who used one of the first transposition[1.11] cipher devices ever devised, called a *skytale*, which consisted of a wooden staff around which a strip of parchment was tightly wrapped, layer upon layer. The secret message was written on the parchment lengthwise down the staff. Then the parchment (which could also be replaced by papyrus or leather) was unwrapped and sent. By itself, the letters on the parchment were disconnected and made no sense until rewrapped around a staff of equal circumference, at which point the letters

---

[1.10] A cipher is often visualized as a table consisting of plaintext symbols together with their ciphertext equivalents, for instance. See Table 1.1 on page 15.

[1.11] For the time being, we will think of a transposition as a permutation of *places* where the plaintext symbols sit. This is distinct from a *substitution* which is a permutation of the actual plaintext symbols themselves. For instance, a transposition of *They hung flags*, via the permutation $(1, 2, 3, 4, 5, 6, 7, 8, 9, 10, 11, 12, 13) \mapsto (1, 2, 3, 4, 10, 7, 8, 9, 5, 6, 11, 12, 13)$ of the places where the plaintext letters sit, yields the ciphertext *They flung hags*. Thus, plaintext letters get moved according to the given place permutation. To decipher, one merely reverses the permutation.

would realign to make sense once again. There are several instances of the Spartans using skytales, mostly to recall recalcitrant generals from the field. One of the most notable such uses occurred around 475 B.C. with the recalling of General Pausanius, who was also a Spartan prince, since he was trying to make alliances with the Persians, upon which the Spartans did not look kindly. Over a hundred years later, General Lysander was recalled, using a skytale, to face charges of sedition.

The first use of substitution ciphers in both domestic and military affairs is due to Julius Caesar, and it is called the *Caesar Cipher* to this day. In *The Lives of the Twelve Caesars*, Suetonius [197, p. 45] notes that Caesar wrote "to Cicero, and others to his friends, concerning domestic affairs; in which, if there was an occasion for secrecy, he wrote in ciphers; that is, he used the alphabet in such a manner, that not a single word could be made out. The way to decipher those epistles was to substitute the fourth for the first letter, as *d* for *a*, and so for the other letters respectively." He also used secret writings in his military efforts, which is documented in his own writing of the *Gallic Wars*.

There were numerous isolated incidents involving the use of cryptography in Egypt, Persia, and Anglo-Saxon Britain, among others. However, with the fall of the Roman empire, Europe descended into the "Dark Ages," with illiteracy rampant, and both art and science a faded memory, including of course any development of cryptography. In the Middle Ages, one of the few authors to discuss cryptography was Roger Bacon.[1.12] In the *Epistle on the Secret Works of Art and the Nullity of Magic*, written around 1250, he describes shorthand, invented characters, and even "magic figures and spells."[1.13]

[1.12]Roger Bacon (ca. 1220–1292) was principally a Franciscan philosopher. He studied at both Oxford and Paris. Among his interests were alchemy, astronomy, languages, optics, and mathematics. He joined the Franciscans in 1257, but his rebellious nature and novel ideas ultimately landed him in prison for an unspecified term in 1277. He was a visionary, whose predictions were close to the modern truth. For instance, Bacon had looked at the possibility of "flying machines," "horseless carriages," "motorboats," "microscopes," and "telescopes," centuries before they were invented. In April of 1921, Romaine Newbold, a professor of philosophy at the University of Pennsylvania, announced that he had cryptanalyzed what has come to be known as the *Voynich manuscript*, which was deemed to have been written by Roger Bacon. His results showed that Bacon had recognized the *Great Nebula in Andromeda* as a spiral galaxy, and identified biological cells and their nuclei. This would have to mean that Bacon not only *speculated* about microscopes and telescopes, but actually *built* them. For instance, Newbold cryptanalyzed a portion that read: "In a concave mirror I saw a star in the form of a snail...between the navel of Pegasus, the girdle of Andromeda, and the head of Cassiopeia." The Newbold theory was later ridiculed, so the manuscript remains an enigma.

[1.13]To this day cryptography still has an air of the occult attached to it, which is partly due to the history of its association with secret spells and incantations that bestowed power upon the "sorcerer" who voiced them. The reader may find that the local bookstore even files books on cryptography under "occult" today. However, the extraction of information by cryptographic techniques has become an objective science, whereas its unfortunate associate *divination*, or insight into the future, usually by "supernatural" means (such as *astrology* and *numerology*, for instance), is subjective and at best an amusing distraction in our modern world. However, the history of cryptography has rendered an association with magic, largely fostered by the perception that the removal of a disguise from a deeply buried secret is somehow miraculous or magical. Thus, through education about cryptography, we can remove the aura of magic attached to it and better understand it as a science with a fascinating history.

Perhaps more famous than Bacon's brief discussion of cryptography, was the use of ciphers by another writer of the Middle Ages, Geoffrey Chaucer.[1.14] In his work, *The Equatorie of the Planetis*, he included six brief, enciphered passages. What sets this apart is that the cryptograms are in Chaucer's own handwriting, thus rendering them to be more illustrious than other enciphered documents in the history of the subject. In his cryptograms, Chaucer described a simplified means for the use of an astronomical instrument called the *equatorie*.

Nowhere in the preceding history do we see a marriage of cryptography and cryptanalysis. In fact, the first to actually record methods of cryptanalysis (and therefore create what we now call cryptology) were the Arabs. The immensely rich civilization created by the Arabs in the seventh century A.D. meant that science and creative writing flourished, and this included the study of secret writing.[1.15] However, it was not until 1412, with the publication of the fourteen-volume work *Ṣubḥ al-a 'sha* that the complete documentation of the Arabic knowledge of cryptology appeared. One of the important features of this great work was that it contained the first systematic explanation of cryptanalysis in recorded history.

Once Europe exited from the Middle Ages, the study of cryptology began in earnest. In Pavia, Italy on July 4, 1474, Cicco Simonetta (a secretary to the Dukes of Sforza, oligarchs of Milan) wrote the first known manuscript devoted solely to cryptanalysis. He wrote thirteen rules for symbol substitution ciphers. Later, another Italian, Giovanni Soro, was appointed *Cipher Secretary* for Venice in 1506. His success at cryptanalysis was so great that by 1542 Soro was given two assistants and an office in the Doge's Palace above the Sala di Segret. There they worked in the highest security deciphering all dispatches from foreign powers that were obtained by the Venetians. Soro died two years later.

Among other authorities who had cryptologic assistants by their side were the popes. Ultimately, the practice became so common and of such importance that the office of *Cipher Secretary* to the pontiff was created in 1555. The first to have the title bestowed upon him was Triphon Bencio de Assisi. In 1557, King Philip II of Spain was warring with Pope Paul IV, who had deep-seated anti-Spanish sentiments, and under the leadership of de Assisi, one of the King's cryptograms was deciphered by the cryptanalysts. Peace was finally made on September 12, 1557.

By the late 1580's the Argentis, a family of cryptologists, took over the cipher

---

[1.14]Geoffrey Chaucer (ca. 1342–1400) was certainly the most prominent English poet before Shakespeare. His father's business as a vintner (wine merchant) contributed to his middle-class upbringing. Chaucer served briefly in the military beginning in 1359. By 1366, he was appointed a court official. Within the decade 1370–80, he traveled on numerous diplomatic missions and was appointed a customs official. During the years 1380–90, he suffered personal problems, but continued to write. In the 1390's, he remained a respected member of court and wrote his best-known work, the unfinished *Canterbury Tales*. He died on October 25, 1400 and was buried at Westminister Abbey.

[1.15]The very term *cipher* comes from Arabic. The goose-egg symbol 0 for the zero is *sifr* in Arabic. It is from these words that our word *zero* is derived. In the thirteenth century A.D., the term *sifr* was introduced into German as *cifra*, from which we get our present word *cipher*. Also, the German word for *digit* is *Ziffer*.

secretariat. Matteo Argenti, for instance, wrote a 135-page leather-bound book on cryptology, which is purported to describe and summarize the height of Renaissance cryptology. The Argentis also were the first to institute certain cryptographic processes, which later became widespread in use, such as using a mnemonic (memory aid) key to mix a cipher alphabet.[1.16]

In Spain, Philip II ascended to the throne in 1556. In that same year, he decided to discard the (much compromised) ciphers used during the reign of his father Charles V. He emulated Soro by dividing his cipher systems into two classes: the *cifra general*, used for correspondence between the king and various ambassadors; and the *cifra particular*, used by an individual messenger and the king. Philip's new general cipher was one of the strongest of its day, and became the template for Spanish cryptography far into the seventeenth century. Cryptography even found its way to the *New World*. The oldest remaining record of this *New World cryptography* is a letter dated June 25, 1532 sent by Cortés from Mexico.

In France, Henry IV was able to decipher letters sent between Philip and his officers in France, who were at war with Henry. He did this with the assistance of François Viète, who was a Huguenot sympathizer with cryptanalytic skills.[1.17] The Spanish, on the other hand, despite their cryptographic skills were sadly lacking in cryptanalytic abilities. For instance, Viète cryptanalyzed a Spanish letter destined for Alessandro Farncsc, the Duke of Parma, who headed the Spanish forces of the *Holy League*, a Catholic faction opposed to the protestant king on the throne of France. When Philip found out about Viète's cryptanalysis of other letters to his commanders in France, he was stunned, having thought they were unbreakable. This failure to understand cryptanalytic techniques was to have disastrous consequences for Philip and his greatest dream, which was to overthrow Queen Elizabeth, establish a marriage with Mary, Queen of Scots, and thereby secure a shared Catholic crown with her.

Philip supported the siege of Paris against the Duke of Mayenne, who headed the French forces of the Holy League. Henry intercepted a Spanish cryptogram from Commander Juan de Moreo to King Philip. Philip van Marnix,[1.18] who

---

[1.16]For now we will think of a *key* as a parameter or set of parameters that determines which cipher we will use. For instance, a key may specify the pattern of moving letters around in a transposition. A *cipher alphabet* is a list of equivalents used to transform the plaintext into secret form.

[1.17]François Viète (1540–1603) was also a mathematician who was known as the father of modern algebraic notation. In fact, his book *In Artem Analyticem Isagoge* or *Introduction to the Analytical Art*, published in 1591, could pass for a modern text in elementary algebra. He also made some advances in the theory of equations.

[1.18]Philip van Marnix van Sint Aldegonde (1540–1598) was a Dutch theologian and poet, whose translation of the Psalms is considered to be a summit of religious literature in sixteenth century Holland. He was also an avid supporter of William of Orange, who was the leader of the Dutch and Flemish revolt against Spain. His anti-Catholicism was evident in both his written words and his deeds. His first major work was *De biencorf der H. Roomsche Kercke* or *The Beehive of the Roman Catholic Church*, published in 1569, a polemical treatise in which the author appears to be defending Catholicism, when in fact he is ridiculing it. Marnix was also an outstanding cryptanalyst, and his efforts resulted in the cryptanalysis of numerous Spanish cryptograms. His efforts got him exiled in the years 1568–72, and he was actually captured by the Catholics in 1573–74. He died December 15, 1598 in Leiden, Holland.

had joined Henry's forces at the siege, was able to decipher it. A report of the decrypted Spanish letter, which revealed Philip's plans for England, reached Sir Francis Walsingham, minister to Queen Elizabeth, who headed a secret intelligence organization in England. Although Philip did not invade England until eleven years later, this cryptanalyzed report prepared them. Moreover, Walsingham employed a man in Paris, named Thomas Phelippes, who became England's first eminent cryptanalyst. This alliance was to prove fatal for Mary. Phelippes was able to cryptanalyze messages sent between herself and one of her pages, which outlined the plot to assassinate Queen Elizabeth. Ultimately, on July 17, 1586, Walsingham had sufficient evidence to submit. Mary, Queen of Scots, was put to death by the axman on February 6, 1587 — an axe set in motion by a cryptanalyst's skills.

The fifteenth and sixteenth centuries also saw the establishment of numerous classical developments in cryptology. The title *Father of Western Cryptology* goes to an architect named Leon Battista Alberti. In 1470, he published his *Trattati in cifra* in which he describes the first *cipher disk*, a mechanical tool for general substitution with shifted mixed alphabets. Thus, Alberti gave rise to the notion of *polyalphabeticity*.[1.19] This is distinct from *homophonic substitution* in which a plaintext letter is always represented by the same ciphertext equivalent, such as 5 for the letter C.

On his disk, Alberti has an outer ring consisting of twenty letters and the numbers $1, 2, 3, 4$, and an inner disc, which was a movable circle, consisting of 24 characters in the Latin alphabet. The plaintext letters were chosen from the inner disc and the ciphertext letters corresponded to the character on the outer ring. With Alberti's disk, the word *hit* might be encrypted as *med* on one setting of the disc, then at a new setting, might be represented by *suf*, for instance. As well as this first polyalphabetic cipher, Alberti is responsible for inventing the *enciphered code*. This was the reason that Alberti put the numbers on the outer ring. In a table he had 336 codegroups corresponding to the use of the numbers 1 to 4 in two-, three- and four-digit groups, from 11 to 4444. (Note that $336 = 4^2 + 4^3 + 4^4$.) The codegroups would have some preassigned meaning for the plaintext such as "We attack at dawn." for the number 434. Hence, 434 would be perhaps enciphered as *rad* in one position and as *bro* in another. This was a concept that was four centuries ahead of its time. In fact, when code enciphering began in earnest at the end of the nineteenth century, the codes used were simpler than his.[1.20]

[1.19]A polyalphabetic cipher is one in which two or more cipher alphabets are used, whereas a *monoalphabetic cipher* is one in which only one cipher alphabet is used. For instance, see Exercises 1.3 –1.6 on pages 15–16.

[1.20]Codes may be viewed as transformations (or simply changes) that replace the words and phrases in the plaintext with alphabetic and numeric *code groups*, such as the "attack at dawn" example above. On the other hand, as we have seen, *ciphers* are transformations that operate on smaller components of a message, such as individual letters, groups of symbols, bits (see page 34) and so on. Nevertheless the distinction between the two, at least historically, is somewhat blurred. What we call *coding theory* today is very different from cryptography in that coding theory has *nothing* to do with secrecy. Today, there is a blurring of coding theory and cryptography. For instance, if a satellite sends data to earth, they are *encoded* using some

In 1553, a small book entitled *La cifra del. Sig. Giovan Batista Belaso*, by Giovan Batista Belaso was published. In this booklet, he introduced the notion of an easily remembered and easily changed key, which he called a *countersign*, for a polyalphabetic cipher. However, Belaso used standard alphabets in his ciphers. The use of mixed alphabets in a polyalphabetic cipher was developed by the next character to enter the stage.

In 1563, a respected work on cryptology entitled *De furtivis Literarum Notis* by Giovanni Battista Porta[1.21] was published. In it he planted the seeds for the modern division of ciphers into transposition and substitution. The major contribution of this work is that it was the first time that polyalphabeticity was fully enunciated, and it contained the first *digraphic* cipher, in which two letters were used as a single symbol. His book went through several editions, culminating in a 1593 edition, published under the title *De Occultis Literarum Notis*, which included the first synoptic (comprehensive overview) tables ever made for cryptology. This outlined the various paths that a cryptanalyst could traverse in the analysis of a cryptogram. Thus, many historians consider Porta to be the most outstanding cryptographer of the Renaissance.

One of the most clever innovations created in sixteenth century cryptological studies was the *autokey*, which was the ingenious idea of using the message as its own key. The inventor of the first such autokey was Girolamo Cardano.[1.22]

---

modern error-correcting/detecting code. If the data are secret, then the *encoded* data are also *encrypted*.

[1.21]Giovanni Battista Porta (1535–1615) was born in Naples. His brilliance became apparent rather early with his composition of essays in Italian and Latin at the age of ten. While still in his twenties, he organized the first known association of scientists, the *Accademia Secretorum Naturae*, and later was vice president of another society of scientists, the *Accademia dei Lincei* or *Academy of Lynxes*, among whose members was Galileo. He wrote a number of books on a variety of topics including astronomy, astrology, physiognomy (the art of supposedly discovering character from outward appearance), meteorology, and pneumatics (the branch of physics that deals with the mechanical properties of gases), among others. As early as the age of twenty-two, he had written a book on oddities and scientific curiosities called *Magia Naturalis*, which was later translated and reprinted over two dozen times.

[1.22]Girolamo Cardano (1501–1576) was born on September 24, 1501 in Pavia, Duchy of Milan, Italy. In the early part of his life, he assisted his father, who was a lawyer and lecturer of mathematics primarily at the Platti foundation in Milan. After his father's death, he went through some unfortunate personal turmoil (see [145, Footnote 5.29, p. 334] for details), but he overcame these difficulties and went on to write and publish 131 books in his lifetime. Among these were two books essentially on cryptology. They were *De Subtilitate* published in 1550, and a continuation of it called *De Rerum Varietate* published in 1556. Among his discussions of the classical cryptographic methods, he inserted his original (albeit flawed) idea for an autokey. The flaws in his idea ensured that he would not be remembered so much for it as for another idea that does indeed bear his name. It is a steganographic device, called the *Cardano grille*, which is a metal (or some other rigid material) sheet consisting of holes the height of a written letter and of varying lengths. The sender of a message places the grille on a piece of paper and writes the message through the holes. Then the grille is removed and the message is "filled in" with some innocent sounding verbiage. The decipherer merely places the grille on the message and gets the intended plaintext. Despite some defects in the use of the device (such as the very clumsy wording giving a hint to the existence of a secret message) diplomats used the Cardano grille well into the seventeenth century. Cardano also contributed to number theory, probability theory, hydrodynamics, and geology. He died on September 21, 1576.

In Section Two of Chapter Two, we will study what is called the *Vigenère Cipher*. However, Blaise de Vigenère,[1.23] after whom it is named, had nothing to do with it. Misattribution has stuck him with a relatively minor system when he was responsible for the invention of a more important system — the first *valid* autokey system. Vigenère's autokey system used the same principle as Cardano's system, namely the plaintext was to be the key. However, his method did not have the inherent flaws attributed to Cardano. Vigenère provided a primer key consisting of a single letter that would be known to the sender and the receiver. This would allow the receiver to decipher the first letter of the cryptogram. Then this now-known first letter of the plaintext could be used to decipher the second cryptogram letter, which in turn could be used to decipher the third cryptogram letter and so on. Unfortunately, his invention was forgotten and reinvented in the late nineteenth century. The cipher with his name attached to it is a very elementary, degenerate version of his original idea, having only one repeating keyletter and a direct standard[1.24] alphabet.

As mentioned earlier, the first use of two-part codes[1.25] began in seventeenth-century France. France's first recognized full-time cryptologist was Antoine Rossignol, who served both Louis XIII and XIV. He also assisted Cardinal Richelieu with his cryptanalytic skills, by ensuring that the Catholic armies under Richelieu prevailed over the Huguenots in southern France in 1628. His cryptographic abilities were also important since he helped develop technical improvements in nomenclatures that were the most vital in over four centuries.

During Rossignol's service, the practice of using two-part nomenclatures went into high gear. This involved a first part called a *tables à chiffer*, consisting of plaintext letters in alphabetical order, and ciphertext symbols in random order, whereas the second part, called the *tables à déchiffer*, consisted of the plaintext letters jumbled, while the ciphertext symbols were in alphabetical or numerical order. Rossignol died in 1682 at the age of eighty-two.

One of the most noteworthy cryptanalysts of the eighteenth century was Edward Willes who was a minister at Oriel College in Oxford in 1716 when he was hired with the title of *Decypherer* for the English crown. He solved a cipher that revealed Sweden's plan to create an uprising in England. For this and other feats, he rose to become Cannon of Westminister in short order, and ultimately in 1742 was made Bishop of St. David's. He died in 1773, and was buried at Westminister Abbey.

[1.23] Blaise de Vigenère (1523–1596) was born in the village of Saint-Pouçain (between Paris and Marseilles) on April 5, 1523. In his mid-twenties, when he was on a diplomatic mission to Rome, he first came into contact with cryptology. He devoured the books by Alberti, Belaso, Cardano, Porta, and Trithemius. Although he published numerous books, only *Traicté des Chiffres*, published in 1586, has escaped oblivion. In this work is the first discussion of plaintext and ciphertext autokey systems. It was not until the seventeenth century that two-part codes were used. The first to actually put them into practice were Kings Louis XIII and Louis XIV of France. Vigenère died of throat cancer in 1596.

[1.24] This means the ordinary alphabet (standard) in its usual order (direct).

[1.25] The first code vocabularies were called *nomenclatures*. These vocabularies consisted of a separate cipher alphabet with homophones and a list of names. These were eventually expanded into what were called *repertories*, then ultimately into codes.

An amusing anecdote concerning cryptanalysis of a polyalphabetic cipher in the eighteenth century occurred in 1757. The central character was the famous Casanova, who received a cryptogram for safekeeping from his wealthy friend, Madame d'Urfé. She believed that the cryptogram could never be broken given that she held the keyword in her memory and had never written it down or disclosed it to anyone. Nevertheless, Casanova was able to cryptanalyze the enciphered manuscript, which contained a description for the transmutation of baser metals into gold. He was also able to recover the key via his calculations. She was incredulous at the revelation. Casanova later wrote in his memoirs: "I could have told her the truth—that the same calculation which had served me for deciphering the manuscript had enabled me to learn the word—but on a caprice it struck me to tell her that a genie had revealed it to me." The keyword?— NEBUCHADNEZZAR, or in Italian *NABUCODONOSOR*.

Meanwhile, in the seventeenth-century American colonies, cryptology was not as sophisticated as that in Europe. Nevertheless, after some preliminary difficulties with secret writing, the Founding Fathers sought to improve their means of secret communication. The chief proponent of that improvement was Thomas Jefferson. Jefferson compiled a nomenclature in 1785 for the purpose of communicating secretly with Madison and Monroe, and this was used until 1793. More importantly for history, just prior to the dawn of the nineteenth century, Jefferson created what he called his *wheel cypher*, which was far ahead of its time. Unfortunately, Jefferson filed away his idea and forgot about it. It was not rediscovered until 1922 among his papers in the Library of Congress. Some departments of government agencies and the military used it thereafter since modern cryptanalysts often could not defeat it! Hence, Jefferson has been rightly termed *The Father of American Cryptography*.

By the time of the American Civil War, the Union Army was using relatively advanced cryptography, while the Confederate Army used the Vigenère Cipher with standard substitution. It gets worse — much worse. On one occasion, a Confederate General, Albert S. Johnson, decided to use a Caesar substitution! Of course, the cryptography used by the Confederate Army was a disastrous failure. Regularly, Confederate troops were being captured with cryptograms that President Lincoln's youngest cryptanalysts could solve. On the other hand, rarely could the Confederates decipher a union cryptogram. A Vigenère Cipher was even found in John Wilkes Booth's hotel room after he was shot. This was used at trial to convict not only Booth, but also eight other Southern sympathizers, all of whom were hanged, even though the connection of the eight with Booth and his cipher was not established.

By the end of the nineteenth century, cryptography was nearing maturity and showed up in some famous cases. In what came to be known as the *Dreyfus Affair* in France, cryptology played a crucial role. On October 15, 1894, Captain Alfred Dreyfus of the French general staff was arrested and charged with high treason. They suspected him of giving military secrets to German or Italian officials. Later, the Italian military attaché, Colonel Alessandro Panizzardi, sent a cryptogram to Rome. The French cryptanalysts, who regularly got copies of such cryptogams, began to work on deciphering it. They deciphered it as:

"If Captain Dreyfus has not had relations with you, it would be wise to have the ambassador deny it officially, to avoid press comment." This suggested that Panizzardi disavowed any contact with Dreyfus. Those who were convinced of Dreyfus's guilt were skeptical. Thus, the French decided to trick Panizzardi into sending a telegram whose contents were known to them. In this way, they would have certain access to decryption. Panizzardi fell for a ruse, enciphered the telegram and sent it to Rome. The French were able to use it to verify the decryption of the original message. This should have exonerated Dreyfus, but again those who were convinced of his guilt, or would rather that an innocent man go to jail than admit an error, refused to let the telegram be admitted at his first trial. Dreyfus was convicted of treason and sent to Devil's Island. Upon appeal, the telegram was admitted into official evidence, but it would take several years before Dreyfus got true justice, which would include reinstatement and the Legion of Honour. The true culprit, Major Ferdinand Walsin Esterhazy, was arrested with several cardboard grilles that implicated him in having secret communication with the German military attaché.

In the dawn of the twentieth century, with a world war brewing, cryptology was headed for a major turning point. In the years 1914–18, World War I saw the use of small codes for low-level communications, and certain complicated cipher systems for high-level communications. For example, the first version of the famous ADFGVX cipher was introduced by the Germans on March 1, 1918. It was named in this fashion since only those six letters were used in the cryptogram. These letters were chosen since the Morse code equivalents were sufficiently dissimilar to minimize errors. The presence of only six letters ensured that the system was quick and easy for the Germans (see Exercises 1.7 – 1.8 on pages 16–17). The Allies could not crack the first cryptograms sent by this cipher system, so it turned out to be the toughest field cipher known to that date. Then the first such messages were brought to the attention of Georges Jean Painvin who was the best cryptanalyst in France's Bureau du Chiffre. Ultimately, he did decrypt them, and later was able to decipher even more. His cryptanalytic efforts ultimately saved French forces and helped to turn the tide for the Allies. However, there were failures that amounted to defeat in battle for some during the war, because of poor cryptography. For instance, the Russians lost the battle of Tannenberg in August 1914, because of failure of their cryptographic communications. The reason for this failure deserves some elucidation as follows.

The Russian Second Army was planning to come up behind the Germans, cut off their retreat, and destroy them. However, Russian communications were, at best, inadequate. They ran out of wire to string, so there was no communication between army headquarters and core headquarters, which were the two highest levels of field command. This was compounded by the fact that inept distribution of military ciphers and their keys meant they did not have the cryptographic tools. Hence, Russian signalmen were sending messages over the radio in the clear, without even attempting to encipher. Thus, the Germans knew the Russian military plans in advance. When it was over, roughly 100,000 Russians were taken prisoner, 30,000 were dead or missing, so the Russian

Second Army had ceased to exist. Hence, Tannenberg became the first battle in history to be determined by cryptographic failure. On the other hand, the deciphering of one of the most famous telegrams in history was accomplished by the British, namely the *Zimmerman Telegram*, of January 16, 1917 that offered Mexico territorial gains if it would enter the war on the side of Germany. This was a major contributing factor to the United States entering the war on April 6, six weeks after President Wilson learned of its contents.

In the postwar years, considerable advances were made in cryptography, most especially with cipher machines, which were used extensively in World War II. Shortly before World War II, the United States was was able to reconstruct the Japanese cipher machine that was used for diplomatic communication. Thus, the American cryptanalysts enjoyed tremendous success in deciphering Japanese cryptograms during the war. One incident that goes down in infamy deserves special mention — Pearl Harbour.

It was November 19, 1941, when the U.S. Navy intercepted a diplomatic radio message sent from Tokyo to Washington. By the twenty-eighth of the month, they had cryptanalyzed the message, which indicated that there would follow an cryptogram announcing intention of hostilities. This resulted in a flurry of activity to intercept Japanese radio traffic for the cue. It came on December seventh, several hours *after* the attack on Pearl Harbour, and the message only mentioned intended hostilities toward Great Britain. However, even this message was intended as a signal to some Japanese outposts, who had not already done so, to burn their codes. Hence, Japan had commenced hostilities without a prior declaration of war, an act that would make up some of the charges laid against Japanese war criminals after the war.

Due to the nature of the Pearl Harbour attack, the Joint Congressional Committee met for an investigation, which concluded that the efforts of the American cryptanalysts had shortened the war, and saved thousands of lives. For instance, cryptanalysts helped to ensure that Japan's lifeline was rapidly cut, and that German U-boats were defeated. Another incident involved the downing of the plane carrying the commander-in-chief of the Combined Fleet of the Japanese Navy, Admiral Isoruko Yamamoto. The American cryptanalysts had been able to decipher a highly secret cryptogram, giving the itinerary of Yamamoto's plane on a tour of the Solomon Islands. Of vital importance was the Battle of Midway, which was a stunning victory by American cryptanalysts since they were able to give complete information on the size and location of the Japanese forces advancing on Midway. This enabled the Navy to concentrate a numerically inferior force in exactly the right place at the right time, and prepare an ambush that turned the tide of the Pacific War.

World War II saw another outstanding achievement in cryptanalysis by the Allies. Essentially, the story begins with the invention of the electric typewriter, which provided the means for the introduction of electromechanical enciphering machines. Credit for inventing the first electric contact rotor machine goes to the American, Edward Hugh Hebern. In 1915, he used two electric typewriters (randomly) connected by twenty-six wires. Hence, a plaintext letter key hit on one typewriter would result in a ciphertext letter to be printed on the other machine. These wire connections provided the seed idea for a rotor, namely a way of varying the monoalphabetic enciphering. By 1918, Hebern had a device that embodied the rotor principle. Also, in that year, the German Arthur Scherbius applied for a patent on a rotor enciphering machine using multiple rotors. In 1923, a corporation was formed to manufacture and sell his machine, which he called the *Enigma*.[1.26] In 1934, the Japanese Navy bought the Enigma for their own use, and it developed into the Japanese cryptosystem[1.27] called *Purple* by the Americans. This was a polyalphabetic cipher cryptanalyzed in August 1940 by the U.S. Signal Intelligence Service.

By the time of Hitler's arrival on the scene, the cryptographers of the Wehrmacht made a (fateful) decision that the Enigma would work well for their security purposes, so the German forces were supplied with it. The German Enigma cryptosystem was cryptanalyzed by the researchers at Bletchley Park, which was a Victorian country mansion in Buckinghamshire, halfway between Oxford and Cambridge, England. This is the place to which the Government Code and Cypher School was seconded in August 1939. Most important among these researchers was Allan Turing (for details on Turing's life, see Footnote 1.69). Turing designed a machine for cryptanalyzing Enigma. Based upon his ideas, a machine was built, called the *BOMBE*, which was operational on March 18, 1939. In early 1940, the researchers at Bletchly Park were routinely breaking the Enigma cryptograms sent by the Luftwaffe. By September of 1941, Field Marshal Rommel's Enigma cryptograms to Berlin were being cryptanalyzed. In 1942, the British had dug deeply into cryptanalyzing Enigma, and the Russians were also deciphering such cryptograms. This played a major role in the Allied victory. World War II saw cryptology reach adulthood, with mathematics as its firm foundation.

In the United States, the National Security Agency (N.S.A.) can be said to have arisen out of the attack on Pearl Harbour. Later, in Canada the agency Communications Security Establishment (C.S.E.), the Canadian version of the N.S.A., came into being. Thus, North America has agencies in place, which are devoted to cryptology, and their activities are classified. Given the official secret work of governments around the globe, the development of cryptographic techniques in modern times is often impossible to ascertain. Much of the detail of cryptanalytic techniques since World War II is cloaked in official secrecy. Nevertheless, there have been astounding cryptographic techniques, which were developed in the public domain, and deserve to be better understood by the greatest number of people.

As we begin a new millennium, the effects of cryptography on our everyday lives will only increase since we are in the midst of a revolution in information processing and telecommunications. To ever increasing depths, our lives are

---

[1.26]Hebern did not receive a patent until 1924, although he filed in 1921, whereas Scherbius filed his patent in 1918. Patents were also filed for rotor enciphering machines by Alexander Koch and by Avrid Gerhard Damm in 1919, the latter for half rotors.

[1.27]We will have to wait until Definition 2.32 on page 78 to get a formal definition of the term *cryptosystem*. For now, we may think of this term as referring to a set of cryptographic tools (primarily) used to encipher data.

impacted on a daily basis by interactions that require our sending of digital messages through cyberspace. This may involve the electronic transfer of digital dollars, the sending of personal "e-mail" messages, or the sending of military secrets. What is common to all these types of message-sending is the need to keep these messages secret, and ensure that nobody tampers with the message. Hence, the importance of cryptography to our information-based society will only deepen in the new millennium. It is essential that we are equipped with the knowledge to understand and deal more effectively with the new reality.

Exercises

1.1. Decipher **NQRZOHGJH LV SRZHU**, which was enciphered using the table below.

Table 1.1 The Caesar Cipher

Plain	A	B	C	D	E	F	G	H	I	J	K	L	M
Cipher	D	E	F	G	H	I	J	K	L	M	N	O	P
Plain	N	O	P	Q	R	S	T	U	V	W	X	Y	Z
Cipher	Q	R	S	T	U	V	W	X	Y	Z	A	B	C

1.2. Decipher **IOHH DW GDZQ**, obtained using the Caesar Cipher.

1.3. The following is a cipher for a monoalphabetic substitution.

Plain	A	B	C	D	E	F	G	H	I	J	K	L	M
Cipher	M	L	O	P	R	Q	T	S	N	W	U	V	Z
Plain	N	O	P	Q	R	S	T	U	V	W	X	Y	Z
Cipher	Y	X	C	B	A	F	E	D	G	K	J	I	H

Using this cipher, decrypt the following: **KMA NF NZZNYRYE**

1.4. Using the cipher in Exercise 1.3, find the plaintext for the following:

EADES OXYBDRAF MVV ESNYTF

1.5. Consider the following Digraph Cipher, where the letters W and X are considered as a single entity.

Table 1.2

A	Z	I	WX	D
E	U	T	G	Y
O	N	K	Q	M
H	F	J	L	S
V	R	P	B	C

Pairs of letters are enciphered according to the following rules.

(a) If two letters are in the same row, then their ciphertext equivalents are immediately to their right. For instance, VC in plaintext is RV in ciphertext. (This means that if one is at the right or bottom edge of the table, then one "wraps around" as indicated in the example.)

(b) If two letters are in the same column, then their cipher equivalents are the letters immediately below them. For example, ZF in plaintext is UR in ciphertext, and XB in plaintext is GW in ciphertext.

(c) If two letters are on the corners of a diagonal of a rectangle formed by them, then their cipher equivalents are the letters in the opposite corners and same row as the plaintext letter. For instance, UL in plaintext becomes GF in ciphertext and SZ in plaintext is FD in ciphertext.

(d) If the same letter occurs as a pair in plaintext, then we agree by convention to put a Z between them and encipher.

(e) If a single letter remains at the end of the plaintext, then an Z is added to it to complete the digraph.

Decipher the following message, which was enciphered using the above digraph cipher.[1.28] **BP DV KT VY FD OE HQ YF SG RTCF TU IC DH JD KU HV XV TK FD**

1.6. Using the digraph cipher in Exercise 1.5, decrypt the following. **UP TG JA CS XT JE EDMZ XW FX JB VU AE XJ**

1.7. The German ADFGVX field cipher used a table such as the following where the twenty-six letters of the alphabet plus the ten digits (with 10 represented by ϕ) populate the six-by-six square, where the coordinates of each letter and digit are uniquely determined by the six letters. For instance, the coordinate of H is FX.

Table 1.3

	A	D	F	G	V	X
A	B	3	M	R	L	I
D	A	6	F	ϕ	8	2
F	C	7	S	E	U	H
G	Z	9	D	X	K	V
V	1	Q	Y	W	5	P
X	N	J	T	4	G	O

Thus, for instance, *The British have landed* would be enciphered as:

[1.28]The idea behind the above digraph cipher was conceived by Sir Charles Wheatstone, and was sponsored at the British Foreign Office by Lord Lyon Playfair. Thus, it has become known as the *Playfair Cipher*.

XF FX FG AA AG AX XF AX FF FX

FX DA GX FG AV DA XA GF FG GF

However, this is only the *transitional* ciphertext, which was then placed in another rectangle to be transposed into the *final* ciphertext using a numerical key as follows. We think of the letters of *GERMAN* as having numerical equivalents according to the alphabetic order of the letters, namely A corresponds to 1 since it is the letter in GERMAN that appears first in the alphabet, then E corresponds to 2, and so on. Then place the above transitional ciphertext by rows into a matrix as follows.

Table 1.4

G	E	R	M	A	N
3	2	6	4	1	5
X	F	F	X	F	G
A	A	A	G	A	X
X	F	A	X	F	F
F	X	F	X	D	A
G	X	F	G	A	V
D	A	X	A	G	F
F	G	G	F		

Now the final ciphertext is obtained by "peeling off" the *columns* in the above rectangle according to the order of the numbers as follows and grouping the letters in convenient five-letter pieces.

FVFDA GFAFX XAGXA XFGDF

XGAXG AFGAF AVFFA VFFXG

Decipher the following, assuming that it was enciphered using the above cipher.

AFAXX AGFAX FFXAA

XGGVX FGXFA XAAFX

1.8. Using the cipher in Exercise 1.7, decrypt the following.

VVFDVVFAA XXXXAGAFG

AXAVFADDV FVFGGGGXA

FGXAXXGGG DFFFFFAXD

☆ 1.9. Assuming that the following is enciphered English text, find the plaintext.
53‡‡ † 305))6∗; 4826)4‡.)4‡); 806∗; 48 † 8¶60))85; ;]8∗; : ‡ ∗ 8 † 83(88)5 ∗ †; 46(; 88 ∗ 96∗?; 8) ∗ ‡(; 485); 5 ∗ †2 : ∗‡(; 4

☆ 1.10. Decipher the following — a continuation of the message in Exercise 1.9.
956∗2(5∗−4)8¶8∗; 4069285);)6†8)4‡‡; 1(‡9; 48081; 8 : 8‡1; 48†85; 4)485†
528806 ∗ 81(‡9; 48; (88; 4(‡?34; 48)4‡; 161; : 188; ‡?;

1.2   A History of Factoring & Primality Testing

In this text, we will be concentrating upon the role played by primality testing and factoring in cryptography. Thus, as was the case for cryptography itself in Section One, we look at a brief history of factoring and primality testing (within the venue of cryptography) to prepare the reader for what is to come.

The definition of a *prime number* (or simply a *prime*) is a natural number[1.29] bigger than 1, which is not divisible by any natural number except itself and 1.[1.30] The first recorded definition of a prime was given by Euclid around 300 B.C. in his *Elements*. However, there is some indirect evidence that the concept of primality must have been known earlier to Aristotle (ca. 384–322 B.C.) for instance, and probably to Pythagoras. If $n \in \mathbb{N}$ and $n > 1$ is *not* prime, then n is called *composite*.

The *Factoring Problem* is the determination of the prime factorization of a given $n \in \mathbb{N}$ guaranteed by The Fundamental Theorem of Arithmetic (see Theorem A.28). This theorem says that the primes in the factorization of a given natural number n are unique to that number n up to order of the factors. Thus, the primes are the fundamental *atoms* or multiplicative building blocks of arithmetic as well as its more elevated relative *the higher arithmetic*, also known as *number theory*.[1.31]

Later, Eratosthenes (ca. 284–204 B.C.) gave us the first notion of a *sieve*, which was what he called his method for finding primes. The following example illustrates the *Sieve*[1.32] *of Eratosthenes*.

Example 1.5 Suppose that we want to find all primes less than 30. First, we write down all natural numbers less than 30 and bigger than 1, and cross out all numbers (bigger than 2) which are multiples of 2, the smallest prime.

$$\{2, 3, \not{4}, 5, \not{6}, 7, \not{8}, 9, \not{10}, 11, \not{12}, 13, \not{14}, 15, \not{16}, 17, \not{18}, 19, \not{20}, 21, \not{22},$$

$$23, \not{24}, 25, \not{26}, 27, \not{28}, 29, \not{30}\}.$$

[1.29]The natural numbers {1,2,3,4,...} are denoted by $\mathbb{N}$ and the *integers* {...,−3,−2,−1,0,1,2,3,...} are denoted by $\mathbb{Z}$. See [144, pp. 1–6] for a brief history of the development of the integers.

[1.30]If $n \in \mathbb{Z}$, then for n to be *divisible* by $m \neq 0$ means that $n/m \in \mathbb{Z}$. The notation for *division* is $m|n$, which means that there is a $t \in \mathbb{Z}$ such that $n = mt$. We also say that m *divides* n, or n is *divisible* by m and use these interchangeably. In the vernacular, it is often said that m *evenly divides* n to describe the preceding. We adopt the convention that the *evenly* be suppressed in view of the given definition. See Appendix A on the basic notions of arithmetic.

[1.31]The Greeks of antiquity used the term *arithmetic* to mean what we consider today to be *number theory*, namely the study of the properties of the natural numbers and the relationships between them. They reserved the word *logistics* for the study of ordinary computations using the standard operations of addition/subtraction and multiplication/division, which we call arithmetic today. The Pythagoreans introduced the term *mathematics*, which to them meant the study of arithmetic, astronomy, geometry, and music. These became known as the *quadrivium* in the middle ages. See Appendix A for the *Fundamental Laws of Arithmetic*.

[1.32]In general, we may think of a *sieve* as any process whereby we find numbers by searching up to a prescribed bound and eliminate candidates as we proceed until only the desired solution set remains.

Next, we cross out all numbers (bigger than 3) which are multiples of 3, the next prime.

$$\{2, 3, 5, 7, \not{9}, 11, 13, \not{15}, 17, 19, \not{21}, 23, 25, \not{27}, 29\}.$$

Then we cross out all numbers (bigger than 5) which are multiples of 5, the next prime.

$$\{2, 3, 5, 7, 11, 13, 17, 19, 23, \not{25}, 29\}.$$

(We need not check any primes bigger than 5 since such primes are larger than $\sqrt{30}$. See below for the historical description of this fact.)

What we have left is the set of primes less than 30.

$$\{2, 3, 5, 7, 11, 13, 17, 19, 23, 29\}.$$

The sieve of Eratosthenes illustrated in Example 1.5 clearly works well, but it is highly inefficient. This sieve represents the only known algorithm from antiquity that could come remotely close to what we call primality testing today. We should agree upon what we mean by *primality testing*. A *primality test* is an algorithm[1.33] the steps of which verify the hypothesis of a theorem the conclusion of which is: "n is prime."

Arab scholars helped to enlighten the exit from Europe's Dark Ages, and they were primarily responsible for preserving much of the mathematics from antiquity, as well as for extending some of the ideas. For instance, Eratosthenes did not address the issue of termination in his algorithm. However, Ibn al-Banna (ca. 1258–1339) appears to have been the first to observe that, in order to find the primes less than n using the sieve of Eratosthenes, one can restrict attention to prime divisors less than $\sqrt{n}$.

One of the chief reasons that the Arabs sought to preserve such works from antiquity was a vision had by Caliph al-Mamun (809–833). His vision included a visit by Aristotle. After this epiphany, he was driven to have all of the Greek classics translated into Arabic, including Euclid's *Elements*. Under the caliphate of al-Mamun lived Mohammed ibn Musa al-Khowarizmi (Mohammed, son of Moses of Kharezm, now Khiva) to whom we owe the introduction of the Hindu-Arabic number system. In around 825 A.D. he completed a book on arithmetic, which was later translated into Latin in the twelfth century under the title *Algorithmi de numero Indorum*. This book is one of the best-known means by which the Hindu-Arabic number system was introduced to Europe after being introduced into the Arab world. This may account for the widespread, although mistaken, belief that our numerals are Arabic in origin. Not long after Latin translations of his book began appearing in Europe, readers began to attribute the new numerals to al-Khowarizmi, and began contracting his name, concerning

---

[1.33] For now, we may think loosely of an algorithm as any methodology following a set of rules to achieve a goal. More precisely, in Section Four, when we discuss complexity theory, we will need the definition of an algorithm as a well-defined (see Footnote A.1) computational procedure, which takes a variable input and halts with an output.

the use of these numerals, to *algorism*, and ultimately to *algorithm*. Also, al-Khowarizmi wrote a book on algebra, *Hisab al-jabr wa'lmuqābala*. The word *algebra* is derived from *al-jabr* or *restoration*. In the Spanish work *Don Quixote*, which came much later, the term *algebrist* is used for a *bone-setter* or *restorer*.

The resurrection of mathematical interest in Europe during the thirteenth century is perhaps best epitomized by the work of Leonardo of Pisa (ca. 1170–1250), better known as Fibonacci. He had an Arab scholar as his tutor while his father served as consul in North Africa. Thus, he was well-educated in the mathematics known to the Arabs. Fibonacci's first, and certainly his best known book is *Liber Abaci* or *Book of the Abacus* first published in 1202, which was one of the means by which the Hindu-Arabic number system was transmitted into Europe. However, only the second edition, published in 1228 has survived. In this work, Fibonacci gave an algorithm to determine if n is prime by dividing n by natural numbers up to $\sqrt{n}$. This represents the first recorded instance of a *Deterministic Algorithm* for primality testing, where *deterministic* means that the algorithm always terminates with either a *yes* answer or a *no* answer.[1.34] Also included in his book was an inspired problem described as follows.

◆ The Rabbit Problem

Suppose that we are given a pair of newborn rabbits on January first, and assume that they reach sexual maturity at two months of age. Suppose further that they give birth to another pair of rabbits on March first and continually produce a new pair the first of every succeeding month. Each newborn pair takes two months to mature and produces a new pair on the first day of the third month of their life, and on the first day of every succeeding month. If no rabbits ever die, how many pairs of rabbits are there on the first day of the nth month?

The answer is given by the *Fibonacci Sequence* $\{F_n\}$:

$$F_1 = F_2 = 1, \quad F_n = F_{n-1} + F_{n-2} \quad (n \geq 3)$$

where F_n is the *nth Fibonacci Number*.[1.35] The answer to the rabbit problem is F_n pairs of rabbits (see Exercise 1.11 on page 29). Later, we will see the influence of Fibonacci Numbers in the history of primality testing.

Another set of numbers also had a deep influence on the development of primality testing, called *perfect numbers*, which are those $n \in \mathbb{N}$ equal to the sum of their *proper divisors* (those $m \in \mathbb{N}$ where $m \mid n$ but $m \neq n$), such as $6 = 3+2+1$.[1.36] Nicomachus of Gerasa (ca. 100 A.D.) knew that $2^{n-1}(2^n-1)$ is

---

[1.34]A Deterministic Algorithm may also be viewed as an algorithm that follows the same sequence of operations each time it is executed with the same input. This is in contrast to *Randomized Algorithms* that make random decisions at certain points in the execution, so that the execution paths may differ each time the algorithm is invoked with the same input.
[1.35]The research journal devoted entirely to the study of such numbers is the *Fibonacci Quarterly*.
[1.36]The Pythagoreans probably knew about these perfect numbers since the idea is founded in mysticism, which was their venue. The notion does explicitly appear in Euclid's *Elements*, so we know that it is a concept deeply rooted in antiquity.

perfect for $n = 2, 3, 5, 7$. These are the first four perfect numbers $6, 28, 496$, and 8128. The ancient Greeks attributed mystical properties to perfect numbers. St. Augustine (354–430 A.D.) is purported to have said that God created the earth in six days since the perfection of the work is signified by the perfect number 6. Also, the moon orbits the earth every twenty-eight days, and 28 is the second perfect number. The first person to actually use an algorithmic approach to primality testing that is the status quo today was Cataldi.[1.37] Cataldi proved that the fifth, sixth, and seventh perfect numbers are;

$$33550336 = 2^{12}(2^{13} - 1),$$

$$8589869056 = 2^{16}(2^{17} - 1),$$

and

$$137438691328 = 2^{18}(2^{19} - 1).$$

It is uncertain whether Cataldi was the first to discover these perfect numbers, but his proofs remain the first known of these facts. Cataldi was also the first to observe that if $2^n - 1$ is prime then n must be prime. In fact, it can be shown that the following result holds.

Theorem 1.6 (Perfect Numbers)
  If $2^n - 1$ is prime, then n is prime and $2^{n-1}(2^n - 1)$ is perfect.

  Proof. Let S_1 be the sum of all divisors of 2^{n-1} and let S_2 be the sum of all the divisors of the prime $2^n - 1$. Then the sum S of all the divisors of $2^{n-1}(2^n - 1)$ is $S_1 S_2$. Also, $S_1 = \sum_{j=0}^{n-1} 2^j$, so by Theorem A.30 on page 283,

$$S_1 = 2^n - 1.$$

Finally, since $2^n - 1$ is prime, then $S_2 = 2^n$. Hence, $S = 2^n(2^n - 1)$, so $2^{n-1}(2^n - 1)$ is perfect.
  Since $(2^{st} - 1) \mid (2^s - 1)$ for any $s, t \in \mathbb{N}$, then n is prime whenever $2^n - 1$ is prime. $\square$

Furthermore, it can be shown that every even perfect number has the form given in Theorem 1.6. We will reserve a proof of this fact for later, when we have developed the machinery to do so. It is unknown if there are any odd perfect numbers and the search for them has exceeded the bound 10^{300}. Moreover, if such a beast exists, then it is known that it must have at least twenty-nine (not

---

[1.37]Pietro Antonio Cataldi (1548–1626) was born on April 15, 1548 in Bologna. He began teaching mathematics at the age of seventeen, and after a few briefly held positions in Italy, outside Bologna, he returned in 1584 to teach at the Studio di Bologna where he remained for the rest of his life. He is probably best known for his work on continued fractions. In particular, his work *Trattato del modo brevissimo di trovar la radice quadra delli numeri* published in 1613, represents a significant contribution to the development of continued fractions. However, his work on perfect numbers was considerable, and among his thirty books, he also wrote on military applications of algebra, even publishing an edition of Euclid's *Elements*. He died in Bologna on February 11, 1626.

necessarily distinct) prime factors (see [94, B1, p. 44]). Theorem 1.6 tells us that the search for even perfect numbers is essentially the search for primes of the form

$$M_p = 2^p - 1, \text{ where } p \text{ is prime.}$$

Such primes are called *Mersenne Primes*.[1.38] They are named after Marin Mersenne.[1.39] Although Mersenne was not a formally trained mathematician, his enthusiasm for number theory was deep-seated. Among his contributions were his multifarious communications with many of the outstanding scholars of the day, including Descartes, Fermat, Frénicle de Bessy, and Pascal. He also published *Cognitata Physica-Mathematica* in 1644 in which he claimed that of all the primes $p \leq 257$, the only M_p that are primes occur for

$$p = 2, 3, 5, 7, 13, 17, 19, 31, 67, 127, 257.$$

It was not until this century that this list was resolved. We now know that Mersenne made five mistakes. For example, M_p is *not* prime for $p = 67$ and $p = 257$, but M_p *is* prime for $p = 61$, $p = 89$ and $p = 107$. It is for this list, and the impact which it had, that these primes were named after him.

Fermat[1.40] kept Mersenne informed of the progress that he was making. In particular, Fermat informed him that he had proved

$$223 \mid (2^{37} - 1) = M_{37}.$$

[1.38] The largest known Mersenne Prime is $2^{6972593} - 1$ (at least at the time of this writing, which could change at any time given the access to high-speed computers and improved algorithms. In fact, for the past several decades it has been held that computing doubles in power every eighteen months, a fact that is called *Moore's Law* (see [64]).) This is the thirty-eighth known Mersenne Prime, and it has $2,098,960$ decimal digits. This was found on June 1, 1999 by Nayan Hajratwala, assisted by George Woltman, Scott Kurowski, and thousands of members of the *Great Internet Mersenne Prime Search* (GIMPS), which was started by Woltman (see: http://www.utm.edu/research/primes/largest.html.)

[1.39] Marin Mersenne (1588–1648) was born in Paris on September 8, 1588. He studied at the new Jesuit college at La Fleche (1604–1609), and at the Sorbonne (1609–1611). He joined the mendicant religious order of the Minims in 1611, and on October 28, 1613, he celebrated his first mass. After teaching philosophy and theology at Nevers, he returned to Paris in 1619 to the Minim Convent de l'Annociade near Place Royale where he was elected Correcteur. This became his home base for the rest of his life. He died on September 1, 1648 in Paris.

[1.40] Pierre Fermat (1607–1665) is most often listed in the historical literature as having been born on August 17, 1601, which was actually the baptismal date of an elder brother, also named Pierre Fermat, born to Fermat's father's first wife, who died shortly thereafter. Fermat, the mathematician, was a son of Fermat's father's second wife. Note also that Fermat's son gave Fermat's age as fifty-seven on his tombstone. Fermat attended the University of Toulouse, and later studied law at the University of Orléans where he received his degree in civil law. By 1631, Fermat was a lawyer as well as a government official in Toulouse. This entitled him to change his name to Pierre *de* Fermat. He was ultimately promoted to the highest chamber of the criminal court in 1652. Throughout his life Fermat had a deep interest in number theory and incisive ability with mathematics. There is little doubt that he is best remembered for Fermat's Last Theorem (FLT). (FLT says that $x^n + y^n = z^n$ has no solutions $x, y, z, n \in \mathbb{N}$ for $n > 2$. This has recently been solved after centuries of struggle by Andrew Wiles. See [162].) However, Fermat published none of his discoveries. It was only after Fermat's son Samuel published an edition of Bachet's translation of Diophantus's *Arithmetica* in 1670 that his father's margin notes, claiming to have had a proof, came to light. The attempts to prove FLT for over three centuries have led to discoveries of numerous results and the creation of new areas of mathematics. Fermat died on January 12, 1665 in Castres, France.

Fermat was able to do this based upon his following result. Readers who require a reminder of the Binomial Theorem should consult page 285 in Appendix A at this juncture.

Theorem 1.7 (Fermat's Divisibility Test)

   Let $b \in \mathbb{N}$ and q a prime such that q does not divide b. Then there exists an $n \in \mathbb{N}$ such that $n \mid (q - 1)$ and $q \mid (b^{(q-1)/n} - 1)$.

   Proof. Since the theorem calls for *any* $n \in \mathbb{N}$, then to prove it for $n = 1$ is sufficient. So we need only show that $q \mid (b^{q-1} - 1)$ for any prime q not dividing b. This is known as *Fermat's Little Theorem*, which we will study later. The result is obvious if $q = 2$, so we assume that $q > 2$. We now use the Binomial Theorem. First, we establish that $q \mid \binom{q}{j}$ for any natural number $j < q$. Since $q > 2$ is prime, then neither j nor $q - j$ divides q for any j with $1 \leq j \leq q - 1$. Therefore, the integer

$$\binom{q}{j} = \frac{q!}{(q-j)!j!}$$

is a multiple of q. Now, the Binomial Theorem in conjunction with this fact tells us that

$$b^q = \left(\sum_{j=1}^{b} 1\right)^q = \sum_{j=1}^{b} 1^q + qa = b + qa \text{ for some } a \in \mathbb{N}.$$

   Hence, $q \mid (b^q - b) = b(b^{q-1} - 1)$, but $q \nmid b$ so $q \mid (b^{q-1} - 1)$.   □

Corollary 1.8 *If $p > 2$ is prime, then any prime divisor q of $2^p - 1$ must be of the form $q = 2mp + 1$ for some $m \in \mathbb{N}$. Also, if $m \in \mathbb{N}$ is the smallest such that $q \mid (b^m - 1)$, then $q \mid (b^t - 1)$ whenever $m \mid t$.*

   Proof. By Theorem 1.7, $p = (q - 1)/n$ for some $n \in \mathbb{N}$, so $q = np + 1$. Also, since $p, q > 2$, then $n = 2m$ for some $m \in \mathbb{N}$. For the second assertion, we merely observe that if $t = ms$ for some $s \in \mathbb{N}$, then $(b^t - 1) = (b^m - 1)\sum_{j=1}^{s} b^{m(s-j)}$.□

   Of course, what Fermat sought was a number $n > 1$ in his trial divisions to test Mersenne Numbers for primality. He saw that Theorem 1.7 is useful in detecting possible primes q such that $q \mid (2^{37} - 1)$. For example, $q = 74n + 1$ may be tested for $1 \leq n \leq 6$ and we find that $223 \mid (2^{37} - 1) = (2^{(q-1)/n} - 1)$ where $n = 6$. Fermat also discovered that

$$47 \mid (2^{23} - 1)$$

using this method since $(47 - 1)/2 = 23$, so he used $q = 47$ and $n = 2$. The reader will also observe that it takes only two trial divisions using this method to prove that

$$233 \mid (2^{29} - 1)$$

since $(233 - 1)/8 = 29$, and by Corollary 1.8 any divisor of $2^{29} - 1$ must be of the form $58m + 1$ for $m = 1, 2, 3, 4$ of which only two are prime, 59 and 233.

Fermat also had correspondence with Frénicle de Bessy,[1.41] from which has arisen some of the most famous mathematics, not the least of which is a letter from Fermat to him dated October 8, 1640, in which (certain special cases of) FLT made its first recorded appearance. Also, after his aforementioned success with Mersenne Primes, he suggested to Frénicle that numbers of the form

$$2^{2^n} + 1$$

should be prime. Today such numbers are called *Fermat Numbers*, denoted by $\mathfrak{F}_n$. Fermat knew that $\mathfrak{F}_n$ for $n = 0, 1, 2, 3, 4$ are primes, called *Fermat Primes*, but could not prove it for $n = 5$. Today we know that $\mathfrak{F}_n$ is composite for $5 \le n \le 24$, and it is suspected that $\mathfrak{F}_n$ is composite for all $n > 24$ as well.[1.42]

Fermat's work not only touched upon primality testing via perfect numbers, but also on factoring, which we now describe. The reader unfamiliar with the *greatest integer function* should solve Exercises 1.12–1.15 before proceeding.

In 1643, Fermat developed a method for factoring that was based upon a simple observation. If $n = rs$ is an odd natural number with $r < \sqrt{n}$, then

$$n = a^2 - b^2 \text{ where } a = (s + r)/2 \text{ and } b = (s - r)/2.$$

Hence, in order to find a factor of n, we need only look at values $x = a^2 - n$ for $a = \lfloor \sqrt{n} \rfloor + 1, \lfloor \sqrt{n} \rfloor + 2, \ldots, (n - 1)/2$ until a perfect square is found. This is called the *difference of squares method* of factoring, and it has been rediscovered numerous times, as we will see later on. We will also see later in the text that the most interesting case, from a cryptographic viewpoint, occurs when

$$n = pq = x^2 - y^2,$$

where $p > q$ are odd primes. In this case, we compute the value $n + y^2$ for $y = 1, 2, \ldots, (p - q)/2$. We cannot get a square value before $y = (p - q)/2$ since Exercise 1.16 tells us that there is exactly one representation $n = x^2 - y^2$ where $x - y > 1$, namely $x = (p + q)/2$ and $y = (p - q)/2$. That exercise also tells us that the only other representation of n as a difference of squares is

$$n = \left(\frac{pq + 1}{2} \right)^2 - \left(\frac{pq - 1}{2} \right)^2,$$

for which $x - y = 1$.

[1.41] Bernard Frénicle de Bessy (1605–1675) was born in Paris in 1605. He was an excellent mathematician (albeit an amateur) who had correspondence with Descartes, Fermat, Huygens, and Mersenne. He was a member of the Académie Royal des Sciences from 1666 and held an official position as counselor at the Court of Monnais. He actually solved several problems posed by Fermat, and posed further queries himself. He died on January 17, 1675 in Paris.

[1.42] On July 25, 1999, F_{382447}, which has over 10^{10^5} decimal digits, was shown to be composite by John Cosgrave (see http://www.spd.dcu.ie/johnbcos/fermat.htm).

Euler[1.43] became interested in Fermat's work in 1730. He found that

$$641 \mid \mathfrak{F}_5,$$

thereby contradicting Fermat's contention to the contrary. The way that he did this was to generalize a result of Fermat as follows.

Theorem 1.9 (Euler's Result on Fermat Numbers)
   If $\mathfrak{F}_n = 2^{2^n} + 1$, then every prime divisor of $\mathfrak{F}_n$ is of the form $2^{n+1}k + 1$ for some $k \in \mathbb{N}$.

Proof. Let p be a prime divisor of $\mathfrak{F}_n$. By Theorem 1.7, $p \mid (2^{p-1} - 1)$. Also, since

$$2^{2^{n+1}} - 1 = (2^{2^n} + 1)(2^{2^n} - 1) = \mathfrak{F}_n(2^{2^n} - 1),$$

Then $p \mid (2^{2^{n+1}} - 1)$. Therefore, we may let $2^{2^{n+1}} - 1 = pv$ for some $v \in \mathbb{N}$. Furthermore, by the Division Algorithm (see page 279), there exist $k, \ell \in \mathbb{Z}$ such that $p - 1 = 2^{n+1}k + \ell$ with $0 \leq \ell < 2^{n+1}$. Hence,

$$p \mid (2^{p-1} - 1) = (2^{2^{n+1}k+\ell}) - 1 = (2^{2^{n+1}})^k 2^\ell - 1 = (pv + 1)^k 2^\ell - 1,$$

and by the Binomial Theorem, this must equal $(pw + 1)2^\ell - 1$ for some $w \in \mathbb{Z}$. We have shown that $p \mid (2^\ell - 1)$, so $\ell = 0$ by Exercise 1.18. Thus, $p - 1 = 2^{n+1}k$, or $p = 2^{n+1}k + 1$. $\square$

In particular, we know from Theorem 1.9 that all divisors of $\mathfrak{F}_5$ must be of the form $64k + 1$. Thus, Euler only needed five trial divisions to find the factor 641, namely for $k = 3, 4, 7, 9, 10$, since the values $64k + 1$ for $k = 2, 5, 8$ are divisible by 3, and those for $k = 1, 6$ are divisible by 5.
   Euler also knew of the seven perfect numbers

$$2^{n-1}(2^n - 1) \text{ for } n = 2, 3, 5, 7, 13, 17, 19.$$

By 1771, he had determined that M_{31} is also prime, the largest known prime to that date, and the record held until 1851. This only touches very briefly upon the contributions that Euler made to primality testing, and number theory in general, which were indeed profound.

[1.43]Leonard Euler (1707–1783) was a Swiss mathematician who studied under Jean Bernoulli (1667–1748). Euler was extremely prolific. He published over five hundred papers during his lifetime, and another three hundred and fifty have appeared posthumously. It took almost fifty years for the Imperial Academy to finish publication of his works after his death. Euler had spent the years 1727–1741 and 1766–1783 at the Imperial Academy in St. Petersburg under the invitation of Peter the Great. Euler lost the sight in his right eye in 1735, and he was totally blind for the last seventeen years of his life. Nevertheless, he had a phenomenal memory, and so his mathematical output remained high. In fact, about half of his works were written in those last seventeen years. He contributed not only to number theory, but also to other areas of mathematics such as graph theory. It may even be argued that he essentially founded that field of mathematics. He died on September 18, 1783.

In 1830, a valuable technique for factoring any odd integer n was discovered by Legendre.[1.44] This method involved the use of the distinguished symbol (about which we will learn later – see Definition 4.26) that bears his name. This allowed him to exclude almost all primes less than $\sqrt{n}$, as potential factors of n. We will illustrate this method later when we have developed the tools to do so. Legendre also proved the result contained in Exercise 1.19, which extended some of Euler's ideas. A version of Legendre's Method for factoring was given by Gauss[1.45] in his influential masterpiece *Disquisitiones Arithmeticae* [80]. Gauss recognized the importance of factoring, as is indicated by the following comment from [80, Art. 329, p. 396]: "The problem of distinguishing prime numbers from composite numbers and of resolving the latter into their prime factors is known to be one of the most important and useful in arithmetic." Other than the elucidation of a version of Legendre's Method, Gauss discussed another factoring method in [80], which may be described as follows. The reader unfamiliar with the notion of a gcd should consult Appendix A before proceeding.

Suppose that we want to factor $n \in \mathbb{N}$. We choose some $m \in \mathbb{N}$ such that $\gcd(m, n) = 1$. Suppose that $x = r, s \in \mathbb{N}$ are two solutions of

$$n \mid (x^2 - m). \tag{1.10}$$

Then, as long as $n \nmid (r \pm s)$, then $\gcd(r - s, n)$ must be a nontrivial divisor of n because $n \mid (r - s)(r + s)$, while $n \nmid (r - s)$ and $n \nmid (r + s)$. Gauss's methods

---

[1.44] Adrien-Marie Legendre (1752–1833) was born on September 18, 1752 in Paris, France. He was educated at the Collège Mazarin in Paris. During the half decade 1775–1780, he taught along with Laplace (1749–1827) at Ecole Militaire. Over the next decade, he published works, not only on number theory, but also celestial mechanics and elliptic functions. He also took a position at the Académie des Sciences, becoming first *adjoint* in 1783, then *associé* in 1785, and his work finally resulted in his election to the Royal Society of London in 1787. In 1793, the Académie was closed due to the Revolution, but Legendre was able to publish his phenomenally successful book *Eléments de Géométrie* in 1794, which remained the leading introductory text in the subject for over a century. In 1795, the Académie was reopened as the *Institut National des Sciences et des Arts*, and met in the Louvre until 1806. In 1808, Legendre published his second edition of *Théorie des Nombres*, which included Gauss's proof of the Quadratic Reciprocity Law (about which we will learn in Chapter Four). Legendre also published his three-volume work *Exercises du Calcul Intégral* in (1811–1819). Then his three-volume work *Traité des Fonctions Elliptiques* was published during the period 1825–1832. Therein he introduced the name "Eulerian Integrals" for beta and gamma functions. This work also provided the fundamental analytic tools for mathematical physics, and today some of these tools bear his name, such as *Legendre Functions*. In 1824, Legendre had refused to vote for the government's candidate for the Institute National and for taking this position his pension was terminated. He died in poverty on January 10, 1833 in Paris.

[1.45] Carl Friederich Gauss (1777–1855) is certainly one of the greatest mathematicians who ever lived. At the age of eight, he astonished his teacher, Büttner, by rapidly adding the integers from 1 to 100 via the observation that the fifty pairs $(j + 1, 100 - j)$ for $j = 0, 1, \ldots, 49$ each sum to 101 for a total of 5050. At the age of eleven Gauss entered a preparatory school for university called a *Gymnasium* in Germany. By the age of fifteen, Gauss entered Brunswick Collegium Carolinum funded by the Duke of Brunswick. In 1795, Gauss entered Göttingen University, and by the age of twenty achieved his doctorate. The reader is referred to [144] and [145] for a discussion of his multitudinous achievements. Gauss was married twice. He married his first wife, Johanna Ostoff on October 9, 1805. She died in 1809 after giving birth to their second son. His second wife was Johanna's best friend Minna, whom he married in 1810. She bore him three children. Gauss remained a professor at Göttingen until the early morning of February 23, 1855 when he died in his sleep.

for finding solutions x to (1.10) will be discussed later when we have developed modular arithmetic in Chapter Two. *Modular arithmetic* is essential not only to an understanding of Gauss's factoring methods, but also as a fundamental tool in modern day factoring methods (such as the quadratic sieve to be studied later).

Once we enter the nineteenth century, the work of several individuals stand out in the development of factoring and primality testing. Among these is C. G. Reuschle.[1.46] Tables compiled by Reuschle included all known prime factors of $b^n - 1$ for

$$b = 2, 3, 5, 7, 11, \text{ and } n \leq 42.^{1.47}$$

He also included some information on factors of $2^n - 1$ for some values of $n \leq 156$, and as we have already seen, this information is valuable in the primality testing arena.

In the late 1860's a Parisian named Fortuné Landry worked on finding factors of M_n. He completely factored $2^n \pm 1$ for all $n \leq 64$ with four exceptions. He even found the largest known prime of the time, namely

$$(2^{53} + 1)/321 = 2805980762433.$$

In 1867, Landry recognized the ubiquitous problem in primality testing, namely if one has factored a number, then the factors can be given and checked,[1.48] but the claim that a given number is prime requires convincing. Unless the party to be convinced has knowledge of primality testing methods, and checks the calculations, then acceptance of the primality can be a matter of faith.

Perhaps the most influential individual in the nineteenth century primality testing arena was Lucas.[1.49] He studied Fibonacci Numbers and by 1877 had

---

[1.46] C. G. Reuschle (1813–1875) was the rector of the Royal Gymnasium in the city of Stuttgart, Germany. In honour of the birthday of King of Württemberg, he often compiled some kind of mathematical discourse. In 1842, he began to compile number theoretic tables, and produced them in 1856 for the king, with a dedication to Jacobi.

[1.47] In 1925, Cunningham and Woodall [56] published tables of factorizations of $b^n \pm 1$ for a small number of bases $b \leq 12$, and some high powers of n. As a consequence, work on extending these tables has come to be known as the *Cunningham Project*. Relatively recent work on the Cunningham Project and related problems may be found in [40]. See also, [144, Appendix D, p. 375].

[1.48] At least, this is *usually* true. There are recent advances, such as the proof that $\mathfrak{F}_{24}$ is composite, which produce no explicit factor. See Footnote 1.50.

[1.49] François Éduard Anatole Lucas (1842–1891) was born on April 4, 1842 in Amiens, France. In 1864, he graduated from École Normale as *Agrégé des sciences mathématiques*, meaning that he had passed the state agrégation examination required for a teaching position at French lycées (high schools). However, his first position was assistant astronomer at the Observatory of Paris. He remained there until the Franco-Prussian war in 1870 in which he served as an auxiliary artillery officer. After the war, he became a mathematics teacher at various high schools in Paris. He had interests in recreational mathematics, which is reflected in his invention of the well-known *Tower of Hanoi* problem (see Exercise 1.36). However, his serious interest was in number theory, especially Diophantine analysis. Although he spent only the years 1875–1878 on the problems of factoring and primality testing, his contribution was impressive. Some of the ideas developed by Lucas may be interpreted today as the beginnings of computer design. His death was untimely and unfortunate. While attending a social function, a plate fell and a chip from it cut his face. Later he died from an infection that developed from that cut.

completely factored the first sixty of them. His work led him to develop results on the divisibility of Fibonacci Numbers, and ultimately to a proof that M_{127} is prime. The significance of this feat is revealed by the fact that this number held the distinction of being the largest known prime for three-quarters of a century.[1.50] After this discovery, Lucas turned his attention to the sequence of numbers that bear his name. Their definition and properties are developed in Exercises 1.20–1.30. Later, when we have studied modular arithmetic, we will learn about a primality test that bears his name. The depth of the influence that Lucas had on modern-day primality testing is given full stage in [208], a recent historical perspective by Hugh Williams, which is a book devoted to a discussion of the work of Lucas, and his influence on the history of primality testing for the thirteen decades following his death. At the end of the book, Williams concludes: "Thus, there is still much work to do on the primality testing problem. The work done by Lucas must be advanced further in order to address concerns about primality testing that go as far back as Landry."

At the end of the nineteenth century, once Lucas went on to other things, the work of Proth and Pocklington on primality testing stands out. We will learn the details of their contributions when we have developed the notion of modular arithmetic. Kraitchik's work, which Lehmer developed into twentieth century primality testing and factorization methods presaged the modern computer age. These contributions also will be studied in detail later. The modern electronic computer suddenly made practical the use of older methods, such as those of Kraitchik, which were impractical by hand. With the dawn of the computer age have arisen numerous factorization and primality testing methods, many of which we will study in this text. The methods that we will investigate in this text have taken this study light years ahead of where they were at the turn of the twentieth century. The implications that this has for cryptography are

---

[1.50]This is unlikely ever to happen again. As mentioned in Footnote 1.38, the access to high speed computers and new algorithms means that records do not stand for long. For instance, this author stated in [144, Footnote 2.3.3, p. 95], published in 1998, that the largest known Mersenne Prime was the thirty-fifth, discovered in February of 1996. However, a disclaimer was given :"...it is not unlikely that there will be a new one by the time of the publication of this text..." This indeed turned out to be the case. In [145, Footnote 2.36, p. 126], published in 1999, the thirty-seventh Mersenne Prime was cited as the largest and a similar disclaimer was made there as has been made above. We are on the doorstep of a new age where records perhaps will be distinguished more by their speedy demise than their longevity. In fact, during the writing of this book, an announcement was made on September 29, 1999, that $\mathfrak{F}_{24} = 2^{2^{24}} + 1$, the twenty-fourth Fermat Number is composite. $\mathfrak{F}_{24}$ has over five million decimal digits. A team consisting of Ernst Mayer, formerly of Case Western Reserve University, Cleveland, Ohio; Jason Papadopoulos of the University of Maryland, College Park, Maryland; and Richard Crandall of the Center for Advanced Computation, were able to conclusively determine that $\mathfrak{F}_{24}$ is composite, although no explicit factor is yet known. Mayer and Papadopoulos used independent, floating-point "wavefront" implementations of the rigorous, classical Pepin primality proof, the runs for which were completed on the twenty-seventh and thirty-first of August 1999, respectively, ending up in complete agreement on the final Pepin Residue, namely that it does not equal -1, where equality is required for primality. During these "wavefront" runs Crandall used a pure-integer convolution scheme in parallel mode, namely running on many computers simultaneously, to check the periodically deposited wavefront residues. This integer verification ensures that the proof is rigorous. In other words, there is no doubt that $\mathfrak{F}_{24}$ is composite.

profound indeed. Now that we have completed the historical motivation for our study, we proceed to learn about the modern development of this important and exciting science.

Exercises

1.11. Prove that the solution to the Rabbit Problem on page 20 is F_n pairs of rabbits.

1.12. If $x \in \mathbb{R}$ (the real numbers), then there is a unique integer n such that $n \leq x < n+1$. We say that n is *the greatest integer less than or equal to x*, sometimes called the *floor of x*, denoted by $\lfloor x \rfloor = n$. Prove that $x - 1 < \lfloor x \rfloor \leq x$.

Exercises 1.13–1.15 are with reference to the floor function defined in Exercise 1.12.

1.13. Prove that $\lfloor x + n \rfloor = \lfloor x \rfloor + n$ for any $n \in \mathbb{Z}$.

1.14. Prove that $\lfloor x \rfloor + \lfloor y \rfloor \leq \lfloor x + y \rfloor \leq \lfloor x \rfloor + \lfloor y \rfloor + 1$.

1.15. Prove that $\lfloor x \rfloor + \lfloor -x \rfloor = \begin{cases} 0 & \text{if } x \in \mathbb{Z}, \\ -1 & \text{otherwise.} \end{cases}$

1.16. Let $n = pq$ where $p > q$ are odd primes. Prove that there are exactly two ordered pairs of natural numbers (x, y) for which $n = x^2 - y^2$, namely

$$(x, y) \in \{((p+q)/2, (p-q)/2), ((pq+1)/2, (pq-1)/2)\}.$$

1.17. Prove a stronger version of the second assertion in Theorem 1.7, namely that $(b^m - 1) \mid (b^t - 1)$ whenever $m \mid t$.

1.18. Let $\mathfrak{F}_n = 2^{2^n} + 1$. Prove that if p is a prime dividing $\mathfrak{F}_n$, then the smallest $m \in \mathbb{N}$ such that $p \mid (2^m - 1)$ is $m = 2^{n+1}$. (*Hint: Use the Division Algorithm and Binomial Theorem as in the proof of Theorem 1.9.*)

1.19. The following is called *Legendre's Divisibility Criterion.*

Let p be a prime and $n \in \mathbb{N}$. Then

(a) If $p \mid (a^n + 1)$, then either $p = 2nm + 1$ for some $m \in \mathbb{N}$, or $p \mid (a^{n/k} + 1)$ where k is an odd divisor of n.

(b) If $p \mid (a^n - 1)$, then $p = mb + 1$ or $p \mid a^k - 1$ where $k \mid n$.

Exercises 1.20–1.36 require that the reader have an understanding of some basic arithmetical concepts, such as the notion of a gcd and knowledge of the Principle of Mathematical Induction. *For those readers unfamiliar with such concepts, a review of Appendix A is suggested at this juncture.*

1.20. Let α, β be roots of
$$x^2 - \sqrt{R}x + Q = 0,$$

where $R, Q \in \mathbb{Z}$ with $\gcd(R, Q) = 1$. Prove each of the following.

(a) $\alpha + \beta = \sqrt{R}$.

(b) $\alpha\beta = Q$.

(c) $\alpha - \beta = \sqrt{R - 4Q}$ for an appropriate choice of α and β.

☆ 1.21. With reference to Exercise 1.20, for $n \in \mathbb{Z}$, $n \geq 0$, set

$$U_n = \frac{\alpha^n - \beta^n}{\alpha - \beta}, \text{ and } V_n = \alpha^n + \beta^n,$$

called the *Lucas Functions*. Prove each of the following.

(a) $2V_{m+n} = V_m V_n + \Delta U_m U_n$, where $\Delta = R - 4Q$ and $m, n \in \mathbb{N}$.

(b) $2^{n-1}U_n = \sum_{j=1}^{\lfloor (n+1)/2 \rfloor} \binom{n}{2j-1} V_1^{n-2j+1} \Delta^{j-1}$, where $n \in \mathbb{N}$.

(c) $2^{n-1}V_n = \sum_{j=0}^{\lfloor n/2 \rfloor} \binom{n}{2j} V_1^{n-2j} \Delta^j$, where $n \in \mathbb{N}$.

Exercises 1.22–1.30 are devoted to Lucas Functions. Prove each of the following facts. Note that when discussing divisibility properties, to avoid confusion, we assume that a factor of $\sqrt{R}$ may be ignored in U_n or V_n. For instance, if $R = 5$, and $Q = -3$, then $U_3 = 8$, and $U_6 = 112\sqrt{5}$. We say that $\gcd(U_3, U_6) = 8$, and U_6 is called even, ignoring $\sqrt{5}$.

1.22. For $n \in \mathbb{N}$, $U_{n+2} = \sqrt{R}U_{n+1} - QU_n$, and $V_{n+2} = \sqrt{R}V_{n+1} - QV_n$.

1.23. For $m, n \in \mathbb{N}$ with $n > m$, $2Q^m V_{n-m} = V_n V_m - \Delta U_n U_m$.

1.24. For $n \in \mathbb{N}$, $V_n^2 - \Delta U_n^2 = 4Q^n$.

1.25. For $m, n \in \mathbb{N}$, $2U_{m+n} = U_n V_m + V_n U_m$.

1.26. For all $m \in \mathbb{N}$,

$$\left(\frac{V_1 + U_1\sqrt{\Delta}}{2} \right)^m = \left(\frac{V_m + U_m\sqrt{\Delta}}{2} \right).$$

1.27. For $m, n \in \mathbb{N}$, with $n > m$, $2Q^m U_{n-m} = U_n V_m - V_n U_m$.

1.28. For $n \in \mathbb{N}$, $\gcd(U_n, Q) = 1 = \gcd(V_n, Q)$, and $\gcd(U_n, V_n)$ divides 2.

1.29. If $m, n \in \mathbb{N}$ with $m \mid n$, then $U_m \mid U_n$.

☆ 1.30. If $\gcd(m, n) = g$, then $\gcd(U_m, U_n) = U_g$.

1.31. Generalize the Fibonacci Sequence (defined on page 20) by putting $g_1 = a \in \mathbb{Z}$, $g_2 = b \in \mathbb{Z}$, and $g_j = g_{j-1} + g_{j-2}$ for $j \geq 3$. Prove that $g_j = aF_{j-2} + bF_{j-1}$.

1.32. Let $\mathfrak{g} = (1 + \sqrt{5})/2$ denote the golden ratio. Prove that

$$\mathfrak{g}^2 = \mathfrak{g} + 1.$$

☆ 1.33. Prove that

$$\mathfrak{g} = \sqrt{1 + \sqrt{1 + \sqrt{1 + \cdots}}}.$$

(*Hint: Use Exercise 1.32.*)

1.34. Let p be prime and define $e_1 = 4$ and $e_{j+1} = e_j^2 - 2$ for all $j \in \mathbb{N}$. The *Generalized Lucas Primality Test* says that the Mersenne Number $M_p = 2^p - 1$ is prime if and only if $M_p \mid e_{p-1}$. Use this test to show that $M_5 = 31$ is prime. Additionally, if the reader has access to a mathematical software package such as *Maple*, then verify, as well, that $M_{13} = 8191$ is prime using this test.[1.51]

1.35. Use induction to prove that the nth Fibonacci Number is given by

$$F_n = \frac{1}{\sqrt{5}} \left[\mathfrak{g}^n - \mathfrak{g}'^n \right],$$

where $\mathfrak{g}' = (1 - \sqrt{5})/2$.

1.36. A recreational problem, called the *Tower of Hanoi* problem, was invented by Lucas in 1883 (see Footnote 1.49). It is described as follows.

Assume that there are three vertical posts and $n \in \mathbb{N}$ rings, all of different sizes, concentrically placed on one of the posts from largest on the bottom to smallest on the top. In other words, no larger ring is placed upon a smaller one. The object of the game is to move all rings from the given post to another post, subject to the following rules:

[1] Only one ring may be moved at a time.

[2] A ring may never be placed over a smaller ring.[1.52]

Determine the number of moves required to transfer n rings from one post to another.

Using the above, answer the following question. Ancient folklore tells us that monks in a temple tower were given 64 rings at the beginning of time. They were told to play the above game, and that the world would end when they were finished.

Assuming that the monks worked in shifts twenty-four hours per day, moving one ring per second without any errors, how long does the world last?

[1.51]This test is also called the *Lucas-Lehmer Test*, which we will study in Chapter Four in detail, including a proof of the assertion made in this exercise. See pages 180–181.

[1.52]For the reader with some knowledge of graph theory, this problem is equivalent to finding a Hamiltonian path on an n-hypercube.

1.3   Computer Arithmetic

As observed at the end of Section Two, the advent of modern day electronic computers has radically altered the nature of cryptography. Therefore, it is imperative that we learn to speak and manipulate the vernacular of computers. That is the purpose of this section.

We are familiar with the Hindu-Arabic numeral system, which is a base 10 (decimal) system. For example, $5146 = 5 \cdot 10^3 + 1 \cdot 10^2 + 4 \cdot 10^1 + 6 \cdot 10^0$. Other civilizations from antiquity used different bases. For instance, the ancient Babylonians used base 60 (sexagesimal) and the ancient Mayans used base 20. Modern day computers use base 2. The reason for the latter has to do with how computers store data internally. In other words, computers, at their very essence, really only understand two possibilities such as "electrical charge or no electrical charge," or "magnetized clockwise or magnetized counterclockwise." Thus, the language that we need to learn begins with an understanding of how digits are represented in a computer. To begin to touch base with this understanding, we need the following elementary result. The reader who is unfamiliar with either the *sigma notation* or the *Division Algorithm* should first consult Appendix A.

Theorem 1.11 (Base Representations of Integers)

If $b > 1$ is an integer, then every $n \in \mathbb{N}$ has a unique representation as

$$n = \sum_{j=0}^{t_n} a_j b^j,$$

where t_n is the smallest nonnegative integer such that $a_j = 0$ for all $j > n$.[1.53]

Proof. We use the Division Algorithm given in Theorem A.18 on page 279. Dividing n by b, we get

$$n = bq_0 + a_0, \quad 0 \le a_0 \le b - 1.$$

If $q_0 \ne 0$, then we apply the Division Algorithm again to b and q_0 to get

$$q_0 = bq_1 + a_1, \quad 0 \le a_1 \le b - 1.$$

Continuing in this fashion, we have

$$q_{j-1} = bq_j + a_j, \quad 0 \le a_j \le b - 1,$$

[1.53] Although we will not be using a more general result than this in the text, it is worth mentioning the following for the sake of completeness of information. In the general case, $r \in \mathbb{R}^+$ the *positive reals*, can be uniquely represented in the form $r = \sum_{j=-\infty}^{\infty} a_j b^j$ for integers $0 \le a_j < b$ and $b > 1$. Then we write $(\ldots a_2 a_1 a_0 . a_{-1} a_{-2} \ldots)$. The dot between the a_0 and a_{-1} is called the *radix point*. For example, when $b = 10$, it is called a *decimal point*, and when $b = 2$ it is called the *binary point*, and so on. In (continental) Europe, the convention is to use a comma instead of a period for the radix point. If there is a $t_n \ge 0$ such that $a_j = 0$ for $j > a_{t_n}$, and $a_{t_n} \ne 0$, then a_{t_n} is called the *most significant digit*. Also, if there is there is a $s_n \in \mathbb{N}$ such that $a_j = 0$ for all $j < -s_n$, and $a_{-s_n} \ne 0$, then a_{-s_n} is called the *least significant digit*, or *trailing digit*.

for $j = 0, 1, \ldots, t_n$, where $q_{-1} = n$ and t_n is the least nonnegative integer such that $q_{t_n} = 0$, and such a q_{t_n} must exist since we have that the sequence $n > q_0 > q_1 > \cdots$ is a decreasing sequence of nonnegative integers, which must terminate with a term equaling 0. Therefore, we may put these together to get:

$$n = bq_0 + a_0 = b(bq_1 + a_1) + a_0 = b^2 q_1 + a_1 b + a_0 = \ldots$$

$$= b^{t_n} q_{t_n - 1} + \sum_{j=0}^{t_n - 1} a_j b^j = \sum_{j=0}^{t_n} a_j b^j,$$

where $0 \le a_j \le b - 1$ for $j = 0, 1, \ldots, t_n$ with $q_{t_n - 1} = a_{t_n} \ne 0$. This establishes existence. We now prove uniqueness.

Let

$$n = \sum_{j=0}^{t_n} a_j b^j = \sum_{j=0}^{t_n} c_j b^j \quad (0 \le a_j, c_j \le b - 1).$$

By subtracting the two representations of n, we get that

$$\sum_{j=0}^{t_n} (a_j - c_j) b^j = 0.$$

Let k be the least nonnegative integer such that $a_k \ne c_k$. Thus,

$$\sum_{j=k}^{t_n} (a_j - c_j) b^j = 0,$$

and multiplying through by b^{-k} we get,

$$\sum_{j=k}^{t_n} (a_j - c_j) b^{j-k} = 0.$$

By rearranging, we get

$$a_k - c_k = \sum_{j=k+1}^{t_n} (c_j - a_j) b^{j-k} = b \sum_{j=k+1}^{t_n} (c_j - a_j) b^{j-k-1}.$$

Thus, $b | (a_k - c_k)$, so there exists an $x \in \mathbb{Z}$ such that $a_k - c_k = bx$. Since $0 \le a_k, c_k \le b - 1$, then $|a_k - c_k| < b$ so $x = 0$. In other words, $a_k = c_k$, a contradiction that establishes uniqueness. $\square$

If $n \in \mathbb{N}$, then by Theorem 1.11, there is a unique representation:

$$(n)_{10} = (a_{t_n} a_{t_n - 1} \ldots a_1 a_0)_b,$$

for the *base b* (or *radix b*) representation of the base 10 integer n, where $j = t_n$ is the largest nonnegative integer such that $a_j \ne 0$. The value $t_n + 1$ is called the *base-b length* of n, and the a_j are called the *base* (radix) b digits of n. For instance, if $b = 2$, then the base 2 length is called the *bitlength* of n.

Important applications of Theorem 1.11 are those representations for $n \in \mathbb{N}$ of the form

$$n = \sum_{j=0}^{t_n} a_j 2^j,$$

where $0 \le a_j \le 1$ $(t_n \ge 0)$. The a_j are called *bits*, which is a contraction of *binary digits*, and $(a_{t_n} a_{t_n-1} \ldots a_0)$ is called a *bitstring of length* $t_n + 1$. In other words, n has bitlength $t_n + 1$. For instance, 1057, a base 10 integer, has a binary representation:

$$1 \cdot 2^{10} + 0 \cdot 2^9 + 0 \cdot 2^8 + 0 \cdot 2^7 + 0 \cdot 2^6 + 1 \cdot 2^5 + 0 \cdot 2^4 + 0 \cdot 2^3 + 0 \cdot 2^2 + 0 \cdot 2^1 + 2^0$$

or simply, $(1057)_{10} = (10000100001)_2$. A *byte*, for future reference is an eight-bit binary integer. The first recorded (but unpublished) appearance of the binary notation occurred in 1605 in a manuscript by Thomas Harriot.[1.54] The first *published* appearance of the binary system was in the 1670 manuscript, *Mathesis Biceps I*, by a Cistercian bishop, Juan de Caramuel Lobkowitz, but his publication contained no discussion or examples of binary arithmetic. The first to contain such a discussion was a paper by Leibniz,[1.55] *Mémoires de l'Académie Royal des Sciences*, which appeared in 1703. Leibniz attributed mystical import to the fact that all numbers could be expressed in terms of zeros and ones. The binary system remained somewhat of a curiosity thereafter until the advent, in the 1930's, of electromechanical and electromagnetic circuitry. By the mid

---

[1.54]Thomas Harriot (1560–1621) was an astronomer and mathematician. He gained fame as the leading scientist surrounding Sir Walter Raleigh, who sent him as a scientific advisor on an expedition in 1585 to Roanoke Island, off the coast of what is now called North Carolina. In Harriot's work *Artis Analyticae Praxis ad Aequationes Algebraicas Resolvendas*, or *The Analytical Arts Applied to Solving Algebraic Equations*, he advanced the theory of equations. He also introduced the *greater than* (>) and the *less than* (<) signs. Among his (unfortunately) unpublished discoveries were the following: sunspots, and the moons of Jupiter (independently of his contemporary Galileo), and the law of refraction (bending of light) before Willebrord van Roijen Snell (1591–1626), who published the discovery, which is basic to modern geometrical optics. He died on July 2, 1621.

[1.55]Gottfried Wilhelm von Leibniz (1646–1716), was born on July 1, 1646 in Leipzig, Saxony (now Germany). By the age of twelve, he had taught himself Latin and Greek in order to be able to read the books of his father who was a philosophy professor at Leipzig. Leibniz studied law at Leipzig from 1661 to 1666 and ultimately received a doctorate in law from the University of Altdorf in 1667. He pursued a career in law at the courts of Mainz from 1667 to 1672. Then he went to Paris from 1672 to 1676, during which time he studied mathematics and physics under Christian Huygens (1629–1695). In 1676, he left for Hannover, where he remained for the balance of his life. Leibniz began looking for a uniform and useful notation for the calculus in 1673. By November of 1676, he discovered the now famous $dx^n = nx^{n-1}dx$ for nonzero $n \in \mathbb{Q}$. In 1684, he published the details of the differential calculus, the year before Newton published his famed *Principia*. Leibniz's formal approach was to have a vital impact upon the development of the calculus. The bitter dispute between Newton and Leibniz concerning priority over the discovery of the calculus is detailed in [144, pp. 234–235]. In 1700, Leibniz founded the Berlin Academy and was its first president. Then he became increasingly reclusive until his death in Hannover on November 14, 1716.

1940's, there was support by major figures in the scientific community, such as von Neumann,[1.56] and since then binary computers have proliferated.

Computers have an upper bound on the size of integers that can be used for its arithmetic operations. This upper bound is called the *word size*, which can be measured in binary as 2^e on an e-bit binary computer, or as 10^e on an e-digit decimal computer, for instance. These considerations are important when talking about large-scale arithmetic in connection with implementation of cryptosystems (see Footnote 1.27 on page 14), for example. *Word length* is defined as the logarithm (taken to the appropriate base) of the word size. When we need to do computer arithmetic with integers bigger than the word size, then we must devote more than one word to each integer. One of the primary functions of this section is to describe how the basic arithmetic operations are formally performed.

Base names other than binary include $b = 3$ or *ternary*, $b = 4$ or *quaternary*, $b = 5$ or *quinary*, and so on.[1.57] In base $b = 8$ we have what we call *octal* representation. For instance, the decimal integer 100 has octal representation $100 = 1 \cdot 8^2 + 4 \cdot 8^1 + 4 \cdot 8^0$, so $(100)_{10} = (144)_8$. Base 16, or *hexadecimal*[1.58] representation uses the numbers 0 through 9 as well as the letters A, B, C, D, E, and F to represent the numbers 10, 11, 12, 13, 14, and 15, respectively. For instance the decimal integer 195951310 has the base 16 representation: $11 \cdot 16^6 + 10 \cdot 16^5 + 13 \cdot 16^4 + 15 \cdot 16^3 + 10 \cdot 16^2 + 12 \cdot 16^1 + 14$, which translates into: $(195951310)_{10} = (BADFACE)_{16}$. See Exercises 1.75–1.76 on page 46.

We have not yet addressed the issue of *negative number representation*. There are numerous ways to do this, among which is the *ones' complement*, which is a term used to describe the following. If $n \in \mathbb{Z}$ and $(|n|)_{10} = (a_{t_n} a_{t_n-1} \ldots a_0)_2$, then for any integer $m > t_n + 1$, the *ones' complement* m-bit representation of n is given by:

[1.56]John von Neumann (1903–1957) was born on December 3, 1903 in Budapest, Hungary as Margiattai Neumann János. After receiving his Ph.D. in mathematics from the University of Budapest, his academic positions included: lecturer at the University of Berlin 1926–29, lecturer at the University of Hamburg 1929–30, visiting lecturer at Princeton University 1930, and ultimately a permanent member, along with Einstein, at the Institute for Advanced Study at Princeton in 1933. By the late 1930's he had published ground-breaking work on rings of operators, now called *Neumann Algebras*. During World War II, he was a consultant for the armed forces, where his accomplishments included the drawing up of a report on computer capabilities. In 1946, he published a paper coauthored with A.W. Burks and H.H. Goldstine that detailed virtually the entire field of "automatic computation," including designs for a stored-program computer. This paper strongly influenced the later design of digital computers. In 1955, von Neumann had been appointed to the Atomic Energy Commission, and in 1956, he received the Enrico Fermi award. Among his varied interests and contributions were: computer design, game theory (in which he set the mathematical cornerstone with the minimax theorem published in 1928), group theory, logic and foundations, meteorology, probability theory, and quantum physics. He died on February 8, 1957 in Washington, D.C.

[1.57]It is even possible to have a negative radix b. This idea can be traced back as far as 1885 in a paper published in an obscure journal by Vittorio Grünwald. The idea resurfaced again in papers published in the mid 1930's and mid 1950's. By the late 1950's, Poland was building experimental computers using -2 as a base for their arithmetic. A nice example is the base $b = -10$ in which system *every real number* can be represented (without sign). See Exercises 1.67–1.70 on pages 46–46 for the integer version.

[1.58]This is a term that has both Greek and Latin roots.

$$\begin{cases} (\overbrace{00\ldots0}^{m-t_n-1\text{copies}} a_{t_n} a_{t_n-1}\ldots a_0)_2 & \text{if } n > 0, \\ (\underbrace{11\ldots1}_{m-t_n-1\text{copies}} 1-a_{t_n} 1-a_{t_n-1}\ldots 1-a_0)_2 & \text{if } n < 0, \end{cases}$$

The $n \in \mathbb{Z}$ which can be represented as m-bit ones' complement binary integers are those in the range $-2^{m-1} - 1 \leq n \leq 2^{m-1} - 1$. The basic idea is to convert n to a binary digit and pack $m - t_n - 1$ zeros to the left. Then if $n > 0$, this is the ones' complement, whereas if $n < 0$, then replace all zeros by ones and all ones by zeros. This is illustrated as follows.

Example 1.12 To represent $(26)_{10}$ and $(-26)_{10}$ as $m = 7$-bit ones' complement integers, we calculate that $(26)_{10} = (11010)_2$, so $(0011010)_2$ is the 7-bit ones' complement representation of the decimal digit 26, whereas for -26 it is $(1100101)_2$. See Exercises 1.39 – 1.40 on page 43.

Notice that in the ones' complement system, $+0$ is represented by $(00)_2$, as an $m = 2$-bit binary integer, and -0 is represented by $(11)_2$. Since -0 and $+0$ are the same, but represented differently, some care in practice is required.

A method of representing negative numbers that does not have this disadvantage is called the *ten's complement* notation, which is described as follows. Assuming that we are working with n-bit numbers, then we always work with 10^n. For instance, if $n = 10$, then a negative number such as -3495657980 is represented as 6504342020 ($= 10^{10} - 3495657980$). Hence, any number with a leading digit bigger than 4 is assumed to be negative, so no explicit sign is attached. This system clearly has limitations on size, but avoids the problem of the representation of $+0$ and -0. See Exercises 1.41–1.46 on pages 43–44.

Now that we know how to represent numbers to various bases in numerous ways, it is time to look at the arithmetic involved in using these bases. We begin with the most basic of arithmetic operations — addition.

◆ **(Addition Using Arbitrary Bases)**
Let $b, m, n \in \mathbb{N}$ and $b > 1$. By Theorem 1.11, we have the unique base b representations

$$m = \sum_{j=0}^{t} a_j b^j \text{ and } n = \sum_{j=0}^{t} c_j b^j, \tag{1.12}$$

where t is the largest integer such that $a_j + c_j \neq 0$. Therefore, by Theorem A.16 on page 280,

$$m + n = \sum_{j=0}^{t} (a_j + c_j) b^j.$$

Also, by the Division Algorithm, Theorem A.18, $a_0 + c_0 = q_0 b + r_0$, where $q_0, r_0 \in \mathbb{Z}$ with $0 \leq r_0 < b$. Given that $0 \leq a_0, c_0 \leq b - 1$, then $0 \leq a_0 + c_0 \leq 2b - 2$, so $q_0 \in \{0, 1\}$. In particular, if $q_0 = 1$, then we call q_0 a *carry* to the next position. Continuing in this fashion, we get,

$$a_1 + c_1 + q_0 = q_1 b + r_1 \text{ with } 0 \leq r_1 \leq b - 1$$

for some $q_1 \in \mathbb{Z}$. Since $0 \le a_1 + b_1 + q_0 \le 2b - 1$, then $q_1 \in \{0, 1\}$. Again if $q_1 = 1$, then it is called a carry to the next position. By induction, there exist $r_j, q_j \in \mathbb{Z}$ for each natural number j such that

$$a_j + c_j + q_{j-1} = q_j b + r_j \text{ with } 0 \le r_j \le b - 1,$$

where $q_j \in \{0, 1\}$. Since t is the largest integer such that $a_j + c_j \ne 0$, then if $q_t = 1$, we set $r_{t+1} = q_t$ and write the base b representation of $m + n$ as $(r_{t+1} r_t \dots r_0)_b$, whereas if $q_t = 0$, then we write it as $(r_t r_{t-1} \dots r_0)_b$.

Example 1.14 Suppose that we wish to calculate the addition of the two binary numbers $(10000011)_2$ and $(11010101)_2$. The following table illustrates the above process. The left-pointing arrows over a given column indicate that there is a carry from that position to the next. The binary addition is on the left and the decimal addition is on the right for easy reference.

j	8	$\overset{\leftarrow}{7}$	6	5	4	3	$\overset{\leftarrow}{2}$	$\overset{\leftarrow}{1}$	$\overset{\leftarrow}{0}$	binary/decimal	2	1	0
a_j		1	0	0	0	0	0	1	1	$\longleftrightarrow$	1	3	1
c_j		1	1	0	1	0	1	0	1	$\longleftrightarrow$	2	1	3
r_j	1	0	1	0	1	1	0	0	0	$\longleftrightarrow$	3	4	4

Thus, the addition of the two binary numbers, each of bitlength 8, is a binary number $(101011000)_2$ of bitlength 9, whereas the corresponding decimal numbers of bitlength 3 each sum to a decimal number $(344)_{10}$ of bitlength 3. Notice that, in the binary addition, $q_0 = q_1 = q_2 = q_7 = 1$, which accounts for the carries from each of those positions. On the other hand $q_0 = q_1 = q_2 = 0$ in the decimal addition, which accounts for the *lack* of carries in that summation.

In order to understand the following example, the reader should first solve Exercise 1.45 on page 44.

Example 1.15 In balanced ternary notation, $\bar{1} + \bar{1} = \bar{1}1$, and $\bar{1} + 1 + 1 = \bar{1}0$, where the leading $\bar{1}$ on the right (in both cases) is a carry. For instance, if we want to add the two balanced ternary numbers $1\,\bar{1}\,\bar{1}\,1$ and $1\,\bar{1}\,\bar{1}\,0$, then we must employ this rule as follows.

1	$\bar{1}$	$\bar{1}$	1
1	$\bar{1}$	$\bar{1}$	0
1	0	1	1

In the second column (from the right) we are adding $\bar{1}$ and $\bar{1}$, so we carry a $\bar{1}$ and leave the 1 in that position. Now in the third column, we have an addition of three $\bar{1}$'s, so we carry a $\bar{1}$ and leave a 0. Thus, the end result is 1011 in balanced ternary. The reader may easily calculate that $1\,\bar{1}\,\bar{1}\,1$ is $(16)_{10}$ and $1\,\bar{1}\,\bar{1}\,0$ is $(15)_{10}$, so we have $16 + 15 = 31$ in decimal, which corresponds to 1011 in balanced ternary.

The reader should now go to Exercises 1.47–1.56 to test understanding of addition for various bases.

We now look at the complementary notion of subtraction of integers to various bases. Of course, properly viewed, subtraction is not new since it is merely the *addition* of a number to a negative number.

◆ **(Subtraction Using Arbitrary Bases)**

Consider the representations given in (1.13) under the assumption that $m > n$. Then

$$m - n = \sum_{j=0}^{t}(a_j - c_j)b^j,$$

where t is the largest integer such that $a_j - c_j \neq 0$. We use the Division Algorithm to get

$$a_0 - c_0 = q_0 b + r_0 \text{ where } q_0, r_0 \in \mathbb{Z} \text{ with } 0 \leq r_0 < b.$$

Since $0 \leq a_0, c_0 < b$, then $-b \leq a_0 - c_0 < b$. Hence, $-1 \leq q_0 \leq 0$. If $q_0 = -1$, then we must borrow from the next position, so q_0 is called a *borrow* in this case. Continuing in this fashion,

$$a_1 - c_1 + q_0 = q_1 b + r_1 \text{ where } q_0, r_0 \in \mathbb{Z} \text{ with } 0 \leq r_1 < b.$$

Since, $-b \leq a_1 - c_1 + q_0 < b$, then $q_1 \in \{-1, 0\}$. Using induction, we see that for any nonnegative integer j, we get

$$a_j - c_j + q_{j-1} = q_j b + r_j \text{ where } q_j, r_j \in \mathbb{Z} \text{ with } 0 \leq r_j < b,$$

and $q_j \in \{-1, 0\}$, with $q_{-1} = 0$. Hence,

$$\sum_{j=0}^{t} r_j b^j = \sum_{j=0}^{t}(a_j - c_j + q_{j-1} - q_j b)b^j = \sum_{j=0}^{t}(a_j - c_j)b^j,$$

since

$$\sum_{j=0}^{t}(q_{j-1} - q_j b)b^j = 0.$$

Since $j = t$ is the largest integer such that $a_j - c_j \neq 0$, the base b representation of $m - n$ is

$$(r_t r_{t-1} \ldots r_0)_b.$$

Example 1.16 Suppose that we wish to calculate the result of subtracting $(100131)_4$ from $(303020)_4$. Consider the following tabular illustration, where the left-pointing arrow designates a borrow from the next position.

j	5	4	3	$\overset{\leftarrow}{2}$	$\overset{\leftarrow}{1}$	$\overset{\leftarrow}{0}$
a_j	3	0	3	0	2	0
c_j	1	0	0	1	3	1
r_j	2	0	2	2	2	3

Hence,
$$(303020)_4 - (100131)_4 = (202223)_4.$$

The actual development in the table that corresponds to the preceding notation is given as follows.

$$a_0 - c_0 = 0 - 1 = -1 \cdot 4 + 3 = q_0 b + r_0,$$

$$a_1 - c_1 + q_0 = 2 - 3 - 1 = -1 \cdot 4 + 2 = q_1 \cdot b + r_1,$$

$$a_2 - c_2 + q_1 = 0 - 1 - 1 = -1 \cdot 4 + 2 = q_2 \cdot b + r_2,$$

$$a_3 - c_3 + q_2 = 3 - 0 - 1 = 0 \cdot 4 + 2 = q_3 \cdot b + r_3,$$

$$a_4 - c_4 + q_3 = 0 - 0 + 0 = 0 \cdot 4 + 0 = q_4 \cdot b + r_4,$$

and
$$a_5 - c_5 + q_4 = 3 - 1 + 0 = 0 \cdot 4 + 2 = q_5 b + r_5 = q_t b + r_t.$$

Example 1.17 The following illustrates the subtraction of the octal integer $(67677)_8$ from $(77501)_8$, where the left-pointing arrow denotes a borrow from the next position.

j	4	$\overleftarrow{3}$	$\overleftarrow{2}$	$\overleftarrow{1}$	0
a_j	7	7	5	0	1
c_j	6	7	6	7	7
r_j	0	7	6	0	2

Thus, $(77501)_8 - (67677)_8 = (7602)_8$.

The reader may now solve Exercises 1.61–1.64 concerning subtraction using various bases.

The next step in the learning of computer arithmetic is multiplication, which is, properly viewed, a sequence of additions.

◆ **(Multiplication Using Arbitrary Bases)**
Consider the representations given in (1.13). We now determine the value of

$$mn = \left(\sum_{j=0}^{t} a_j b^j \right) \left(\sum_{j=0}^{t} c_j b^j \right),$$

where t is the largest integer such that $a_j c_j \neq 0$. We may simplify our task by observing that

$$mn = \sum_{j=0}^{t} (mc_j) b^j.$$

In other words, our task is simplified to the task of understanding how to find mc_j in base b for each j, then determining how to find $(mc_j)b^j$, and finally adding up the terms. For simplicity of explanation, we let $c_j = c$ for now.

By the Division Algorithm, $ca_0 = q_0 b + s_0$ for some integers s_0, q_0 with $0 \leq s_0 \leq b - 1$. Also, since $0 \leq ca_0 \leq (b-1)^2$, then $0 \leq q_0 \leq b - 2$. Continuing in this fashion, we get

$$a_1 c + q_0 = q_1 b + s_1 \text{ where } s_1, q_1 \in \mathbb{Z} \text{ with } 0 \leq s_1, q_1 \leq b - 1.$$

By induction we get,

$$a_j c + q_{j-1} = q_j b + s_j \text{ where } s_j, q_j \in \mathbb{Z} \text{ with } 0 \leq s_j, q_j \leq b - 1$$

for $j = 0, 1, \ldots, t - 1$ with $q_{-1} = 0$, and $s_t = q_{t-1}$, where $j = t$ is the largest integer such that $mc \neq 0$. Thus, if $q_t = 0$, then mc is

$$(a_t \ldots a_0)_b (c)_b = (s_t \ldots s_0)_b,$$

and if $q_t = 1$, then mc is

$$(a_{t+1} a_t \ldots a_0)_b (c)_b = (s_{t+1} s_t \ldots s_0)_b.$$

Now we can achieve $(mc_i)b^i$ by what is a *shift* (to the left). For instance, $(1301)_4$ multiplied by 4^3 is just a shift three places to the left of the digits of $(1301)_4$ and a fill-up of the original three places with zeros. In other words, 4^3 times $(1301)_4$ is $(1301000)_4$. Now we just add up the results of our efforts, and we have the base b representation of mn.[1.59]

Example 1.18 We wish to multiply the two ternary digits $(222)_3$ and $(121)_3$. The following diagram illustrates the process.

```
              2   2   2
              1   2   1
          ─────────────
              2   2   2
      1   2   2   1
      2   2   2
  ─────────────────────
  1   2   0   1   0   2
```

In terms of the notation preceding this example, we have the following. Let

$$m = (222)_3 = (a_2 a_1 a_0)_3$$

and

$$n = (121)_3 = (c_2 c_1 c_0)_3.$$

For $c = c_0 = 1$, we clearly have

$$(a_2 a_1 a_0)_3 c_0 = (222)_3 = (s_2 s_1 s_0)_3.$$

[1.59] Note that for $s, t \in \mathbb{N}$ an s-bit integer multiplied by a t-bit integer yields an $(s + t)$-bit integer.

For $c = c_1 = 2$,

$$ca_0 = 4 = 1 \cdot 3 + 1 = q_0 b + s_0,$$
$$ca_1 + q_0 = 5 = 1 \cdot 3 + 2 = q_1 b + s_1,$$
$$ca_2 + q_1 = 5 = 1 \cdot 3 + 2 = q_2 b + s_2,$$

and $s_3 = 1$. Therefore,

$$(a_2 a_1 a_0)_3 c_1 = (1221)_3 = (s_3 s_2 s_1 s_0)_3.$$

For $c = c_2 = 1$, we have

$$(a_2 a_1 a_0)_3 c_2 = (222)_3 = (s_2 s_1 s_0)_3.$$

Now we perform shifting on each mc_j. We have

$$(222)_3 \cdot 3^0 = (222)_3,$$

$$(1221)_3 \cdot 3^1 = (12210)_3,$$

and

$$(222)_3 \cdot 3^2 = (22200)_3.$$

When we add up these amounts, we get

$$(222)_3 + (12210)_3 + (22200)_3 = (120102)_3,$$

which is what the above diagram told us.

The reader may now test knowledge of multiplication by solving Exercises 1.73–1.76 on page 46.

Now we turn our attention to division in various bases. As we will see, the division process amounts to a sequence of subtractions.

◆ **(Division Using Arbitrary Bases)**

Suppose that $m, n \in \mathbb{N}$, with $m \leq n$, then by the Division Algorithm,

$$n = mq + r \text{ for some } q \in \mathbb{N}, r \in \mathbb{Z} \text{ with } 0 \leq r \leq m - 1.$$

First we determine the base b representation of q as follows. Suppose that

$$q = (d_t d_{t-1} \ldots d_0)_b, \quad t \geq 0.$$

Set

$$s_i = m \sum_{j=0}^{t-i} d_j b^j + r \geq 0$$

for any nonnegative integer $i \leq t$. Since $d_j \leq b - 1$ for all nonnegative integers $j \leq t$, we have

$$s_i \leq m \sum_{j=0}^{t-i} (b-1) b^j + r = m \left(\sum_{j=0}^{t-i} b^{j+1} - \sum_{j=0}^{t-i} b^j \right) + r = m(b^{t-i+1} - 1) + r.$$

Since $r < m$, it follows that $s_i < mb^{t-i+1}$. Also, since

$$s_i = s_{i-1} - md_{t-i+1}b^{t-i+1},$$

for $1 \leq i \leq t+1$ where $s_{t+1} = r$, we have

$$\frac{s_{i-1}}{mb^{t-i+1}} = d_{t-i+1} + \frac{s_i}{mb^{t-i+1}} \geq d_{t-i+1} = \frac{s_{i-1} - s_i}{mb^{t-i+1}} > \frac{s_{i-1}}{mb^{t-i+1}} - 1.$$

In other words,

$$d_{t-i+1} = \left\lfloor \frac{s_{i-1}}{mb^{t-i+1}} \right\rfloor,$$

for $1 \leq i \leq t+1$.

The above is essentially an algorithm for finding each digit in the base b representation of q; namely, we subtract mb^{t-i+1} from s_{i-1} enough times until the result is negative. Then d_{t-i+1} is one less than the number of subtractions. We also observe that, since

$$d_t = \lfloor s_0/(mb^t) \rfloor = \lfloor n/(mb^t) \rfloor,$$

then this is our starting point.

Example 1.19 Suppose we want to divide $n = (101)_2$ by $m = (11)_2$. Set

$$(101)_2 = (11)_2(d_2d_1d_0)_2 + r = mq + r.$$

Then $t = 2$, $b = 2$ and $q = (d_2d_1d_0)$. Since

$$mb^t = (11)_2 \cdot 2^2 = (1100)_2,$$

then subtracting this from $s_0 = (101)_2 = n$ yields a negative number, so $d_2 = d_t = 0$. Since

$$mb^{t-1} = (11)_2 \cdot 2 = (110)_2,$$

then subtracting this from

$$s_1 = s_0 - md_tb^t = (101)_2 - (11)_2 \cdot 0 \cdot 2^2 = (101)_2,$$

yields a negative number, so $d_1 = d_{t-1} = 0$. Since $mb^{t-2} = (11)_2$, and the subtraction of this from $s_2 = (101)_2$ yields $(10)_2$, whereas subtracting $(11)_2$ from $(10)_2$ yields a negative number, then $d_0 = d_{t-2} = 1$. We have shown that $q = (1)_2$. To get r, we look at

$$r = s_{t+1} = s_3 = s_2 - md_{t-2}b^{t-2} = (101)_2 - (11)_2 \cdot 1 \cdot 1 = (10)_2.$$

Hence,

$$n = (101)_2 = (11)_2(1)_2 + (10)_2 = mq + r.$$

To master the above technique, the reader should now solve Exercises 1.77–1.80 on page 46. In the next section, we will learn about the amount of time that a computer takes to perform operations such as those which we learned about in this section. This will be a special case of a more general notion called *computational complexity*.

Exercises

1.37. Let $R_{n,b}$ denote a number, in base $b > 1$, which is a string of n 1's. ($R_{n,b}$ *is called a* repunit, *which is a term introduced in 1964 by Beiler* [13].) Prove that
$$R_{n,b} = (b^n - 1)/(b - 1)$$
in base b.

1.38. With reference to Exercise 1.37, express $R_{19,10}$ in binary. ($R_{19,10}$ *is an example of a* prime repunit.)

1.39. Determine the ones' complement representation of each of the following.
   (a) -10. (b) -21.
   (c) 30. (d) 33.

1.40. Assuming that each of the following is the ones' complement of a seven bit number, find each of the corresponding base 10 representations.
   (a) 111101. (b) 101111.
   (c) 100001. (d) 101010.

1.41. Find the ten's complement representation of the following decimal digits.
   (a) $-1,234,554,321$. (b) $-1,010,101,010$.
   (c) $-9,999,999,999$. (d) $-5,000,111,111$.

1.42. Find the ten's complement representation of each of the following decimal digits.
   (a) $-5,432,112,345$. (b) $-5,550,000,001$.
   (c) $-8,888,888,888$. (d) -0.

1.43. A *nines' complement* representation of a negative number is achieved by replacing each digit x of the given number by $9 - x$. Thus, -1234554321 for example, would be 8765445678 in nines' complement notation. Find the nines' complement notation of each of the following decimal digits.
   (a) $-8,888,888,888$. (b) $-1,010,101,010$.
   (c) $-9,999,999,990$. (d) $-0,000,000,000$.

1.44. Find the nines' complement notation for each of the following decimal digits.
   (a) $-8,888,888,111$. (b) $-1,233,333,333$.
   (c) $-1,111,111,110$. (d) $-9,999,999,999$.

1.45. Assume that we are working with base 3 (ternary) notation. However, instead of using $0, 1, 2$ for our ternary representations, we will use $-1, 0, 1$, called *trits* or *ternary digits*. Furthermore, we will use the symbol $\overline{1}$ to denote -1. Using these symbols is called *balanced ternary notation*.[1.60] For instance, the decimal digit 242 is represented by $1000\overline{1}$ in balanced ternary notation. Find the representation in balanced ternary notation of each of the following.

(a) 332.   (b) -222.

(c) 111.   (d) 0.

1.46. With reference to Exercise 1.45, show that the negative of a number in balanced ternary notation is achieved by interchanging each 1 and $\overline{1}$.

1.47. Add $(1111100111)_2$ and $(1011000)_2$.

1.48. Add $(10001010111)_2$ and $(100010101110)_2$.

1.49. Convert $(999)_{10}$ and $(88)_{10}$ to binary and add the resulting binary digits.

1.50. Convert $(1111)_{10}$ and $(2222)_{10}$ to binary and add the results.

1.51. Add the two hexadecimal numbers $(FEED)_{16}$ and $(BEEF)_{16}$.

1.52. Add the two hexadecimal numbers $(FADE)_{16}$ and $(BEAD)_{16}$.

1.53. Add the two balanced ternary numbers $10\overline{1}$ and $\overline{1}10\overline{1}1$. (See Exercise 1.45.)

1.54. Add the two balanced ternary numbers $11\overline{1}010\overline{1}$ and $10\overline{1}011$.

1.55. Add the two base six numbers $(123443215)_6$ and $(543223411)_6$.

1.56. Add the two hexadecimal numbers $(E8BEEF)_{16}$ and $(B9CAF)_{16}$.

1.57. Convert $(43210A)_{16}$ to binary.

1.58. Convert $(10000110010000100001010)_2$ to quaternary.

1.59. Convert $(21012)_3$ to base nine.

1.60. Prove that if the base b representation of n is given by

$$n = (a_{t_n} \ldots a_2 a_1 a_0)_b,$$

[1.60] Although the seeds of balanced ternary notation may be found as far back as ancient Persia, as well as later in Fibonacci's work, and rediscovered by others including Cauchy, the first complete description of a balanced ternary notation was given in a paper published in 1840 by Léon Lalanne, a designer of mechanical devices for doing arithmetical computations. However, it was not until the advent of electronic computers over a century later that balanced ternary was considered, along with binary, as a replacement for decimal in computer design. In the early 1960's the Russians actually used balanced ternary in an experimental computer called SETUN. Nevertheless, binary has won the upper hand. (See [104].)

then for $k \in \mathbb{N}$, the base b^k representation of n is given by

$$n = (A_\ell \ldots A_2 A_1 A_0)_{b^k},$$

where

$$A_j = (a_{kj+k-1} \ldots a_{kj+1} a_{kj})_b$$

and

$$\ell = \lfloor t_n / k \rfloor.$$

(*Now the reader may go back to Exercises 1.57–1.59 and solve them from a different perspective.*)

1.61. Subtract the balanced ternary number $\overline{1}110$ from the balanced ternary number $1\overline{1}\overline{1}0$. (See Exercises 1.45–1.46.)

1.62. Subtract the hexadecimal number $(BEEF)_{16}$ from $(FADE)_{16}$.

1.63. Convert $(2222)_3$ to base 9 and subtract it from $(8888)_9$.

1.64. Convert the hexadecimal number $(DAB)_{16}$ to binary and subtract from $(110111100100)_2$.

1.65. Find the binary representations of the (decimal) primes 17, 19, 23, 29, 31 and 37.

☆ 1.66. Prove that any $n \in \mathbb{N}$ can be uniquely represented in the form

$$n = \sum_{j=1}^{m} a_j j!$$

for some $m \in \mathbb{N}$, $a_m \neq 0$, $a_j \in \mathbb{Z}$, and $0 \leq a_j \leq j$ for all $j = 1, 2, \ldots, m$, called the *combinatorial number system*. (*This is an example of what is called a* mixed-radix *number system, namely where there is more than one* radix *(base) in the representation.*)

1.67. Prove that every integer $n \neq 0$ is uniquely representable in base $b < -1$. In other words show that every nonzero $n \in \mathbb{Z}$ has a unique representation

$$n = \sum_{j=0}^{t_n} a_j b^j,$$

where $0 \leq a_j < |b|$ for $j = 0, 1, \ldots, t_n$ and $a_{t_n} \neq 0$.

1.68. With reference to Exercise 1.67, find base -10 representations for each of the following decimal digits.

(a) -110.   (b) -15.

(c) 16.   (d) -17.

1.69. With reference to Exercise 1.67, find base -2 representations for each of the following decimal digits.

(a) -1.   (b) -10.

(c) -17.   (d) -100.

☆ 1.70. Let $b > 1$ be a fixed natural number, and let $r \in \mathbb{R}$ be arbitrary but fixed. Show that there exist $a_j \in \mathbb{Z}$ with $0 \le a_j < b$ such that

$$r = \lfloor r \rfloor + a_1 b^{-1} + a_2 b^{-2} + a_3 b^{-3} + \cdots.$$

Also, prove that for each $j \in \mathbb{N}$, we have

$$a_j = \lfloor b^j r \rfloor - b \lfloor b^{j-1} r \rfloor.$$

1.71. Prove that a representation of r as given in Exercise 1.70 terminates (namely, there is an $N \in \mathbb{N}$ such that $a_j = 0$ for all $j \ge N$) if and only if

$$r = n/b^m$$

for some $n, m \in \mathbb{Z}$ with $m \ge 0$.

☆ 1.72. Prove that if

$$r = n/b^m$$

for some $n, m \in \mathbb{Z}$ with $m > 0$, then there is a representation of r as given in Exercise 1.70 which does not terminate. (*This shows that the representations as given in Exercise 1.70 of such r are not unique.*)

1.73. Multiply $(101010)_2$ and $(1101010)_2$.

1.74. Multiply $(765)_8$ and $(565)_8$.

1.75. Multiply the two hexadecimal numbers $(F)_{16}$ and $(10FDA5)_{16}$.

1.76. Multiply the two hexadecimal numbers $(1, 1)_{16}$ and $(AFE1D)_{16}$.

1.77. Divide $(276)_8$ by $(167)_8$ giving quotient and remainder.

1.78. Divide the hexadecimal number $(FED12)_{16}$ by $(BAD2)_{16}$, giving quotient and remainder.

1.79. Divide the binary number $(101010)_2$ by $(1111)_2$, showing the quotient and remainder.

1.80. Divide $(11211)_3$ by $(212)_3$.

1.81. Devise a recursive algorithm for multiplying two n-bit integers that is faster than the usual method requiring n^2 operations.

1.82. Estimate an upper bound on the number of operations required for the algorithm created in Exercise 1.81. (*In the next section, we will be dealing with such issues, which are in the realm of complexity.*)

1.4 Complexity

The amount of time required for the execution of an algorithm on a computer is measured in terms of *bit operations*, which are defined as follows: addition, subtraction, or multiplication of two binary digits; the division of a two-bit integer by a one-bit integer; or the shifting of a binary digit by one position. The number of bit operations necessary to complete the performance of an algorithm is called its *computational complexity* or simply its *complexity*. This method of estimating the amount of time taken to execute a calculation does not take into account such things as memory access or time to execute an instruction. However, these executions are very fast compared with a large number of bit operations, so we can safely ignore them. These comments are made more precise by the introduction of the following notation introduced by Edmund Landau.[1.61]

Definition 1.20 (Big O Notation) *Suppose that f and g are positive real-valued functions. If there exists a positive real number c such that*

$$f(x) < cg(x) \tag{1.21}$$

for all sufficiently large x, then we write[1.62]

$$f(x) = O(g(x)) \text{ or simply } f = O(g). \tag{1.22}$$

(Mathematicians also write $f << g$ to denote $f = O(g)$.)[1.63]

Big O is *the order of magnitude of the complexity*, an *upper bound* on the number of bit operations required for execution of an algorithm in the *worst-case scenario*, namely in the case where even the trickiest or the nastiest inputs are given. It is possible that most often, for a given algorithm, even less time will be used, but we must always account for the worst-case scenario.

The comments made before Definition 1.20 may now be put into perspective. The definition of the time taken to perform a given algorithm does not take

---

[1.61]Edmund Georg Hermann Landau (1877–1938) was born on February 14, 1877 in Berlin, Germany. He attended the French Lycée in Berlin, then entered the University of Berlin to study mathematics at the age of sixteen. He received his doctorate in 1899 in number theory, having studied under Frobenius. In 1901, he submitted his Habilitation on analytic number theory. Then he taught at the University of Berlin from 1899 until 1909, when he was appointed to Göttingen as a successor to Minkowski (1864–1909). In 1909, he published the first systematic presentation of analytic number theory. In 1933, he was forced out of Göttingen by the National Socialist regime. After this he lectured only outside Germany. He died on February 19, 1938 in Berlin.

[1.62]Here sufficiently large means that there exists some bound $B \in \mathbb{R}^+$ such that $f(x) < cg(x)$ for all $x > B$. We just may not know explicitly the value of B. Often f is defined on $\mathbb{N}$ rather than $\mathbb{R}$, and occasionally over any subset of $\mathbb{R}$.

[1.63]The notation $<<$ was introduced by I.M. Vinogradov, a Russian mathematician who proved, in 1937, that every sufficiently large positive integer is the sum of at most four primes. This is related to Goldbach's Conjecture, which says that every even $n \in \mathbb{N}$ with $n > 2$ is a sum of two primes. The "=" in (1.22) should be considered as a $<$ and the "O" should be considered as *a constant multiple*. The equality is a means of saying that f is a member of the family satisfying (1.21).

into consideration of time spent *reading and writing* such as memory access, timings of instructions, even the speed or amount of memory of a computer, all of which are negligible in comparison with the order of magnitude complexity. The greatest merit of this method for estimating execution time is that it is machine-independent. In other words, it does not rely upon the specifics of a given computer, so the order of magnitude complexity remains the same, irrespective of the computer being used. In the analysis of the complexity of an algorithm, we need not know *exactly* how long it takes (namely, the *exact* number of bit operations required to execute the algorithm), but rather it suffices to compare with other objects, and these comparisons need not be immediate, but rather long term. In other words, what Definition 1.20 says is that if f is $O(g)$, then *eventually* $f(x)$ is bounded by *some* constant multiple $cg(x)$ of $g(x)$. We do not know exactly *what* c happens to be or just *how big* x must be before (1.21) occurs. However, for reasons given above, it is enough to account for the efficiency of the given algorithm in the worst-case scenario.

Example 1.23 A simple illustration of the use of Big O is to determine the number of bits in a base b integer. If n is a t_n-bit base b integer, then

$$b^{t_n-1} \leq n < b^{t_n}.$$

Therefore, $t_n = \lfloor \log_b n \rfloor + 1$, so an estimate on the size of t_n is, in general, $t_n = O(\log_b n)$. Shortly, we will demonstrate that the base b of the logarithm is irrelevant in determining complexity.

Another simple illustration of the use of the Big O notation is to refer to Section Three, where we introduced the algorithms for adding, subtracting, multiplying and dividing two s-bit integers. Review of these algorithms shows us that addition or subtraction take $O(s)$ bit operations, which is also the number of bit operations required to compare them (determine which is larger, or whether they are equal). On the other hand, the multiplication of an s-bit integer with an t-bit integer requires $O(st)$ bit operations (see Exercise 1.83 on page 56).[1.64] By Exercise 1.85, division of an s-bit integer by an t-bit integer, with $s \leq t$, takes $O(st)$ bit operations.

If a number n has no more than s bits, then $n \leq 2^s$, so if we wish to describe complexity in terms of the numbers themselves rather than their respective bit sizes, then we can rephrase the above as follows. The addition, subtraction or comparison of two integers less than n takes $O(\log_2(n))$ bit operations, and

---

[1.64]However, it is possible to perform the multiplication of two s-bit integers in $O(s)$ steps on what is called a *pointer machine*, which has no built-in facilities for doing arithmetic (see [104, pp. 311–317], where there is even a discussion of bringing multiplication of s-bit numbers down to just s steps). Furthermore, it is possible to multiply to n-bit numbers in $O(n\ln(n)\ln(\ln(n)))$ steps on a standard computer using fast Fourier multiplication (see [182, Theorem S, p. 237]). Here, $\ln n$ means $\log_e n$, the logarithm to the base e, the *natural* or *canonical* base, where we often use $\exp(x)$ in place of e^x for convenience. In the mathematical literature, $\log x$ is often used for $\log_e x$. We remind the reader that $\log_b x = \ln x / \ln b$.

the multiplication of two such integers takes $O(\log_2^2(n))$ bit operations, while division of n by $m \leq n$ takes $O(\log_2 m \log_2 n)$ bit operations.

The amount of time taken by a computer to perform a task is (essentially) *proportional* [1.65] to the number of bit operations. In the simplest possible terms, the constant of proportionality, which is the number of nanoseconds[1.66] per bit operation, depends upon the computer being used. This accounts for the machine-independence of the Big O method of estimating complexity since the constant of proportionality is of no consequence in the determination of Big O.

A fundamental *time estimate* in executing an algorithm is *polynomial time* (or simply *polynomial*)[1.67]. In other words, an algorithm is polynomial when its complexity is $O(n^c)$ for some constant $c \in \mathbb{R}^+$, where n is the bitlength of the input to the algorithm, and c is independent of n. (Observe that any polynomial of degree c is $O(n^c)$.) In general, these are the desirable algorithms, since they are the fastest. Therefore, roughly speaking, the polynomial-time algorithms are the *good* or *efficient* algorithms. For instance, the algorithm is constant if $c = 0$; if $c = 1$, it is linear; if $c = 2$, it is quadratic, and so on. Examples of polynomial time algorithms are those for the ordinary arithmetic operations of addition, subtraction, multiplication, and division. On the other hand, those algorithms with complexity $O(c^{f(n)})$ where c is constant and f is a polynomial on $n \in \mathbb{N}$ are *exponential time algorithms* or simply *exponential*. A *subexponential* time algorithm is one for which the complexity for input $n \in \mathbb{N}$ is $O(\exp((c + o(1))(\ln n)^r((\ln n)(\ln \ln n))^{1-r})$ where $r \in \mathbb{R}$ with $0 < r < 1$ and c is a constant (see Footnote 5.6 on page 200 for a nice example of such an algorithm), where $o(1)$ denotes a function $f(n)$ such that $\lim_{n \to \infty} f(n) = 0$.[1.68] Subexponential time algorithms are faster than exponential time algorithms but slower than polynomial time algorithms. These are, again roughly speaking, the *inefficient* algorithms. For instance, the method of trial-division as a test for primality of $n \in \mathbb{N}$ uses $\sqrt{n}$ steps to prove that n is prime, if indeed it is. If we take the maximum bitlength $N = \log_2 n$ as input, then

$$\sqrt{n} = 2^{(\log_2 n)/2} = 2^{N/2},$$

which is exponential. Algorithms with complexity $O(c^{f(n)})$ where c is constant and $f(n)$ is more than constant but less than linear are called *superpolynomial*. It is generally accepted that modern-day cryptanalytic techniques for breaking known ciphers are of superpolynomial time complexity, but nobody has been able to prove that polynomial time algorithms for cryptanalyzing ciphers do not exist.

In calculating complexity using the Big O notation, the following properties are essential.

[1.65]To say that a is proportional to b means that $a/b = c$, a constant, called the *constant of proportionality*. This relationship is often written as $a \propto b$ in the literature.

[1.66]A nanosecond is $1/10^9$ of a second — a billionth of a second.

[1.67]Recall that a (nonconstant) polynomial is a function of the form $\sum_{i=0}^n a_i x^i$ for $n \in \mathbb{N}$, where the a_i are the coefficients (see page 289).

[1.68]In general, $f(n) = o(g(n))$ means that $\lim_{n \to \infty} f(n)/g(n) = 0$. Thus, $o(1)$ is used to symbolize a function whose limit as n approaches infinity is 0.

Theorem 1.24 (Properties of the Big O Notation)
    Suppose that f, g are positive real-valued functions.

(a) *If $c \in \mathbb{R}^+$, then $cO(g) = O(g)$.*

(b) *$O(\max\{f, g\}) = O(f) + O(g)$.*

(c) *$O(fg) = O(f)O(g)$.*

    Proof. (a) If $f = O(g)$, then there is a constant $k \in \mathbb{R}^+$ such that $f(x) < kg(x)$ for all sufficiently large x. Therefore,

$$cf(x) < (ck)g(x),$$

from which we get $cf = O(g)$. In other words,

$$O(g) = cO(g).$$

    (b) Let $h_1 = O(f)$, and $h_2 = O(g)$, then there exist $c_1, c_2 \in \mathbb{R}^+$ such that $h_1(x) < c_1 f(x)$ and $h_2(x) < c_2 g(x)$ for sufficiently large x. Therefore,

$$h_1(x) + h_2(x) < \max\{c_1 f(x), c_2 g(x)\}.$$

Hence, $O(f) + O(g) = O(\max\{f, g\})$.
    (c) Let $h_1 = O(f)$, and $h_2 = O(g)$, then $h_1(x) < c_1 f(x)$ and $h_2(x) < c_2 g(x)$, for some $c_1, c_2 \in \mathbb{R}^+$, and for sufficiently large x. Therefore,

$$h_1(x)h_2(x) < c_1 c_2 f(x)g(x),$$

for sufficiently large x. This implies that

$$O(f)O(g) = h_1 h_2 = O(fg),$$

which completes the proof. (*Note that part* (a) *is now a special case of part* (c) *with $f = 1$. Also, note that if $f = g$, then this provides the induction step for the more general fact that $O(f^n) = O(f)^n$ for any $n \in \mathbb{N}$.*) $\square$

    The following illustration involves the factorial notation (see Appendix A).

Example 1.25 Suppose that we wish to calculate the number of bit operations required to evaluate $n!$ for $n \in \mathbb{N}$ using only standard techniques. Assume that each natural number less than n has at most t bits. Then $n!$ has at most $n(t+1)$ bits and $n(t+1) = O(nt)$. Therefore, in the $n-1 = O(n)$ multiplications involved in computing $n!$, we multiply an integer with at most t bits by an integer with $O(nt)$ bits. This requires $O(nt^2)$ bit operations. Since we do this $O(n)$ times, the total number of bit operations required is $O(nt^2)O(n) = O(n^2 t^2)$, by part (c) of Theorem 1.24. However, we know that $t = O(\log_2 n)$ from above, so the number of bit operations to compute $n!$ is

$$O(n^2 t^2) = O(n^2 \log_2^2 n),$$

by part (c) of the theorem again. This is exponential in the number of bits of n.

Note that Theorem 1.24 shows us that in the complexity analysis of such algorithms as the division of two integers discussed above, where division of n by $m \le n$ takes $O(\log_2 m \log_2 n)$ bit operations, it is irrelevant which logarithm we use, since it does not change the Big O estimate. To see this we note that

$$O(\log_b(n)) = O(\ln n / \ln b) = O(\ln n) / \ln b = O(\ln n), \qquad (1.26)$$

by Theorem 1.24. For this reason, we omit any subscripts on logarithms in Big O notation henceforth, unless specified otherwise for a given reason.

To get some idea of what the various classes of complexity analysis mean in "real-world" terms, let's look at times related to some of these classes. Suppose that the unit of time on the computer at our disposal is a microsecond (a millionth $(1/10^6)$ of a second). Assuming an input of $n = 10^6$ bits, then a constant algorithm (complexity $O(1)$) would take a microsecond to execute, since the number of bit operations is one. A linear algorithm (complexity $O(n)$) would take a second, since the number of bit operations is 10^6. A quadratic algorithm (complexity $O(n^2)$) would take $11.5741 = 10^{12}/(10^6 \cdot 24 \cdot 3600)$ days, since there are 10^{12} bit operations, and a cubic algorithm (complexity $O(n^3)$) would take $31,709 = 10^{18}/(10^6 \cdot 24 \cdot 3600 \cdot 365)$ years, since the number of bit operations is 10^{18}. By the time we get to exponential algorithms, we are looking at times astronomically larger than the age of the known universe. Hence, a problem is called *intractable* if no polynomial time algorithm could possibly solve it, whereas one that can be solved using a polynomial time algorithm is called *tractable*. (By a *problem*, we mean a general question to be answered. A *decision problem* is one whose solution is "yes" or "no." A problem may possess *parameters* whose values are left unspecified, and an *instance* of a problem is achieved by specifying values for those parameters.)

To understand how complexity theory divides problems into classes, we must imagine a theoretical computer, called a *Turing Machine*,[1.69] which is a finite

---

[1.69] Alan Mathison Turing (1912–1954) was born on June 23, 1912 in London, England. He studied under Alonzo Church (1903–1995), at Princeton and received his doctorate in 1938 for his thesis entitled *Systems of Logic Based on Ordinals*. During World War II, he worked in the British Foreign Office, and was a major player in cryptanalyzing enemy codes. In 1945, he began work at the National Physical Laboratory in London, where he helped to design the *Automatic Computing Engine*, which led the world at the time as a design for a modern computer. In 1948, Turing became the deputy director of the Computing Laboratory at Manchester, where the first running example of a computer using electronically stored programs was being built. His contributions include pioneering efforts in artificial intelligence. In 1952, he was arrested for violation of the British homosexuality statutes. His death from potassium cyanide poisoning occurred while he was doing experiments in electrolysis, and it is uncertain whether this was an accident or self-inflicted.

For the reader interested in more detail and background, a (deterministic one-tape) Turing Machine has an infinitely long magnetic tape on which instructions can be written and erased. It also has a single bit register of memory and a processor that carries out the instructions: (1) move the tape right, (2) move the tape left, (3) change the state of the register based upon its current value and a value on the tape, and write or erase on the tape. The Turing Machine runs until it reaches a desired state causing it to halt. A famous problem in theoretical computer science is to determine when a Turing Machine will halt for a given set of input and rules. This is called the *Halting Problem*. Turing proved that this problem is *undecidable*, meaning that it is neither formally provable nor unprovable.

state machine having an infinite read-write tape. In other words, our theoretical computer has infinite memory and the ability to search for and retrieve any data from memory. *Church's Thesis* essentially says that the Turing Machine as a model of computation is equivalent to any other model for computation.[1.70] Therefore, Turing Machines are realistic models for simulating the running of algorithms, and they provide a powerful computational model.[1.71] Complexity theory designates a decision problem to be in class **P** if it can be solved in polynomial time, whereas a decision problem is said to be in class **NP** if it can be solved in polynomial time on a *nondeterministic* Turing Machine, which is a variant of the normal Turing Machine in that it *guesses* solutions to a given problem and checks its guess in polynomial time. Another way to look at the class **NP** is to think of these problems as those for which the *correctness of a guess* at an answer to a question can be proven in polynomial time.[1.72] Those problems that can be *disproved* or for which a *no* answer can be guessed in polynomial time are said to be in **Co − NP**. In other words, **Co − NP** consists of the problems whose complement is in **NP**. For example, the complement of: "Is $n \in \mathbb{N}$ composite?" is "Is $n \in \mathbb{N}$ prime?." Note that the complement of problems in **P** are also in **P**. However, this is not known for **NP**. It is generally held that **NP** $\neq$ **Co − NP**, but this has not been proved.

The class **P** is a subset of the class **NP** since a problem that can be solved in polynomial time on a *deterministic* machine can also be solved, by eliminating the guessing stage, on a nondeterministic Turing Machine. It is an open problem in complexity theory to resolve whether or not **P** = **NP**. However, virtually everyone believes that they are unequal. It is generally held that most modern ciphers can be cryptanalyzed in nondeterministic polynomial time. However, in practice it is the deterministic polynomial-time algorithm that is the end-goal of modern-day cryptanalysis. Defining what it means to be a "computationally hard" problem is a *hard problem*. One may say that problems in **P** are *easy*, and those not in **P** are considered to be *hard* [1.73] However, there are problems that are regarded as computationally easy, yet are not known to be in **P**.[1.74] A practical (but mathematically less satisfying) way to define "hard" problems is to view them as those which have continued to resist solutions after a concerted

---

[1.70] Here we may think of a "model" naively as a simplified mathematical description of a computer system.

[1.71] However, a Turing Machine is not meant to be a practical design for any actual machine, but rather is a sufficiently simple model to allow us to prove theorems about its computational capabilities while at the same time being sufficiently complex to include any digital computer irrespective of implementation.

[1.72] Another equivalent way to define the class **NP** is the class of those problems for which a "yes" answer can be verified in polynomial time using some extra information, called a *certificate* (see Definition 4.56 on page 186). For instance, the problem of answering whether or not a given $n \in \mathbb{N}$ is composite is a problem in **NP** since, we can verify in polynomial time if it is composite given the certificate of a nontrivial divisor a of n. However, it is not known if this problem is in **P**. See [134].

[1.73] For instance, see [173, pp. 195–196].

[1.74] For instance, the Miller-Rabin-Selfridge Test, which we will study in Section Three of Chapter Four, is such a problem. It is in the class **RP**, called *randomized polynomial time* or *probabilistic polynomial time*. Here, **P** $\subseteq$ **RP** $\subseteq$ **NP**.

attack by competent investigators for a long time up to the present.

Another aspect of problem classification in complexity theory is the **NP**-complete problem, which is a problem in the class **NP** that can be proved to be as difficult as any problem in the class. Should an **NP**-complete problem be discovered to have a deterministic polynomial time algorithm for its solution, this would prove that $\mathbf{NP} \subseteq \mathbf{P}$, so $\mathbf{P} = \mathbf{NP}$. Hence, we are in the position that there is no proof that there are *any* hard problems in this sense, namely those in **NP** but not in **P**. Nevertheless, this has not prevented the flourishing of research into complexity theory.[1.75] There are other distinctions up to and including the set **EXPTIME** of problems that can be solved in exponential time. However, the focus of this book will not need to address the finer distinctions.

We now look at some applications of complexity theory and the use of the Big O symbol.

Theorem 1.27 (Lamé)[1.76]

If $a, b \in \mathbb{N}$ such that $a > b$, and it takes $n + 1$ iterations (divisions) to find $\gcd(a, b)$ *using the Euclidean Algorithm, then $n < \log_{\mathfrak{g}} b$, where $\mathfrak{g} = (1 + \sqrt{5})/2$ is the golden ratio.*

Proof. If $a = r_{-1}$, $b = r_0$, then by Euclid's Algorithm,

$$r_{j-1} = r_j q_{j+1} + r_{j+1} \quad (0 < r_{j+1} < r_j)$$

for all nonnegative integers $j < n$, where n is the smallest value such that $r_{n+1} = 0$ (namely, we have $n + 1$ iterations). Therefore, if F_j denotes the jth Fibonacci Number, then

$$r_n \geq 1 = F_2$$

$$r_{n-1} = r_n q_{n+1} \geq 2 = F_3$$

$$r_{n-2} \geq r_{n-1} + r_n \geq F_3 + F_2 = F_4$$

$$\vdots$$

$$b = r_0 \geq r_1 + r_2 \geq F_{n+1} + F_n = F_{n+2}$$

Therefore, $b \geq F_{n+2}$. By Theorem A.31 on page 284, $F_{n+2} \geq \mathfrak{g}^n$, so $b \geq \mathfrak{g}^n$. Hence, $n < \log_{\mathfrak{g}} b$, so we have established the theorem of Lamé. $\qquad \square$

[1.75]The classical **NP**-complete problem is the *Travelling Salesman Problem*: A travelling salesman wants to visit $n \in \mathbb{N}$ cities. Is there a round trip that he can map out, which allows him to visit each city exactly once? This has been shown to be equivalent to the *Knapsack Problem*, which we will study in Section Four of Chapter Three.

[1.76]Gabriel Lamé (1795–1870) was born on July 22, 1795 in Tours, France. He both studied at École Polytechnique and was later a professor there. His primary contributions were to mathematical physics, but he also worked in differential geometry, diffusion in crystalline material, and elasticity. In fact, two elastic constants are named after him. His contributions to number theory were this result, and his proof of Fermat's Last Theorem for the exponent $n = 7$, which he gave in 1839. He died on May 1, 1870 in Paris, where there is now a street named after him.

Corollary 1.28 *With the hypothesis of Lamé's Theorem holding,*

$$n < 5 \log_{10} b.$$

Proof. An easy check shows that $\log_{10} \mathfrak{g} > 1/5$, so by Lamé's Theorem,

$$\frac{n}{5} < \log_{\mathfrak{g}} b \log_{10} \mathfrak{g} = \log_{10} b.$$

(For the equality, see Footnote 1.64.) Hence, $n < 5 \log_{10} b$.   □

In common language, what Corollary 1.28 says is the following. The number of iterations required to find $\gcd(a, b)$ is at most five times the number of decimal digits in the smaller value b. To see this, suppose that b has s decimal digits, so $b < 10^s$, namely $\log_{10} b < s$. Therefore, $5s > n$ by Corollary 1.28, so $5s \geq n+1$, which is the number of iterations required to find the $\gcd(a, b)$. What is implicit in this discussion is that the complexity of the given algorithm depends on b and not on a. Moreover, in computing the gcd of two consecutive Fibonacci numbers, the upper bound on complexity is reached. In other words, the worst case scenario does indeed occur.

Now we use Lamé's Theorem to find the computational complexity of the gcd.

Theorem 1.29 (Computational Complexity of the GCD)

If $a, b \in \mathbb{N}$ such that $a > b$, then the number of bit operations required to find $\gcd(a, b)$ using Euclid's Algorithm, is $O(\ln^3 a)$.[1.77]

Proof. From the proof of Theorem 1.27, $r_j \geq F_{n-j+2}$. In particular, we have

$$b \geq F_{n+2} \text{ and } a \geq F_{n+3}.$$

Since $F_{n+3} \geq \mathfrak{g}^{n+1}$, then

$$\log_{\mathfrak{g}} a \geq n + 1.$$

By the Euclidean Algorithm, Theorem A.21, the number of divisions required to find $\gcd(a, b)$ is $n + 1$, and

$$n + 1 = O(\ln a)$$

by (1.26). This is the number of bit operations required to perform the $n + 1$ divisions. Since each iteration of the algorithm computes a quotient and remainder involving numbers no bigger than a, then each iteration can be done in time $O(\ln^2 a)$. Therefore, by part 3 of Theorem 1.24, $\gcd(a, b)$ may be found using $O(\ln^3 a)$ bit operations.   □

The following is a variant of the Euclidean Algorithm.

[1.77]This can be improved by a more refined analysis to show that the *running time* (defined as the number of bit operations executed for a given input) of the Euclidean Algorithm is $O(\ln^2 a)$. See [134, 2.105 Fact, p. 66].

◆ **The Least Remainder Algorithm**

Let $a, b \in \mathbb{Z}$ with $a \geq b > 0$ and set $a = s_{-1}$, $b = s_0$. As with the Euclidean Algorithm, we repeatedly apply the Division Algorithm according to the recursive formula for each $j \geq 0$:

$$|s_{j-1}| = |s_j|t_{j+1} + s_{j+1} \text{ where } -|s_j|/2 < s_{j+1} \leq |s_j|/2.$$

Note that an analogue of the Division Algorithm (Theorem A.18) holds for the least remainder algorithm. This guarantees the existence and uniqueness of the s_j and t_j.

Example 1.30 If $1001 = s_{-1}$ and $s_0 = 221$, then to apply the Least Remainder Algorithm, we proceed as follows.

$$s_{-1} = 1001 = 221 \cdot 5 - 104 = s_0 t_1 + s_1,$$

$$s_0 = 221 = 104 \cdot 2 + 13 = |s_1|t_2 + s_2,$$

and $|s_1| = 104 = 13 \cdot 8 = s_2 t_3$, so $\gcd(1001, 221) = 13$.

If we now compare this with the ordinary Euclidean Algorithm, we get,

$$r_{-1} = 1001 = 221 \cdot 4 + 117 = r_0 q_1 + r_1,$$

$$r_0 = 221 = 117 \cdot 1 + 104 = r_1 q_2 + r_2,$$

$$r_1 = 117 = 104 \cdot 1 + 13 = r_2 q_3 + r_3,$$

and $r_2 = 104 = 13 \cdot 8 + 0 = r_3 q_4 + r_4$, so $\gcd(1001, 221) = 13 = r_3 = r_{n-1}$.

In general, the Euclidean Algorithm is less efficient than the Least Remainder Algorithm. It can be shown that we *save* approximately $1 - \log_2 \mathfrak{g} \approx 0.306$ of the division steps by using the Least Remainder Algorithm over the Euclidean Algorithm (see [104, Exercise 30, p. 376]). The Least Remainder Algorithm allows for negative remainders, the *least* in absolute value, which accounts for its increased efficiency. In Chapter Two, we will see that what underlies this algorithm is some modular arithmetic.

This completes the introductory material and thus we conclude Chapter One with some general remarks concerning complexity theory.

Roughly speaking, complexity theory can be subdivided into two categories: (a) structural complexity theory, and (b) the design and analysis of algorithms. Essentially, category (a) is concerned with lower bounds, and category (b) deals with upper bounds. Basically, the primary goal of structural complexity theory is to classify problems into classes determined by their intrinsic computational difficulty. In other words, how much computing time (and resources) does it take to solve a given problem? As we have seen in this section, the fundamental question in structural complexity theory remains unanswered, namely does $\mathbf{P} = \mathbf{NP}$? In this section, we have been primarily concerned with the analysis of algorithms, which is of the most practical importance to cryptography.

The foundations of complexity theory were laid by the work done starting in the 1930's by Turing and Church, among others (see Footnote 1.69). As we have seen in this section, the first goal was to formalize the notion of a computer (or realistic model thereof such as the Turing Machine). Then the goal was whether such devices could solve various mathematical problems. One of the outcomes of this research, again as we have seen, is that there are problems that cannot be solved by a computer. This dashed the program, set out by Hilbert[1.78] at the turn of the twentieth century, which sought to show that all mathematical problems could, at least in principle, be answered in some deterministic or mechanical way. Although the design of better and more efficient algorithms has been a goal of mathematicians and scientists in general for some time, it was not until the late 1960's that complexity theory began to be recognized as a formal discipline. The establishment of the theory may be credited to the pioneering work of Stephen Cook, Richard Karp, Donald Knuth, and Michael Rabin. Each of these individuals have since been awarded the highest honour in computer science research — the Turing Award.

Exercises

1.83. Prove that the multiplication of an m-bit integer with an n-bit integer, using the algorithms of Section Three, takes $O(mn)$ bit operations (see Footnote 4.12 on page 186).

1.84. Prove that for any $n \in \mathbb{N}$, $n! = O(n^n)$.

1.85. Let $n, m \in \mathbb{N}$ with $m \leq n$. Prove that the number of bit operations required to divide n by m, using the algorithm in Section Three, is $O(mn)$.

1.86. If $n, k \in \mathbb{N}$, find the number of bit operations required to compute $\binom{n}{k}$.

1.87. Show that if $f(x) = \sum_{j=0}^{n} a_j x^j$, where $a_j \in \mathbb{N}$ for each $j \geq 0$, then $O(f) = O(x^n)$.

1.88. Show that $O(c) = O(1) = O(|\cos x|)$, where $c \in \mathbb{R}^+$.

1.89. Estimate the number of bit operations required to compute $\sum_{j=1}^{n} j^2$.

[1.78]David Hilbert (1862–1943) was born in Königsberg, Prussia (now Kaliningrad, Russia). In 1895, Hilbert was appointed to the chair of mathematics at the University of Göttingen where he remained until his retirement in 1930. Among his students were Hermann Weyl (1885–1955) and E.F.E. Zermelo (1871–1953). Hilbert's contributions to twentieth-century mathematics were deep indeed. Perhaps this is best epitomized in the speech he delivered to the Second International Congress of Mathematicians held in 1900 in Paris, where he presented his now-famous list of twenty-three problems, many of which remain unsolved. Among these problems was the aforementioned one, namely that a finite number of logical steps based upon the axioms of arithmetic can never lead to contradictory results. However, the work of a mathematician named Kurt Gödel (1906–1978) destroyed any hope of that in 1931, when he proved that, given the axioms of arithmetic, statements can be made that can neither be proved nor disproved, namely they are *undecidable*. Hilbert died on February 14, 1943. See [144, p. 290] and [145, p. 10] for more information.

Chapter 2

Symmetric-Key Cryptosystems

2.1 An Introduction to Congruences

A new concept will be introduced in this chapter, *congruences*, invented by Gauss (see Footnote 1.45). The stage is set by the discussion of divisibility given in Appendix A. Gauss sought a convenient tool for abbreviating the family of expressions $a = b+nk$, called an *arithmetic progression* with modulus n, wherein k varies over all natural numbers, $n \in \mathbb{N}$ is fixed, as are $a, b \in \mathbb{Z}$. He did this as follows.

Definition 2.1 (Congruences)

If $n \in \mathbb{N}$, then we say that a is congruent *to b modulo n if $n|(a-b)$, denoted by*

$$a \equiv b \,(\text{mod } n).$$

On the other hand, if $n \nmid (a - b)$, then we write

$$a \not\equiv b \,(\text{mod } n),$$

and say that a and b are incongruent *modulo n, or that a is* not congruent *to b modulo n. The integer n is the* modulus *of the congruence. The set of all integers that are congruent to a given integer m modulo n, denoted by $\overline{m}$, is called the* congruence class *or* residue class *of m modulo n.*[2.1]

The reader may now solve the first exercise of this chapter, Exercise 2.1, which shows that $\equiv$ is an equivalence relation.

Example 2.2   (a) Since $3|(82 - 1)$, $82 \equiv 1 \,(\text{mod } 3)$.

[2.1]Note that since the notation $\overline{m}$ does not specify the modulus n, then the bar notation will always be taken in context.

(b) Since $11|(16 - (-6))$, $16 \equiv -6 \pmod{11}$.

(c) Since $7 \nmid (10 - 2)$, $10 \not\equiv 2 \pmod{7}$.

(d) For any $a, b \in \mathbb{Z}$, $a \equiv b \pmod{1}$, since $1|(a - b)$.

Gauss used the congruence notation to replace the assertion: *a and b are in the same arithmetic progression with difference a multiple of n* by the statement: *a is congruent to b modulo n*, and he denoted this by $a \equiv b \pmod{n}$, which is the content of Definition 2.1. In other words, $a \equiv b \pmod{n}$, if and only if $a = b + nk$ for some $k \in \mathbb{Z}$. Thus, $a \equiv b \pmod{n}$ if and only if $\bar{a} = \bar{b}$ with modulus n. Therefore, it makes sense to have a canonical representative.

Definition 2.3 (Least Residues)
   If $n \in \mathbb{N}$, $a \in \mathbb{Z}$, and $a = nq + r$ where $0 \leq r < n$ is the remainder when a is divided by n, given by Theorem A.18, the Division Algorithm, then r is called the least (nonnegative) *residue of a modulo n, and the set $\{0, 1, 2, \ldots, n-1\}$ is called the set of* least *nonnegative residues modulo n.*

We now show that for all $n \in \mathbb{N}$, congruence modulo n partitions[2.2] the integers $\mathbb{Z}$ into disjoint subsets. We need to show that every $m \in \mathbb{Z}$ is in *exactly one* residue class modulo n. (Note that Definition 2.3 justifies the use of the term *residue* class, given in Definition 2.1.) Since $m \in \bar{m}$, then m is in *some* congruence class, so we need show that it is in *no more than one* such class. If $m \in \bar{m}_1$ and $m \in \bar{m}_2$, then both $m \equiv m_1 \pmod{n}$ and $m \equiv m_2 \pmod{n}$. Thus, $m_1 \equiv m_2 \pmod{n}$ by Exercise 2.1, so $\bar{m}_1 = \bar{m}_2$, and we are done.
   As well as the above being true, it is also true that, for any $n \in \mathbb{N}$, and $0 \leq i \leq j \leq n - 1$, $i \equiv j \pmod{n}$ if and only if $i = j$. To see this, we observe that $j - i = mn$ for some $m \in \mathbb{Z}$ by definition, so $n \mid (j - i)$. If $j - i > 0$, then $n \leq (j - i)$. Since $j < n$ and $-i \leq 0$, it follows that $j - i = j + (-i) < n$, contradicting that $n \leq (j - i)$. Hence, $i = j$. We have shown that there are exactly n congruence classes for each $n \in \mathbb{N}$.

Example 2.4 There are four congruence classes modulo 4, namely

$$\bar{0} = \{\ldots, -4, 0, 4, \ldots\},$$
$$\bar{1} = \{\ldots, -3, 1, 5, \ldots\},$$
$$\bar{2} = \{\ldots, -2, 2, 6, \ldots\},$$

and

$$\bar{3} = \{\ldots, -1, 3, 7, \ldots\},$$

since each element of $\mathbb{Z}$ is in exactly one of these sets.

[2.2]For the definition of a partition, see Definition A.4 on page 272.

In order to motivate the next notion we let $r \in \mathbb{Z}$, $n \in \mathbb{N}$, and consider the set $\{r, r+1, \ldots, r+n-1\}$. If $r+i \equiv r+j \pmod{n}$ for $0 \le i \le j \le n-1$, then $i \equiv j \pmod{n}$, so by the same argument as above $i = j$. This shows that the $\overline{r+j}$ for $0 \le j \le n-1$ are n distinct congruences classes. Moreover, if $m \in \mathbb{Z}$, then by the argument given after Definition 2.3, m must be in exactly one of the n congruence classes. In other words, $m \equiv r+j \pmod{n}$ for some nonnegative integer $j < n$. This motivates the following.

Definition 2.5 (Complete Residue System)
Suppose that $n \in \mathbb{N}$ is a modulus. A set of integers

$$\mathcal{T} = \{r_1, r_2, \ldots, r_n\}$$

such that every integer is congruent to exactly one element of $\mathcal{T}$ modulo n is called a complete residue system modulo n. *In other words for any $a \in \mathbb{Z}$, there exists a unique $r_i \in \mathcal{T}$ such that $a \equiv r_i \pmod{n}$. The set $\{0, 1, \ldots, n-1\}$ is a complete residue system, called the* least residue system modulo n.

For example, $\mathcal{T} = \{-4, -3, -2, -1\}$ is a complete residue system modulo 4. Also, $\mathcal{T} = \{0, 1, 2, 3\}$ is the least residue system modulo 4. In fact, as proved in the discussion preceding Definition 2.5, any set of n consecutive integers forms a complete residue system modulo n. By choosing $r = 0$ in that discussion, we get the least residues.

In Exercises 2.1–2.6 on page 72 are the beginnings of the properties of what we call *modular arithmetic*, namely an arithmetic for congruences.

Example 2.6 The least residue system modulo 4 is $\mathcal{T} = \{0, 1, 2, 3\}$. Suppose that we want to calculate the addition of $\overline{3}$ and $\overline{2}$ in $\{\overline{0}, \overline{1}, \overline{2}, \overline{3}\}$. First, we must define what we mean by this addition. Let $\overline{a} \oplus \overline{b} = \overline{a+b}$ where $+$ is the ordinary addition of integers. Since $\overline{3}$ represents all integers of the form $3 + 4k$, $k \in \mathbb{Z}$, and $\overline{2}$ represents all integers of the form $2 + 4\ell$, $\ell \in \mathbb{Z}$,

$$3 + 4k + 2 + 4\ell = 5 + 4(k+\ell) = 1 + 4(1+k+\ell).$$

Hence, $\overline{3} \oplus \overline{2} = \overline{1} = \overline{3+2}$. Similarly, we may define $\overline{a} \otimes \overline{b} = \overline{a \cdot b}$, where $\cdot$ is the ordinary multiplication of integers. The reader may verify that $\overline{2} \otimes \overline{3} = \overline{2} = \overline{2 \cdot 3}$. Notice as well that since $\overline{a-b} = \overline{a+(-b)} = \overline{a} \oplus \overline{-b}$, then $\overline{2} \oplus \overline{-3} = \overline{3} = \overline{2-3}$, for instance.

Example 2.6 illustrates the basic operations of addition and multiplication in $\{\overline{0}, \overline{1}, \ldots, \overline{n-1}\}$ for any $n \in \mathbb{N}$, namely

$$\overline{a} \oplus \overline{b} = \overline{a+b} \text{ and } \overline{a} \otimes \overline{b} = \overline{a \cdot b},$$

where $\oplus$ and $\otimes$ are well-defined since $+$ and $\cdot$ are well-defined. Since it would be cumbersome to use the notations of $\oplus$, and $\otimes$ in general, we maintain the usage of $+$ for $\oplus$ and $\cdot$ for $\otimes$, where we will understand that the the result of the given operation is in the appropriate residue class. The following result formalizes this for us in general. The reader is encouraged to review the fundamental laws for arithmetic beginning on page 274, so that we will see that these seemingly trivial laws have a generalization to the following important scenario.

Theorem 2.7 (Modular Arithmetic)

Let $n \in \mathbb{N}$ and suppose that for any $x \in \mathbb{Z}$, $\overline{x}$ denotes the congruence class of x modulo n. Then for any $a, b, c \in \mathbb{Z}$ the following hold.

(a) $\overline{a} \pm \overline{b} = \overline{a \pm b}$.   (Modular additive closure)

(b) $\overline{a}\overline{b} = \overline{ab}$.   (Modular multiplicative closure)

(c) $\overline{a} + \overline{b} = \overline{b} + \overline{a}$.   (Commutativity of modular addition)

(d) $(\overline{a} + \overline{b}) + \overline{c} = \overline{a} + (\overline{b} + \overline{c})$.   (Associativity of modular addition)

(e) $\overline{0} + \overline{a} = \overline{a} + \overline{0} = \overline{a}$.   (Additive modular identity)

(f) $\overline{a} + \overline{-a} = \overline{-a} + \overline{a} = \overline{0}$.   (Additive modular inverse)

(g) $\overline{a}\overline{b} = \overline{b}\overline{a}$.   (Commutativity of modular multiplication)

(h) $(\overline{a}\overline{b})c = \overline{a}(\overline{b}\overline{c})$.   (Associativity of modular multiplication)

(i) $\overline{1} \cdot \overline{a} = \overline{a} \cdot \overline{1} = \overline{a}$.   (Multiplicative modular identity)

(j) $\overline{a}(\overline{b} + \overline{c}) = \overline{a}\overline{b} + \overline{a}\overline{c}$.   (Modular Distributivity)

Proof. Part (a) is a consequence of Exercise 2.2 on page 72, and part (b) is a consequence of Exercise 2.3. Each of the remaining parts is an easy exercise for the reader. We prove part (c) as an example. By part (a),

$$\overline{a} + \overline{b} = \overline{a + b} = \overline{b + a} = \overline{b} + \overline{a}.$$

In other words, the commutativity property is inherited from the integers $\mathbb{Z}$. The other parts follow in a similar manner. $\square$

Parts (a)–(b) of Theorem 2.7 tell us that the bar operation is well defined under addition and multiplication (see Appendix A). The remaining properties of this theorem tell us that there is an underlying structure. Any set that satisfies the (named) properties (a)–(j) of Theorem 2.7 is called a *commutative ring with identity*. In particular, we have the following.

Definition 2.8 (The Ring $\mathbb{Z}/n\mathbb{Z}$)

For $n \in \mathbb{N}$, the set

$$\mathbb{Z}/n\mathbb{Z} = \{\overline{0}, \overline{1}, \overline{2}, \ldots, \overline{n-1}\}$$

is called the Ring of Integers Modulo n, *where $\overline{m}$ denotes the congruence class of m modulo n.*[2.3]

[2.3]Occasionally, when the context is clear and no confusion can arise when talking about elements of $\mathbb{Z}/n\mathbb{Z}$, we will eliminate the *overline bars*.

Notice that since $\{0, \ldots, n - 1\}$ is the least residue system modulo n, then every $z \in \mathbb{Z}$ has a *representative* in the ring of integers modulo n, namely an element $j \in \{0, \ldots, n - 1\}$ such that $z \equiv j \pmod{n}$. The ring $\mathbb{Z}/n\mathbb{Z}$ will play an important role in the cryptographic applications that we study later in the text. There are other structures hidden within the properties listed in Theorem 2.7, which are worth mentioning, since we will also encounter them in our cryptographic travels. Any set satisfying the properties (a), (d)–(f) is called an *additive group*, and if additionally it satisfies (c), then it is called an *additive abelian group*. A fortiori, $\mathbb{Z}/n\mathbb{Z}$ is an additive abelian group as is $\mathbb{Z}$. Any set satisfying (a)–(f), (h) and (j) is called a *ring*, and if in addition it satisfies (g), then it is a *commutative ring*. As we have seen, any set satisfying all of the conditions (a)–(j) is a commutative ring with identity. In general, we would use different symbols than the bar operation, and possibly different binary symbols than the multiplication and addition symbols, but the listed properties in Theorem 2.7 would remain essentially the same for the algebraic structures defined above.

There is a multiplicative property of $\mathbb{Z}$ that $\mathbb{Z}/n\mathbb{Z}$ does not have. On page 274, the Cancellation Law for $\mathbb{Z}$ is listed. This is not the case for $\mathbb{Z}/n\mathbb{Z}$ in general. For instance, $2 \cdot 3 \equiv 2 \cdot 8 \pmod{10}$, but $3 \not\equiv 8 \pmod{10}$. In other words, $2 \cdot 3 = 2 \cdot 8$ in $\mathbb{Z}/10\mathbb{Z}$, but $3 \neq 8$ in $\mathbb{Z}/10\mathbb{Z}$. We may ask for conditions on n under which a modular law for cancellation would hold. In other words, for which $n \in \mathbb{N}$ does it hold that:

$$\text{for any } a, b, c \in \mathbb{Z}/n\mathbb{Z} \text{ with } a \neq 0, \ ab = ac \text{ if and only if } b = c? \qquad (2.9)$$

By Exercise 2.10, (2.9) cannot hold if $\gcd(a, n) > 1$, but if $\gcd(a, n) = 1$, then there is a solution $x \in \mathbb{Z}$ to $ax \equiv 1 \pmod{n}$. This motivates the following.

Definition 2.10 (Modular Multiplicative Inverses)

Suppose that $a \in \mathbb{Z}$, and $n \in \mathbb{N}$. A multiplicative inverse *of the integer a modulo n is an integer x such that $ax \equiv 1 \pmod{n}$. If x is the least positive such inverse, then we call it the* least multiplicative inverse *of the integer a modulo n, denoted $x = a^{-1}$.*

Example 2.11 Consider $n = 11$, $a = -3$, and suppose that we want to find the least multiplicative inverse of a modulo n. Since $-3 \cdot 7 \equiv 1 \pmod{11}$ and no smaller natural number than 7 satisfies this congruence, then $a^{-1} = 7$ modulo 11.

Example 2.12 If $n = 22$ and $a = 6$, then no multiplicative inverse of a modulo n exists since $\gcd(a, n) = 2$. Asking for a multiplicative inverse of such a value a modulo n is similar to asking for division by 0 with ordinary division of integers. In other words, this is undefined.

Since any composite n has a prime $p < n$ dividing it, then this means that (2.9) holds for all $c \in \mathbb{Z}/n\mathbb{Z}$ if and only if n is prime. Another way of stating this is as follows. Every nonzero $z \in \mathbb{Z}/n\mathbb{Z}$ has a multiplicative inverse if and only if n is prime.

If the existence of multiplicative inverses is satisfied for any given element along with (b), (h)–(i) of Theorem 2.7 for a given set, then that set is called a *multiplicative group*. In addition, if the set satisfies (g) of Theorem 2.7, then it is called an *abelian multiplicative group*. Hence, $\mathbb{Z}/n\mathbb{Z}$ is a multiplicative abelian group if and only if n is prime. Notice that $\mathbb{Z}$ is *not* a multiplicative group since any nonzero $a \in \mathbb{Z}$ with $a \neq \pm 1$ has no multiplicative inverse.

There is one property that is held by $\mathbb{Z}$, which is of particular importance to the ring $\mathbb{Z}/n\mathbb{Z}$. In Footnote A.2 on page 279, we mentioned *zero divisors*, namely those elements of a given set which satisfy the property that $a \cdot b = 0$, but $a \neq 0$ and $b \neq 0$. The integers $\mathbb{Z}$ have no zero divisors. What is the situation for $\mathbb{Z}/n\mathbb{Z}$ with respect to zero divisors? If n is composite, then there are natural numbers $n > n_1 > 1$ and $n > n_2 > 1$ such that $n = n_1 n_2$. Hence, $n_1 n_2 = 0$ in $\mathbb{Z}/n\mathbb{Z}$. Therefore, $\mathbb{Z}/n\mathbb{Z}$ has *no zero divisors* if and only if n is prime. Any set that satisfies all the conditions (a)–(j) of Theorem 2.7, together with having no zero divisors, and having multiplicative inverses for all of its nonzero elements is called a *field*. Hence, we have established the following.

Theorem 2.13 (The Field $\mathbb{Z}/p\mathbb{Z}$)
If $n \in \mathbb{N}$, then $\mathbb{Z}/n\mathbb{Z}$ is a field if and only if n is prime.

In Theorem A.47 on page 292, we employed the notation F^* to denote the multiplicative group of nonzero elements of a given field F. In particular, when we have a finite field $\mathbb{Z}/p\mathbb{Z} = \mathbb{F}_p$ of p elements for a given prime p, then $(\mathbb{Z}/p\mathbb{Z})^*$ denotes the multiplicative group of nonzero elements of $\mathbb{F}_p$. This is tantamount to saying that $(\mathbb{Z}/p\mathbb{Z})^*$ is the group of units in $\mathbb{F}_p$, and $(\mathbb{Z}/p\mathbb{Z})^*$ is cyclic by Theorem A.47. Thus, this notation and notion may be generalized as follows. Let $n \in \mathbb{N}$ and let the group of units of $\mathbb{Z}/n\mathbb{Z}$ be denoted by $(\mathbb{Z}/n\mathbb{Z})^*$. Then

$$(\mathbb{Z}/n\mathbb{Z})^* = \{\overline{a} \in \mathbb{Z}/n\mathbb{Z} : 0 \leq a < n \text{ and } \gcd(a, n) = 1\}. \tag{2.14}$$

The structure of $(\mathbb{Z}/n\mathbb{Z})^*$ is going to be of vital importance as we move through the text. Moreover, we will only be interested in *finite* groups, rings and fields, except for the obvious infinite cases such as $\mathbb{Z}$ and $\mathbb{Q}$.

Now we go on to look at some of the consequences of this notion of modular division, which is implicit in the above. Definition 2.10 gives us the means to do modular division since multiplication by a^{-1} is equivalent to division.

A classic example in the use of multiplicative inverses is the following.

◆ The Coconut Problem

Three sailors and a monkey are shipwrecked on an island. The sailors pick n coconuts as a food supply, and place them in a pile. During the night, one of the sailors wakes up and goes to the pile to get his *fair share*. He divides the pile into three, and there is a coconut left over, which he gives to the monkey. He then hides his third and goes back to sleep. Each of the other two sailors does the exact same thing, by dividing the remaining pile into three, giving the left over coconut to the monkey and hiding his third. In the morning, the sailors divide the remaining pile into three and give the monkey its fourth coconut. What is the *minimum* number of coconuts that could have been in the original pile?

We begin by observing that the first sailor began with a pile $n \equiv 1 \pmod 3$ coconuts. The second sailor began with a pile of

$$m_1 = \frac{2(n-1)}{3} \equiv 1 \pmod 3$$

coconuts, and the third sailor began with a pile of

$$m_2 = \frac{2(m_1 - 1)}{3} \equiv 1 \pmod 3,$$

coconuts, after which the three of them divided up the remaining pile of

$$m_3 = \frac{2(m_2 - 1)}{3} \equiv 1 \pmod 3$$

coconuts. We calculate m_3 and get

$$m_3 = \frac{8}{27} n - \frac{38}{27} \equiv 1 \pmod 3.$$

We now solve for n by multiplying through both sides and the modulus by 27, then simplifying to get

$$8n \equiv 65 \pmod{81}.$$

(Note that each of n, m_1, m_2, and m_3 must be natural numbers.) Since the multiplicative inverse of 8 modulo 81 is 71, namely $8^{-1} \equiv 71 \pmod{81}$, then

$$n = 8^{-1} \cdot 65 \equiv 71 \cdot 65 \equiv 79 \pmod{81},$$

and the smallest solution is 79.

The reader may now solve Exercise 2.27 for another version of the coconut problem. Also, Exercise 2.8 explores the notion of *self-multiplicative inverses*. It is used in the proof of the following famous result, which was conjectured by John Wilson (1741–1793) on the basis of some heuristic evidence. See Exercise 2.17 for Wilson's other claim to fame.

Theorem 2.15 (Wilson's Theorem)
   If p is a prime, then

$$(p-1)! \equiv -1 \pmod p.$$

Proof. The result is trivial if $p \leq 3$, so we assume that $p > 3$. By Definition 2.10, any $a \in \mathbb{Z}$ such that $2 \leq a \leq p - 2$ has a unique $a^{-1} \in \mathbb{Z}$ with $2 \leq a^{-1} \leq p - 2$ such that $aa^{-1} \equiv 1 \pmod p$. By Exercise 2.8, $a = a^{-1}$ if and only if $a = 1$ or $a = p - 1$. Hence, the product of the values a^{-1} for $2 \leq a^{-1} \leq p - 2$ is just the product of the integers $a = 2, 3, \ldots, p - 2$ in some order. Therefore, after possible rearrangement of those values of a, we must have $(p-2)! \equiv 1 \pmod{p}$. Thus, $(p-1)! \equiv (p-1) \equiv -1 \pmod{p}$.   □

The first to actually prove Theorem 2.15 was Lagrange.[2.4]

Example 2.16 If $p = 17$, then for each $a \in \mathbb{N}$ with $2 \le a \le 15$, we have

$$2 \cdot 9 \equiv 3 \cdot 6 \equiv 4 \cdot 13 \equiv 5 \cdot 7 \equiv 8 \cdot 15 \equiv 10 \cdot 12 \equiv 11 \cdot 14 \equiv 1 \pmod{17}.$$

Therefore, $16! \equiv 16 \equiv -1 \pmod{17}$.

Lagrange also proved the following.

Theorem 2.17 (The Converse of Wilson's Theorem)
If $n \in \mathbb{N}$ and $(n-1)! \equiv -1 \pmod{n}$, then n is a prime.

Proof. If $p \mid n$ is a prime and $p < n$, then $p \mid (n-1)!$. Thus, given that $(n-1)! \equiv -1 \pmod{n}$, we have

$$0 \equiv (n-1)! \equiv -1 \pmod{p},$$

a contradiction.   □

Another famous result that is linked to Exercise 2.8 is the following.

Theorem 2.18 (Fermat's Little Theorem)
If $a \in \mathbb{Z}$, and p is a prime such that $\gcd(a,p) = 1$, then $a^{p-1} \equiv 1 \pmod{p}$.

Proof. See the proof of Theorem 1.7.   □

When $p > 2$ and $p \nmid a \in \mathbb{Z}$, we see that $b = a^{(p-1)/2}$ is its own multiplicative inverse modulo p, since $b^2 = a^{p-1} \equiv 1 \pmod{p}$.

Notice that Theorem 2.18 tells us that $a^{-1} \equiv a^{p-2} \pmod{p}$, when $p \nmid a$, so this provides a means for computing inverses in $\mathbb{Z}/p\mathbb{Z}$.

Fermat's Little Theorem, which is worthy of the description *a gem*, was generalized by Euler (see Footnote 1.43). In order to understand how he did this, we need to introduce another concept that bears Euler's name.

[2.4]Joseph-Louis Lagrange (1736–1813) was born on January 25, 1736 in Turin, Sardinia-Piedmont (now Italy). Although Lagrange's primary interests as a young student were in classical studies, his reading of an essay by Edmund Halley (1656–1743) on the calculus converted him to mathematics. While still in his teens, Lagrange became a professor at the Royal Artillery School in Turin in 1755 and remained there until 1766 when he succeeded Euler as director of mathematics at the Berlin Academy of Science. Lagrange left Berlin in 1787 to become a member of the Paris Academy of Science where he remained for the rest of his professional life. In 1788 he published his masterpiece *Mécanique Analytique*, which may be viewed as both a summary of the entire field of mechanics to that time, and an establishment of mechanics as a branch of analysis, mainly through the use of the theory of differential equations. When he was fifty-six, he married a young woman, almost forty years younger than he, the daughter of the astronomer Lemonnier. She became his devoted companion until his death in the early morning of April 10, 1813 in Paris.

Definition 2.19 (Euler's ϕ-Function[2.5]—the Totient[2.6])

For any $n \in \mathbb{N}$ the Euler ϕ-function, also known as Euler's Totient $\phi(n)$ is defined to be the number of $m \in \mathbb{N}$ such that $m < n$ and $\gcd(m, n) = 1$.

Example 2.20 If p is prime, then by Exercise 2.16, $\phi(p) = p - 1$.

Example 2.21 Let $n \in \mathbb{N}$. Then the cardinality of $(\mathbb{Z}/n\mathbb{Z})^*$ is $\phi(n)$. See (2.14) and Exercise 2.35.

Theorem 2.22 (The Arithmetic of the Totient)

If $n = \prod_{j=1}^{k} p_j^{a_j}$ where the p_j are distinct primes, then

$$\phi(n) = \prod_{j=1}^{k} (p_j^{a_j} - p_j^{a_j - 1}) = \prod_{j=1}^{k} \phi(p_j^{a_j}).$$

Proof. Let $n = \prod_{j=1}^{k} p_j^{a_j}$, and perform induction on k. If $k = 1$, then this is Exercise 2.16. Assume that

$$\phi(M) = \prod_{j=1}^{k-1} (p_j^{a_j} - p_j^{a_j - 1}),$$

where

$$M = \prod_{j=1}^{k-1} p_j^{a_j}.$$

Claim 2.23 *If $n \in \mathbb{N}$ and p is prime, then*

$$\phi(pn) = \begin{cases} p\phi(n) & \text{if } p \mid n, \\ (p-1)\phi(n) & \text{otherwise.} \end{cases}$$

[2.5]Gauss introduced the symbol $\phi(n)$ (see [80, Articles 38–39, pp. 20–21]). Euler used the symbol πn to denote $\phi(n)$.

[2.6]The name *totient* was given to the function $\phi(n)$ by James Joseph Sylvester (1814–1897), who defined the *totatives* of n to be the natural numbers $m < n$ relatively prime to n. Sylvester was born in London, England on September 3, 1814. He taught at University of London from 1838 to 1841 with his former teacher Augustus De Morgan (1806–1871). Later he left mathematics to work as an actuary and a lawyer. This brought him into contact with Arthur Cayley (1821–1895) who also worked the courts of Lincoln's Inn in London, and thereafter they remained friends. Sylvester returned to mathematics, being appointed professor of mathematics at the Military Academy at Woolrich in 1854. In 1876 he accepted a position at the newly established Johns Hopkins University. He founded the first mathematical journal in the U.S.A., the *American Journal of Mathematics*. In 1883, he was offered a professorship at Oxford University. This position was to fill the chair left vacant by the death of the Irish number theorist Henry John Stephen Smith (1826–1883). When his eyesight began to deteriorate in 1893, he retired to live in London. Nevertheless, his enthusiasm for mathematics remained until the end as evidenced by the fact that he began work on Goldbach's Conjecture (see footnote 1.63 on page 47) in 1896. He died in London on March 15, 1897 from complications involving a stroke.

In order to calculate the value $\phi(pn)$, we look at each of the range of numbers

$$in + 1, in + 2, \ldots, in + n,$$

for $i = 0, 1, \ldots, p - 1$. If we eliminate all of the values j from these intervals that satisfy $\gcd(n, j) > 1$, then we have $p\phi(n)$ integers left. If $p|n$, then this is all of those values relatively prime to pn. However, if $p \nmid n$, then we must also eliminate all those values kp for $k = 1, 2, \ldots, n$. Of these, those kp with $\gcd(k, n) > 1$ have already been eliminated. Hence, there are just $\phi(n)$ more to eliminate, namely

$$\phi(pn) = p\phi(n) - \phi(n) = (p - 1)\phi(n),$$

and we have Claim 2.23. Therefore, it follows that

$$\phi(p_k^{a_k} M) = p_k \phi(p_k^{a_k - 1} M) = p_k^2 \phi(p_k^{a_k - 2} M) = \cdots$$

$$= p_k^{a_k - 1} \phi(p_k M) = p_k^{a_k - 1}(p_k - 1)\phi(M)$$

and by the induction hypothesis, this equals

$$p_k^{a_k - 1}(p_k - 1) \prod_{j=1}^{k-1} (p_j^{a_j} - p_j^{a_j - 1}) = \prod_{j=1}^{k} (p_j^{a_j} - p_j^{a_j - 1}) = \prod_{j=1}^{k} \phi(p_j^{a_j}).$$

This completes the induction and secures the result. $\square$

In order to get Euler's generalization of Fermat's Little Theorem, we need another concept.

Definition 2.24 (Reduced Residue Systems)
  If $n \in \mathbb{N}$, then the set

$$\mathcal{R} = \{m_j \in \mathbb{N} : \gcd(m_j, n) = 1 \text{ and } m_j \not\equiv m_k \,(\text{mod } n) \text{ where } 1 \le j \ne k \le \phi(n)\}$$

is called a reduced residue system modulo n.

If $\mathcal{R} = \{r_1, \ldots, r_{\phi(n)}\}$ is a reduced residue system modulo n, then so is $\mathfrak{R} = \{mr_1, \ldots, mr_{\phi(n)}\}$. To see this, note that since $\gcd(m, n) = \gcd(r_j, n) = 1$, then $\gcd(mr_j, n) = 1$ for all natural numbers $j \le \phi(n)$. If $mr_j \equiv mr_k \,(\text{mod } n)$ for some $j \ne k$ with $1 \le j, k \le \phi(n)$, then $r_j \equiv r_k \,(\text{mod } n)$, by part (e) of Theorem A.24, a contradiction.

Theorem 2.25 (Euler's Generalization of Fermat's Little Theorem)
  If $n \in \mathbb{N}$ and $m \in \mathbb{Z}$ such that $\gcd(m, n) = 1$, then $m^{\phi(n)} \equiv 1 \,(\text{mod } n)$.

Proof. By the discussion immediately preceding the theorem, each element in $\mathcal{R}$ is congruent to a unique element in $\mathfrak{R}$ modulo n. Hence,

$$\prod_{j=1}^{\phi(n)} r_j \equiv \prod_{j=1}^{\phi(n)} mr_j \equiv m^{\phi(n)} \prod_{j=1}^{\phi(n)} r_j \,(\text{mod } n),$$

and $\gcd(\prod_{j=1}^{\phi(n)} r_j, n) = 1$, so

$$m^{\phi(n)} \equiv 1 \,(\text{mod } n),$$

by part (e) of Theorem A.24. $\qquad\qquad\qquad\qquad\qquad\qquad\qquad\qquad\qquad$ □

Example 2.26 By Euler's Theorem, $3^{\phi(7)-1} \equiv 3^{6-1} = 3^5 \equiv 5 \,(\text{mod } 7)$, and 5 is a (least) multiplicative inverse of 3 modulo 7.

Example 2.26 is a special case of a result that is the content of Exercise 2.21, which is in turn a simple application of Theorem 2.25.

Some knowledge of the arithmetic of Euler's Function is necessary in order to do the calculations required later in our applications to cryptography. We develop these properties now.

Theorem 2.22 may now be used to solve Exercise 2.22, which establishes the multiplicativity of the totient.

In Exercises 2.10–2.14, we explore solutions of *linear congruences*, namely those of the form $ax \equiv b \,(\text{mod } n)$ for given $a, b \in \mathbb{Z}$ and $n \in \mathbb{N}$. As an application of the totient, we now look at the solution of *simultaneous* linear congruences, which has a long history beginning with Chinese mathematicians of antiquity. In particular, in a Chinese work entitled *Suangching* from the first century A.D., Sun Tsǔ provided a method for determining integers having remainders $2, 3, 2$ when divided by $3, 5, 7$, respectively. In the modern terminology of congruences, he determined how to solve the simultaneous congruences

$$x \equiv 2 \,(\text{mod } 3), x \equiv 3 \,(\text{mod } 5), x \equiv 2 \,(\text{mod } 7).$$

He calculated that, since $70 \equiv 0 \,(\text{mod } 5{\cdot}7)$, $21 \equiv 0 \,(\text{mod } 3{\cdot}7)$, $15 \equiv 0 \,(\text{mod } 3{\cdot}5)$, and,

$$70 \equiv 1 \,(\text{mod } 3), 21 \equiv 1 \,(\text{mod } 5), 15 \equiv 1 \,(\text{mod } 7),$$

then

$$233 - 2 \cdot 70 + 3 \cdot 21 + 2 \cdot 15$$

is a solution to the above simultaneous congruences. By reducing 233 modulo $3 \cdot 5 \cdot 7 = 105$, we get $x = 23$ as the unique solution that is the least residue modulo 105. Gauss [80, Article 36, p. 16] also recognized this method in his proof of the multiplicativity of the totient. He observed that if a natural number $a < \phi(m)$ with $\gcd(a, m) = 1$, and $b < \phi(n)$ with $\gcd(b, n) = 1$, then there is a unique $x < mn$ with $\gcd(x, mn) = 1$ where $x \equiv a \,(\text{mod } m)$ and $x \equiv b \,(\text{mod } n)$. This is formalized in the following, which takes its name from the work of Sun Tsǔ and other Chinese mathematicians of antiquity.

Theorem 2.27 (Chinese Remainder Theorem)
Let $n_i \in \mathbb{N}$ for natural numbers $i \le k \in \mathbb{N}$ be pairwise relatively prime, set

$$n = \prod_{j=1}^{k} n_j$$

and let $r_i \in \mathbb{Z}$ for $i \leq k$. Then the system of k simultaneous linear congruences given by:

$$x \equiv r_1 \,(\text{mod } n_1),$$

$$x \equiv r_2 \,(\text{mod } n_2),$$

$$\vdots$$

$$x \equiv r_k \,(\text{mod } n_k),$$

has a unique solution modulo n.

Proof. If $N_j = n/n_j$ $(1 \leq j \leq k)$, then $\gcd(N_j, n_j) = 1$. Also, by Definition 2.10 on page 61, there is a multiplicative inverse M_j of N_j modulo n_j. Therefore,

$$M_j N_j \equiv 1 \,(\text{mod } n_j).$$

Hence for any $m \leq k$,

$$x \equiv \sum_{j=1}^{k} r_j M_j N_j \equiv r_m \,(\text{mod } n_m),$$

which means that x is a solution of the system of linear congruences modulo n. Furthermore, if x_1 and x_2 are solutions of this system, then for each $j \leq k$, $x_1 \equiv x_2 \equiv r_j \,(\text{mod } n_j)$. It follows that $x_1 \equiv x_2 \,(\text{mod } n)$. Therefore, the simultaneous solution is unique modulo n. $\square$

Example 2.28 In the example given by Sun Tsŭ, in the discussion preceding Theorem 2.27, we set $n = n_1 n_2 n_3 = 105$ with $n_1 = 3$, $n_2 = 5$, and $n_3 = 7$. Also, let $r_1 = 2$, $r_2 = 3$, and $r_3 = 2$. Then the least multiplicative inverse of $N_1 = n/n_1 = 35$ modulo $n_1 = 3$ is $M_1 = 2$. The least multiplicative inverse of $N_2 = n/n_2 = 21$ modulo $n_2 = 5$ is $M_2 = 1$, and the least multiplicative inverse of $N_3 = n/n_3 = 15$ modulo $n_3 = 7$ is $M_3 = 1$. Hence,

$$x = \sum_{j=1}^{3} r_j M_j N_j = 2 \cdot 2 \cdot 35 + 3 \cdot 1 \cdot 21 + 2 \cdot 1 \cdot 15 = 233,$$

as calculated by Sun Tsŭ. By reducing $x = 233$ modulo $n = 105$, we get $x_0 = 23$, the unique solution modulo n.

One may wonder about the situation where the moduli are *not* relatively prime. In 717 A.D. a priest named Yih-hing generalized Theorem 2.27 in his book *t'ai-yen-lei-schu* as follows.

Theorem 2.29 (Generalized Chinese Remainder Theorem)

Let $n_j \in \mathbb{N}$, set $\ell = \mathrm{lcm}(n_1, n_2, \ldots, n_k)$, and let $r_j \in \mathbb{Z}$ be any integers for $j = 1, 2, \ldots, k$. Then the system of k simultaneous linear congruences given by:

$$x \equiv r_1 \pmod{n_1},$$

$$x \equiv r_2 \pmod{n_2},$$

$$\vdots$$

$$x \equiv r_k \pmod{n_k},$$

has a solution if and only if $\gcd(n_i, n_j) \mid (r_i - r_j)$ for each natural number $i, j \leq k$. Moreover, if a solution exists, then it is unique modulo ℓ. Additionally, if there exist integer divisors $m_j \geq 1$ of r_j with $\ell = m_1 \cdot m_2 \cdots m_k$ such that the m_j are pairwise relatively prime, and there exist integers $s_j \equiv 0 \pmod{\ell/m_j}$ and $s_j \equiv 1 \pmod{m_j}$ for $1 \leq j \leq k$, then

$$x = \sum_{j=1}^{k} s_j r_j$$

is a solution of the above congruence system.

Proof. In view of Exercise 2.28 and induction, we need only prove the result for $k = 2$. If $x \equiv r_j \pmod{n_j}$, for $j = 1, 2$, then $x = r_j + u_j n_j$ $(j = 1, 2)$. Therefore, $r_1 - r_2 = u_2 n_2 - u_1 n_1$. Thus, if $g = \gcd(n_1, n_2)$, then

$$r_1 - r_2 = g(u_2 n_2/g - u_1 n_1/g).$$

We have shown that if a solution exists, then $g \mid (r_1 - r_2)$. Conversely if $g \mid (r_1 - r_2)$, then there is an integer z such that $r_1 = r_2 + gz$. Also, by Theorem A.22 on page 280, there are $a, b \in \mathbb{Z}$ such that $g = an_1 + bn_2$. Thus,

$$r_1 = r_2 + z(an_1 + bn_2) = r_2 + zan_1 + zbn_2.$$

If we set

$$x = r_1 - zan_1 = r_2 + zbn_2,$$

then

$$r_j \equiv x \pmod{n_j} \text{ for } j = 1, 2.$$

This establishes the necessary and sufficient condition for existence. We now establish uniqueness.

Suppose that $x \equiv r_j \pmod{n_j}$ and $y \equiv r_j \pmod{n_j}$ for $1 \leq j \leq k$. Then $x - y \equiv 0 \pmod{n_j}$ for each such j. This means that $\ell \mid (x - y)$. Hence, any solution x is unique modulo ℓ.

The last statement of the theorem is clear since if such m_j and s_j exist, then $x = \sum_{j=1}^{k} s_j r_j \equiv r_j \pmod{m_j}$ for $1 \leq j \leq k$ has a unique solution modulo ℓ by Theorem 2.27, so by the above the proof is completed. $\square$

Yin-hing designed Theorem 2.29 to solve the following problem.

◆ **The Units of Work Problem**
Determine the number of completed units of work, when the same number x of units to be performed by each of four sets of $2, 3, 6$, and 12 workers performing their duties for certain numbers of whole days such that there remain $1, 2, 5$, and 5 units of work not completed by the respective sets. We assume further that no set of workers is lazy, namely each completes a nonzero number of units of work.

Here we are looking to solve

$$x \equiv 1 \,(\text{mod } 2),\ x \equiv 2 \,(\text{mod } 3),\ x \equiv 5 \,(\text{mod } 6),\ \text{and } x \equiv 5 \,(\text{mod } 12).$$

Since $\ell = \text{lcm}(2, 3, 6, 12) = 12$, then we let

$$m_1 = m_2 = 1,\ m_3 = 3,\ \text{and } m_4 = 4.$$

Thus, $s_1 = s_2 = 0$ since $m_1 = m_2 = 1$. Also, $s_3 = 4$ since $s_3 \equiv 0 \,(\text{mod } 4)$ and $s_3 \equiv 1 \,(\text{mod } 3)$; and $s_4 = 9$, since $s_4 \equiv 0 \,(\text{mod } 3)$ and $s_4 \equiv 1 \,(\text{mod } 4)$. Since $(r_1, r_2, r_3, r_4) = (1, 2, 5, 5)$, then

$$x = \sum_{j=1}^{4} r_j s_j = 5 \cdot 4 + 5 \cdot 9 = 65 \equiv 17 \,(\text{mod } 12).$$

Note that we cannot choose $x = 5$ since this would mean that no units of work had been completed by the last two sets of workers. For $x = 17$, the completed units of work must be $8 \cdot 2 = 16$ for the first set since they do not complete one unit; $5 \cdot 3 = 15$ for the second set since they do not complete two units; $2 \cdot 6 = 12$ for the third set since they do not complete five units; and $1 \cdot 12 = 12$ for the fourth set for the same reason. Hence, the total completed units of work is 55, and Yin-hing's problem is solved.

Another classic illustration of Theorem 2.29 is the following, which is due to the Hindu mathematician Brahmagupta.[2.7]

◆ **The Egg Basket Problem**
Suppose that a basket has n eggs in it. If the eggs are taken from the basket $2, 3, 4, 5$, and 6 at a time, there remain $1, 2, 3, 4$, and 5 eggs in the basket, respectively. If the eggs are removed from the basket 7 at a time, then no eggs remain in the basket. What is the smallest value of n such that the above could occur?

Essentially, this problem asks for a value of x such that $x \equiv j = r_j \,(\text{mod } j+1)$ for $j = 1, 2, 3, 4, 5$ and $x \equiv 0 \,(\text{mod } 7)$. Since $\ell = \text{lcm}(2, 3, 4, 5, 6, 7) = 420$, then we may choose

$$m_1 = 1,\ m_2 = 3,\ m_3 = 4,\ m_4 = 5,\ m_5 = 1,\ \text{and } m_6 = 7.$$

[2.7]Brahmagupta was considered to be the greatest of the Hindu mathematicians. In 628 he wrote his masterpiece on astronomy *Brahma-sphuta-siddhanta* or *The revised system of Brahma*, which had two chapters devoted to mathematics. He is also credited with first studying the equation $x^2 - py^2 = 1$ for a prime p. The Arab mathematician al-Khowarizmi based some of his work on the Arabic translation of Brahmagupta's work (see page 20).

Thus, $s_1 = s_5 = 0$, since $m_1 = m_5 = 1$. Also,

$$s_2 = 280 \text{ since } s_2 \equiv 0 \,(\text{mod } 140) \text{ and } s_2 \equiv 1 \,(\text{mod } 3).$$

Similarly, we calculate that

$$s_3 = 105 \text{ since } s_3 \equiv 0 \,(\text{mod } 105) \text{ and } s_3 \equiv 1 \,(\text{mod } 4).$$

and

$$s_4 = 336 \text{ since } s_4 \equiv 0 \,(\text{mod } 84) \text{ and } s_4 \equiv 1 \,(\text{mod } 5).$$

We need not calculate s_6 since $r_6 = 0$ given that $x \equiv 0 \,(\text{mod } 7)$. Hence, by Theorem 2.29,

$$x_0 = \sum_{j=1}^{6} r_j s_j = 2 \cdot 280 + 3 \cdot 105 + 4 \cdot 336 = 2219.$$

To get the smallest value modulo ℓ, we reduce 2219 modulo 420 to get

$$2219 - 420\lfloor 2219/420 \rfloor = 2219 - 420 \cdot 5 = 119,$$

which is the solution to Brahmagupta's Problem.

In the above, the linear congruences all have coefficient 1 for x. However, by using modular multiplicative inverses, we can solve more general systems of linear congruences.

Example 2.30 Suppose that we wish to solve the system of linear congruences

$$2x \equiv 1 \,(\text{mod } 3), \ 3x \equiv 1 \,(\text{mod } 5), \text{ and } 3x \equiv 2 \,(\text{mod } 7).$$

Since $2^{-1} \equiv 2 \,(\text{mod } 3)$, $3^{-1} \equiv 2 \,(\text{mod } 5)$, and $3^{-1} \equiv 5 \,(\text{mod } 7)$, then the system of congruences becomes $x \equiv 2 \,(\text{mod } 3)$, $x \equiv 2 \,(\text{mod } 5)$, and $x \equiv 3 \,(\text{mod } 7)$, for which $x = 17$ is clearly seen to be the least nonnegative solution modulo 105.

The reader may now go to Exercises 2.29–2.32 to test understanding of the solutions of systems of linear congruences.

The next aspect of modular arithmetic that we will need later in the text is called *modular exponentiation*. For $b, r \in \mathbb{N}$, this involves the finding of a least nonnegative residue of b^r modulo a given $n \in \mathbb{N}$, especially when the given natural numbers r and n are large. There is an algorithm for doing this that is far more efficient than repeated multiplication of b by itself.

◆ **The Repeated Squaring Method for Modular Exponentiation**

Suppose that we want to calculate the rth power of $b \in \mathbb{N}$ for a given $r \in \mathbb{N}$, by repeated squaring. Begin by setting $c = 1$ if $a_0 = 0$, $c = b$ if $a_0 = 1$, and letting $b_0 = b$, where $r = \sum_{j=0}^{k} a_j 2^j$ is the binary representation of r (see Section

Three of Chapter One for an algorithm to compute binary representations). For each natural number $j \le k$, perform the following step, called the jth step.

Calculate the least nonnegative residue b_j of b_{j-1}^2 modulo n. If $a_j = 1$, then replace c by $c \cdot b_j$, and reduce modulo n. If $a_j = 0$ leave c unchanged. What is achieved at the jth step is the computation of $b_j \equiv b^{r_j} \pmod{n}$, where b_j is the least nonnegative residue of b^{r_j} modulo n, and $r_j = \sum_{i=0}^{j} a_i 2^i$. Hence, at the kth step, we have calculated $c \equiv b^r \pmod{n}$.

The above algorithm will not only be valuable later where we look at the elliptic curve factoring method, but also in Chapter Three when we talk about *Exponentiation Ciphers*. The reader may conclude this section by looking at Exercise 2.34, which is a practical application of the repeated squaring method for modular exponentiation.

Exercises

2.1. Let $n \in \mathbb{N}$. Prove that each of the following holds.

   (a) For each $a \in \mathbb{Z}$, $a \equiv a \pmod{n}$, called the *reflexive property*.

   (b) For any $a, b \in \mathbb{Z}$, if $a \equiv b \pmod{n}$, then $b \equiv a \pmod{n}$, called the *symmetric property*.

   (c) For any $a, b, c \in \mathbb{Z}$, if $a \equiv b \pmod{n}$, and $b \equiv c \pmod{n}$, then we have $a \equiv c \pmod{n}$, called the *transitive property*.[2.8]

2.2. Prove that if $a \equiv b \pmod{n}$ and $c \equiv d \pmod{n}$, then $a \pm c \equiv b \pm d \pmod{n}$.

2.3. Let $a, b \in \mathbb{Z}$, $n \in \mathbb{N}$, and $a \equiv b \pmod{n}$. Prove that $am \equiv bm \pmod{mn}$.

2.4. Prove that if $a \equiv b \pmod{n}$ and $c \equiv d \pmod{n}$, then $ac \equiv bd \pmod{n}$. Then deduce that if $a \equiv b \pmod{n}$, then $a^m \equiv b^m \pmod{n}$ for any $m \in \mathbb{N}$.

2.5. Let $a, b \in \mathbb{Z}$, $m, n \in \mathbb{N}$ with m dividing n. Prove that if $a \equiv b \pmod{n}$, then $a \equiv b \pmod{m}$.

2.6. Prove that if $a, b, c \in \mathbb{Z}$, $n \in \mathbb{N}$ and $\gcd(c, n) = g$, then $ac \equiv bc \pmod{n}$ if and only if $a \equiv b \pmod{n/g}$.

2.7. Let $n \in \mathbb{N}$ such that $n \equiv 3 \pmod{4}$. Prove that $x^2 \equiv -1 \pmod{n}$ is not solvable for any $x \in \mathbb{Z}$.

2.8. Let $a \in \mathbb{Z}$, $n \in \mathbb{N}$, and $p > 2$ be a prime. Prove that a is its own inverse modulo p^n if and only if $a \equiv \pm 1 \pmod{p^n}$.

[2.8]This shows that congruence modulo n is an *equivalence relation*, which is defined to be a set R of ordered pairs on $S \times S$ for a given set S satisfying the reflexive, symmetric, and transitive properties. Moreover, the set $\{x : (x, a) \in R\}$ is called the *equivalence class* containing a. In the case of congruences, this latter notion coincides with that of a *congruence class*.

2.9. Let $a \in \mathbb{Z}$ and $n \in \mathbb{N}$. Exactly one of the following holds. Prove it and provide counterexamples for the other two.

(a) If $a \equiv \pm 1 \pmod{p}$ for all primes p dividing n, then $a^2 \equiv 1 \pmod{n}$.

(b) If $a^2 \equiv 1 \pmod{n}$, then $a \equiv \pm 1 \pmod{p}$ for all primes p dividing n.

(c) The congruence $a \equiv \pm 1 \pmod{p}$ for all primes p dividing n holds if and only if $a^2 \equiv 1 \pmod{n}$.

2.10. Let $a, b \in \mathbb{Z}$, $n \in \mathbb{N}$. Prove that $ax \equiv b \pmod{n}$ has a solution if and only if $\gcd(a, n) \mid b$. (*Hint: Use Theorem A.22.*)

2.11. Let p be an odd prime. Establish the binomial coefficient congruence,

$$\binom{p}{j} \equiv 0 \pmod{p}$$

for all natural numbers $j \leq p - 1$.

2.12. With reference to Exercise 2.10, let $g = \gcd(a, n)$, and suppose that $x = x_0$ is a solution of $ax \equiv b \pmod{n}$. Prove that a given $y \in \mathbb{Z}$ is a solution of

$$ay \equiv b \pmod{n}$$

if and only if

$$x_0 \equiv y \pmod{n/g}.$$

2.13. Assume that there is a solution to $ax \equiv b \pmod{n}$ (see Exercise 2.12). Show that exactly one solution x_0 of the congruence is in the least residue system modulo n/g, and that x_0 is the unique solution modulo n/g of that congruence. (*Exercises 2.10, 2.12–2.13 provide necessary and sufficient conditions for the existence of solutions to congruences $ax \equiv b \pmod{n}$, called* linear congruences. *The next exercise will test the reader's practical understanding of this concept.*)

2.14. Find all incongruent solutions modulo the given modulus in each of the following.

(a) $3x \equiv 1 \pmod{7}$.　　　(b) $3x \equiv 9 \pmod{21}$.

(c) $11x \equiv 5 \pmod{13}$.　　(d) $-x \equiv 7 \pmod{25}$.

(e) $5x \equiv 27 \pmod{101}$.　　(f) $101x \equiv 3 \pmod{217}$.

2.15. Prove that

$$\sum_{j=1}^{p-1} j^{p-1} \equiv -1 \pmod{p}$$

for any prime p. (*Is it true that* $\sum_{j=1}^{n-1} j^{n-1} \equiv -1 \pmod{n}$ *for a given* $n \in \mathbb{N}$ *implies that* n *is prime? This is open. However, it has been verified up to* 10^{1700}. *See* [94, p. 37]).

2.16. Prove that if p is prime and $a \in \mathbb{N}$, then $\phi(p^a) = p^{a-1}(p-1)$. (See Definition 2.19.)

2.17. A prime p is called a Wilson prime if $(p-1)! \equiv -1 \,(\mathrm{mod}\; p^2)$. Find all Wilson primes less than 564. (*Other than the ones that the reader will find in the solution to this exercise, there are no other known ones less than* $5 \cdot 10^8$. *See* [55]. *It is conjectured that there are infinitely many Wilson primes. A famous quote on the infinitude of Wilson primes is given by Vandiver who is purported to have said:* "This question seems to be of such a character that if I should come to life after my death and some mathematician were to tell me that it had definitely been settled, I think that I would immediately drop dead again." [2.9] *Other special types of primes for which few are known are those of the form* $n^n + 1$, *for which only* $n = 1, 2, 4$ *are known to be prime. There are also the four Fermat primes given on page 24, as well as the repunit primes* $R_{n,10}$, *defined in Exercise 1.38 on page 43, for which only* $n = 2, 19, 23, 317, 1031$ *are known to be prime. See* [144, pp. 373–374].)

☆ 2.18. Let $b \in \mathbb{N}$ and let m be the product of all natural numbers less than b and relatively prime to b. (For instance if $b = p$ is prime, then $m = (p-1)! = (b-1)!$.) Prove that if b is of one of the forms: 4, p^t, or $2p^t$ where $t \in \mathbb{N}$ and $p > 2$ is prime, then $m \equiv -1 \,(\mathrm{mod}\; b)$, whereas if b is not one of these three forms, then $m \equiv 1 \,(\mathrm{mod}\; b)$.[2.10] (*Hint: Use Euler's Totient.*)

If $b \in \mathbb{N}$ is composite and $m \equiv \pm 1 \,(\mathrm{mod}\; b^2)$, then b is called a *Wilson composite*. The only Wilson composite less than $5 \cdot 10^4$ is 5971. Find a Wilson composite bigger than $5 \cdot 10^5$.

2.19. Prove that if $r \mid \mathfrak{F}_n$, then $r = 2^{n+1}m + 1$ for some $m \in \mathbb{N}$. (*Hint: Use Fermat's Little Theorem and the Division Algorithm.*)

2.20. Suppose that $p \equiv 3 \,(\mathrm{mod}\; 4)$ is prime. Prove that

$$\left(\frac{p-1}{2}\right)! \equiv \pm 1 \,(\mathrm{mod}\; p).$$

(*Hint: Use Wilson's Theorem and Exercise 2.8.*)[2.11]

[2.9]Harry Schultz Vandiver (1882–1973) was born in Philadelphia, Pennsylvania on October 21, 1882. His mathematical collaborations included work with G.D. Birkhoff (1884–1944) and with L.E. Dickson (1874–1954). In 1924, he took a position at the University of Texas at Austin. In 1931, he won the Cole prize for his papers on Fermat's Last Theorem, and it is for his work on FLT that he is perhaps best known. He verified FLT for exponents less than two thousand on a computer in 1952. One of his idiosyncrasies was that he lived with his wife in the Alamo Hotel in Austin rather than ever choosing to own a house. He died in Austin on January 9, 1973.

[2.10]Gauss stated this generalization of Wilson's Theorem in [80, Article 78, p. 51], and he suggested a method of proof there.

[2.11]The reader with some background in algebraic number theory may be interested in the fact (proved by Mordell — see Footnote 6.8 on page 231) that if $p \equiv 3 \,(\mathrm{mod}\; 4)$ is prime and $h(-p)$ is the class number of the quadratic field $\mathbb{Q}(\sqrt{-p})$, then $[(p-1)/2]! \equiv (-1)^a \,(\mathrm{mod}\; p)$ where $a \equiv (1 + h(-p))/2 \,(\mathrm{mod}\; 2)$.

2.21. Suppose that $m \in \mathbb{Z}$, $n \in \mathbb{N}$ and $\gcd(m, n) = 1$. Prove that $m^{\phi(n)-1}$ is a multiplicative inverse of m modulo n.

2.22. Prove that $\phi(mn) = \phi(m)\phi(n)$ for any relatively prime $m, n \in \mathbb{N}$.

2.23. Use Exercise 2.22 to prove the following. If $m, n \in \mathbb{N}$ with $g = \gcd(m, n)$, then
$$\phi(mn) = g\phi(m)\phi(n)/\phi(g).$$

☆ 2.24. Prove that for all $n > 1$, and $a, b \in \mathbb{N}$,
$$\gcd(n^a - 1, n^b - 1) = n^{\gcd(a,b)} - 1.$$

2.25. Prove that if $d \mid n \in \mathbb{N}$, then $\phi(d) \mid \phi(n)$

2.26. Prove that if $b^n + 1$ is an odd prime for some $b, n \in \mathbb{N}$, then b is even and $n = 2^k$ for some nonnegative $k \in \mathbb{Z}$.

2.27. Solve for minimum $n \in \mathbb{N}$ in the coconut problem on page 62 for the case of five sailors who subdivide into five piles, each time giving the monkey one coconut.

2.28. Suppose that $a, b, n_i \in \mathbb{N}$ for $i = 1, 2, \ldots, k$, and $\ell = \operatorname{lcm}(n_1, n_2, \ldots, n_k)$. Prove that $a \equiv b \,(\operatorname{mod} n_i)$ for each i if and only if $a \equiv b \,(\operatorname{mod} \ell)$.

In Exercises 2.29–2.32, find the unique solution for each modulus in the given systems of linear congruences.

2.29. $x \equiv 2 \,(\operatorname{mod} 3)$, $x \equiv 3 \,(\operatorname{mod} 5)$, $x \equiv 1 \,(\operatorname{mod} 11)$.

2.30. $x \equiv 1 \,(\operatorname{mod} 7)$, $x \equiv 2 \,(\operatorname{mod} 13)$, $x \equiv 5 \,(\operatorname{mod} 17)$.

2.31. $x \equiv 1 \,(\operatorname{mod} 3)$, $x \equiv 4 \,(\operatorname{mod} 6)$, $x \equiv 1 \,(\operatorname{mod} 10)$.

2.32. $2x \equiv 1 \,(\operatorname{mod} 3)$, $4x \equiv 5 \,(\operatorname{mod} 9)$, and $5x \equiv 1 \,(\operatorname{mod} 7)$.

2.33. Calculate the number of bit operations required to execute the algorithm for modular exponentiation given on page 71.

2.34. Use the repeated squaring method given on page 71 to find the least nonnegative residue of 3^{61} modulo 101.

2.35. For given pairwise relatively prime natural numbers $n_1, n_2, \ldots, n_\ell$, prove that
$$\mathbb{Z}/n\mathbb{Z} \cong \mathbb{Z}/n_1\mathbb{Z} \oplus \cdots \oplus \mathbb{Z}/n_\ell\mathbb{Z},$$
where $n = n_1 \cdot n_2 \cdots n_\ell$. (See Appendix A for a discussion of the abstract algebra needed for this exercise. This exercise is a restatement of the Chinese Remainder Theorem 2.27, which may be used to prove this result.)

2.36. For given pairwise relatively prime natural numbers $n_1, n_2, \ldots, n_\ell$, with $n_j > 1$ for all natural numbers $j \leq \ell$, and integers $m_1, m_2, \ldots, m_\ell$, prove that we can find a solution to the congruences $x \equiv m_j \,(\operatorname{mod} n_j)$ for $1 \leq j \leq \ell$ using $O(\log_2^2 n)$ bit operations where $n = \prod_{j=1}^{\ell} n_j$.

2.2   Block Ciphers

In order to discuss even the most basic ciphers in depth, we need to use a
rigorous mathematical language in which to carry on this discussion. In Section
One of Chapter One, we defined certain basic notions such as the plaintext
and ciphertext. Both the plaintext and the ciphertext are written in terms of
elements from a finite set A, called an *alphabet of definition*. The alphabet of
definition may consist of numbers, letters from an alphabet such as the English,
Greek, or Russian alphabets, or symbols such as !, @, *, or any other symbols
that we choose to use when sending messages. The alphabet of definition for the
plaintext and ciphertext may differ, but the usual convention is to use the same
for both. For instance, as we have already seen in Section Three of Chapter One,
a commonly used one is $A = \{0, 1\}$, called the *binary alphabet* of definition,
in terms of which any given alphabet may be given binary equivalents.   An
example is the English alphabet, in which each letter may be assigned a unique
binary string of length five since there are $2^5 = 32$ binary strings of length
five. Another example, which we have also seen in Section Three of Chapter
One, is $A = \{0, 1, 2\}$, called the ternary alphabet, in which each letter of the
English alphabet may be replaced by a unique ternary string of length three
since there exist $3^3 = 27$ ternary strings of length three. Once we have agreed
upon an alphabet of definition, we choose a *message space*, M which is defined
to be a finite set consisting of strings of symbols from the alphabet of definition.
Elements of M , which may be anything from binary strings to English text, are
called *plaintext message units*. For instance, in Section One of Chapter One, we
saw examples of the use of digraphs, or pairs of letters. Of course any block of
$n \in \mathbb{N}$ letters may be used. A finite set C, consisting of strings of symbols from
an alphabet of definition for the ciphertext is called the *ciphertext space*, and
elements from C are called *ciphertext message units*. Most often, it is convenient
to let M be the message space consisting of *all possible* plaintext message units
and C be the set of *all possible* ciphertext message units. It is within this context
that we will choose to work below.

To make cryptanalysis more difficult, we need a set of parameters K, called
the *keyspace*, whose elements are called *keys*. For instance, in Section One of
Chapter One, we learned about the Caesar Cipher. With our newly acquired
knowledge of congruences, we may now restate the cipher as follows. Given the
alphabet of definition as the numbers 0 through 25, corresponding to the letters
A through Z, respectively, any $m \in M$ is enciphered as $c \in C$, where

$$c = m + 3 \in \mathbb{Z}/26\mathbb{Z}.$$

Thus, the (enciphering) key is $k = 3 \in K$, since we are using the parameter 3
as the shift from $m \in M$ to achieve $c \in C$. Also, the (deciphering) key is also
the parameter 3 since we achieve $m \in M$ from $c \in C$ by

$$c - 3 = m \in \mathbb{Z}/26\mathbb{Z}.$$

We formalize the above in the following.

Definition 2.31 (Enciphering and Deciphering Transformations)

An enciphering transformation (*also called an* enciphering function) *is a bijective function*

$$\mathfrak{E}_e : \mathcal{M} \mapsto \mathcal{C},$$

where the key $e \in \mathcal{K}$ *uniquely determines* $\mathfrak{E}_e$ *acting upon plaintext message units* $m \in \mathcal{M}$ *to get ciphertext message units*

$$\mathfrak{E}_e(m) = c \in \mathcal{C}.$$

*A deciphering transformation (*or *deciphering function) is a bijective function*

$$\mathfrak{D}_d : \mathcal{C} \mapsto \mathcal{M},$$

which is uniquely determined by a given key $d \in \mathcal{K}$, *acting upon ciphertext message units* $c \in \mathcal{C}$ *to get plaintext message units*

$$\mathfrak{D}_d(c) = m.^{2.12}$$

The application of $\mathfrak{E}_e$ *to* m, *namely the operation* $\mathfrak{E}_e(m)$, *is called* enciphering, encoding, *or* encrypting $m \in \mathcal{M}$, *whereas the application of* $\mathfrak{D}_d$ *to* c *is called* deciphering, decoding, *or* decrypting $c \in \mathcal{C}$.

In Definition 2.31 we have mathematically formalized the two notions of enciphering and deciphering, which we informally discussed in Chapter One, as a motivator for this formal setup. For instance, returning to the Caesar Cipher, it may be defined as that transformation $\mathfrak{E}_e$ uniquely determined by the key e, which is addition of 3 modulo 26. Thus,

$$\mathfrak{E}_e(m) = c \equiv m + 3 \,(\mathrm{mod}\ 26),$$

or simply $\mathfrak{E}_e(m) = c = m + 3 \in \mathcal{C} = \mathbb{Z}/26\mathbb{Z}$. Also, $m \in \mathcal{M} = \mathbb{Z}/26\mathbb{Z}$ is the numerical equivalent of the plaintext letter as described above. Similarly, $\mathfrak{D}_d(c)$ is that deciphering transformation uniquely defined by the key d, which is modular subtraction of 3 modulo 26. In other words,

$$\mathfrak{D}_d(c) = m \equiv c - 3 \,(\mathrm{mod}\ 26),$$

or simply $\mathfrak{D}_d(c) = m = c - 3 \in \mathbb{Z}/26\mathbb{Z}$, and $c \in \mathcal{C} = \mathbb{Z}/26\mathbb{Z}$ is the numerical equivalent of the ciphertext letter. Notice that

$$\mathfrak{D}_d(\mathfrak{E}_e(m)) = m$$

[2.12]To be absolutely precise, one only needs a *left* inverse for enciphering since $\mathfrak{D}_d(\mathfrak{E}_e(m)) = m$, and one only needs a *right* inverse for deciphering by the same reasoning. However, for our purposes, we will maintain the notion of a bijective function for both, since this is the only instance that we will encounter in the text. Moreover, we could simply define an enciphering transformation as a bijective function and a deciphering transformation as its inverse. However, we want to tie the notion of a *key* as a parameter that actually uniquely determines the function. This makes Definition 2.31 more cryptographically oriented than the simpler formal mathematical definition would indicate. For instance, see the discussion of the Caesar Cipher after the definition.

for each $m \in \mathcal{M}$. In other words, $\mathfrak{D}_d = \mathfrak{E}_e^{-1}$ is the inverse function of $\mathfrak{E}_e$. This is formalized as follows.

Definition 2.32 (Cryptosystems)

A cryptosystem *is comprised of a set* $\{\mathfrak{E}_e : e \in \mathcal{K}\}$ *consisting of enciphering transformations, and the corresponding set* $\{\mathfrak{E}_e^{-1} : e \in \mathcal{K}\} = \{\mathfrak{D}_d : d \in \mathcal{K}\}$ *of deciphering transformations. In other words, for each $e \in \mathcal{K}$, there exists a unique $d \in \mathcal{K}$ such that $\mathfrak{D}_d = \mathfrak{E}_e^{-1}$, so that $\mathfrak{D}_d(\mathfrak{E}_e(m)) = m$ for all $m \in \mathcal{M}$. The keys (e, d) are called a* key pair *where possibly $e = d$. A cryptosystem is often called a* cipher. *We reserve the term* Cipher Table *for the pairs of plaintext symbols and their ciphertext equivalents* $\{(m, \mathfrak{E}_e(m)) : m \in \mathcal{M}\}$.[2.13]

Definition 2.32 mathematically formalizes the notions of the terms *cipher* and *key*, which were informally discussed in Chapter One. The case where $e = d$ or where one of them may be "easily" determined from the other in the key pair has a special name, which is the simplest of the possibilities for cryptosystems, and so has the longest history.

Definition 2.33 (Symmetric-Key Ciphers)

A cryptosystem is called symmetric-key (*also called* single-key, one-key, *and* conventional) *if for each key pair (e, d), the key d is "computationally easy" to determine knowing only e and to similarly determine e knowing only d.* [2.14]

[2.13]The reader is cautioned that the *mathematical* use of the term *cipher* is not uniform throughout the literature and, as we have noted in Chapter One, is historically blurred with the use of the term "code" (see footnote 1.20). For instance, in [134, p. 12] and [173, p.4], the term cipher is used for what we have called a cryptosystem and in the former, they also call such a setup an "encryption scheme." Also, in [193], although the term is not explicitly defined in the text, it is clear from the usage that the meaning is equivalent to that used in Definition 2.32. In [12, p. 34], [64, pp. 13–14], and [100, pp. xiv–xv], the term cipher is used in agreement with what we call a Cipher Table, which is also what we informally called a cipher in Footnote 1.10. In *practice*, the term cipher and Cipher Table are often synonymously used. We will use the terms "cipher" and "cryptosystem" interchangeably to mean the notion given in Definition 2.32. Thus, the Cipher Table for the Caesar Cipher is Table 1.1, if we take the alphabet of definition to be the English alphabet. The Caesar Cipher is the system of enciphering and deciphering transformations discussed above. We have separated the terms "cipher" and "Cipher Table" to clarify the otherwise overlapping and possibly confusing notions.

[2.14]We will use the term "computationally easy problem" to mean one that can be solved in expected (see Footnote 2.30 on page 105) polynomial time and can be attacked using available resources. (The reason for adding the latter caveat is to preclude problems that are of polynomial time complexity but for which the degree is "large.") The antithesis of this would be a *computationally infeasible problem*, which means that, given the enormous amount of computer time that would be required to solve the problem, this task cannot be carried out in *realistic* computational time. Thus, "computationally infeasible" means that, although there (theoretically) exists a unique answer to our problem, we cannot find it even if we devoted every scintilla of the time and resources available. This is distinct from a problem that is unsolvable in any amount of time or resources. For example, an unsolvable problem would be to cryptanalyze XYZ assuming that it was enciphered using a monoalphabetic substitution. There is simply no unique verifiable answer without more information. However, it should be stressed here that there is no proved example of a computationally infeasible problem.

Usually $e = d$ with practical symmetric-key ciphers, thereby justifying the use of the term symmetric-key.[2.15]

There are two kinds of symmetric-key cryptosystems. The first is defined as follows.

Definition 2.34 (Block Ciphers)

A Block Cipher is a cryptosystem which separates the plaintext message into strings, called blocks, of fixed length $k \in \mathbb{N}$, called the blocklength and enciphers one block at a time.

We now look at some special cases of these Block Ciphers, starting with one familiar to us, an example of the simplest kind of encryption.

The Caesar Cipher is a special case of a *Shift Cipher*, defined as follows. Let $b, n \in \mathbb{N}$ and for each nonnegative $j < n$, define the enciphering transformation by:

$$\mathfrak{E}_e(m_j) = c_j \equiv m_j + b \,(\text{mod } n),$$

for $m_j \in \mathcal{M}$ and $c_j \in \mathcal{C}$, or simply $\mathfrak{E}_e(m_j) = c_j = m_j + b \in \mathbb{Z}/n\mathbb{Z} = \mathcal{C}$. The deciphering transformation is given by:

$$\mathfrak{D}_d(c_j) = m_j \equiv c_j - b \,(\text{mod } n),$$

or simply $\mathfrak{D}_d(c_j) = m_j = c_j - b \in \mathbb{Z}/n\mathbb{Z} = \mathcal{M}$. The Shift Cipher is a Symmetric-key Cipher with $d = -e$, since e is addition of b modulo n and $-e = d$ is subtraction of b modulo n, the additive inverse of e. This is an example of a Block Cipher where the blocklengths are $k = 1$. The Caesar Cipher is the special case obtained by taking $b = 3$ and $n = 26$. Also, for fans of the Stanley Kubrick film *2001: A Space Odyssey*, take $b = 1$ and $n = 26$, wherein the message HAL is enciphered as IBM. Shift Ciphers are relatively easy to cryptanalyze since there are only $|\mathcal{A}|$ keys to exhaustively search, where $\mathcal{A}$ is the alphabet of definition. For instance, with the Caesar Cipher, $|\mathcal{A}| = 26$.

In turn, the Shift Cipher is a special case of a more general cryptosystem, for which we now set the stage. Let $a, b, n \in \mathbb{N}$ and for $m \in \mathbb{Z}$ define

$$\mathfrak{E}_e(m) = am + b \in \mathbb{Z}/n\mathbb{Z},$$

where the key e is the ordered pair (a, b). Notice that for $a = 1$ we are back to the Shift Cipher where the key is b. Such a transformation is called an *Affine function*. In order to guarantee that the deciphering transformation exists, we need to know that the inverse of the affine function exists. By Exercise 2.37 on page 100, this means that $f^{-1}(c) \equiv a^{-1}(c - b) \,(\text{mod } n)$ must exist. By the preamble to Definition 2.10, this can only happen if $\gcd(a, n) = 1$. Also, by Definition 2.19, there are $\phi(n)$ natural numbers less than n and relatively prime

---

[2.15]In a symmetric-key cryptosystem e must be kept secret since d can be deduced from it. This is different from the *public-key* cryptosystems that we will study in Chapter Three.

to it. Hence, since b can be any of the choices of natural numbers less than n, we have shown that there are exactly $n\phi(n)$ possible Affine Ciphers, the product of the possible choices for a with the number for b, since this is the total number of possible keys. We have motivated the following.

Definition 2.35 (Affine Ciphers)
Let $\mathcal{M} = \mathcal{C} = \mathbb{Z}/n\mathbb{Z}$, $n \in \mathbb{N}$, $\mathcal{K} = \{(a,b) : a,b \in \mathbb{Z}/n\mathbb{Z} \text{ and } \gcd(a,n) = 1\}$, and for $e, d \in \mathcal{K}$, and $m, c \in \mathbb{Z}/n\mathbb{Z}$, set

$$\mathcal{E}_e(m) \equiv am + b \,(\mathrm{mod}\ n), \text{ and } \mathcal{D}_d(c) = a^{-1}(c - b)\,(\mathrm{mod}\ n).$$

Thus, as with the Shift Cipher of which the Affine Cipher is a generalization, $e = (a,b)$ since e is multiplication by a followed by addition of b modulo n, and $d = (a^{-1}, -b)$ is subtraction of b followed by multiplication with a^{-1}. In the case of the Shift Cipher, the inverse is additive and in the case of the Affine Cipher, the inverse is multiplicative. Of course, these coincide precisely when $a = 1$. In either case, knowing e or d allows us to easily determine the other, so they are symmetric-key cryptosystems. They are also Block Ciphers with the trivial blocklengths of $k = 1$.

Example 2.36 Let $n = 26$, and let $\mathcal{M} = \mathcal{C} = \mathbb{Z}/26\mathbb{Z}$. Define an Affine Cipher as follows.

$$\mathcal{E}_e(m) = 3m + 7 = c \in \mathbb{Z}/26\mathbb{Z},$$

and since $3^{-1} \equiv 9\,(\mathrm{mod}\ 26)$,

$$\mathcal{D}_d(c) = 9(c - 7) = 9c - 11 \in \mathbb{Z}/26\mathbb{Z} = \mathcal{M}.$$

Let the following table give the numerical equivalents in the alphabet of definition for each element in $\mathcal{M} = \mathcal{C}$.

Table 2.37

A	B	C	D	E	F	G	H	I	J	K	L	M
0	1	2	3	4	5	6	7	8	9	10	11	12
N	O	P	Q	R	S	T	U	V	W	X	Y	Z
13	14	15	16	17	18	19	20	21	22	23	24	25

Using the above, we wish to encipher the following message and provide a cryptogram:

SEND IN THE TROOPS

To do this, we first translate each letter into the numerical equivalent in the alphabet of definition, via Table 2.37 as follows.

18   4   13   3   8   13   19   7   4   19   17   14   14   15   18.

Then we apply $\mathcal{E}_e(m)$ to each of these numerical equivalents m to get the following.

9   19   20   16   5   20   12   2   19   12   6   23   23   0   9.

Finally, we use Table 2.37 to translate back into the English alphabet to get the cryptogram:

JTUQ FU MCT MGXXAJ

which we send.

Affine Ciphers are special cases of a more general class of cryptosystems. In order to define these, we first need to know about the following special type of mapping (see Definition A.6).

Definition 2.38 (Permutations)

If S is a finite set, then a permutation on S is a bijection $\sigma : S \mapsto S$.

Often one can visualize a permutation as a table with the elements of S listed in the first row and the permuted elements in the second row. For instance, the Cipher Table given by Table 1.1 is a simple permutation where the letters are shifted three units to the right. For an important group associated with the notion of a permutation, the reader is referred to Exercise 2.38 on page 100.

The notion in Definition 2.38 allows us to define a more general cryptosystem that encompasses Affine Ciphers among others.

Definition 2.39 (Substitution Ciphers)

Let A be an alphabet of definition consisting of n symbols, and let M be the set of all blocks of length r over A. The keyspace K will consist of all ordered r-tuples $e = (\sigma_1, \sigma_2, \ldots, \sigma_r)$ of permutations σ_j on A. For each $e \in K$, and $m = (m_1 m_2 \ldots m_r) \in M$ let

$$\mathfrak{E}_e(m) = (\sigma_1(m_1), \sigma_2(m_2), \ldots, \sigma_r(m_r)) = (c_1, c_2, \ldots, c_r) = c \in \mathcal{C},$$

and for $d = (d_1, d_2, \ldots, d_r) = (\sigma_1^{-1}, \sigma_2^{-1}, \ldots, \sigma_r^{-1}) = \sigma^{-1}$,

$$\mathfrak{D}_d(c) = (d_1(c_1), d_2(c_2), \ldots, d_r(c_r)) = (\sigma_1^{-1}(c_1), \sigma_2^{-1}(c_2), \ldots, \sigma_r^{-1}(c_r)) = m.$$

This type of cryptosystem is called a Substitution Cipher. *If all keys are the same, namely, $\sigma_1 = \sigma_2 = \cdots = \sigma_r$, then this cryptosystem is called a* Simple Substitution Cipher *or* Monoalphabebetic Substitution Cipher. [2.16] *If the keys differ, then it is called a* Polyalphabetic Substitution Cipher.

Simple Substitution Ciphers suffer from the inherent weakness that a so-called *frequency analysis* can be done on the ciphertext. In other words, a Block Cipher does not change the number of times that a letter appears in the plaintext. For instance, suppose that the letter E is the most frequently occurring letter in the English language, and we encipher the letter as F. Then the letter that occurs most often in the ciphertext will be F. Thus, by looking at a relatively small amount of ciphertext, a cryptanalyst can recover the key. Polyalphabetic Ciphers do not suffer from this weakness since frequencies of symbols are not preserved in the ciphertext. However, Polyalphabetic Ciphers are not that much more difficult to cryptanalyze since, once the block size r

---

[2.16]Sometimes the term *Simple Substitution Cipher* (for large r) is used synonymously with the term *Block Cipher*. For instance, see [134, p. 224] and [193, p. 20]. However, in the latter, a Substitution Cipher is considered to be a more elementary subset of Block Ciphers, since it is defined differently therein.

is determined, the ciphertext symbols can be separated into r groups and a frequency analysis can be performed on each group.

Thus far, our examples have been monoalphabetic. Let us now consider a Polyalphabetic Cipher to which we alluded in Chapter One.

Definition 2.40 (Vigenère Ciphers)

Fix $r, n \in \mathbb{N}$, *and let* $\mathcal{M} = \mathcal{C} = (\mathbb{Z}/n\mathbb{Z})^s$, *the elements of which are ordered* s-*tuples from* $\mathbb{Z}/n\mathbb{Z}$, *and* $\mathcal{K} = \mathbb{Z}^r$ *where* $s \geq r$. *For* $e = (e_1, e_2, \ldots, e_r) \in \mathcal{K}$, *and* $m = (m_1, m_2, \ldots, m_s) \in \mathcal{M}$, *let*

$$\mathfrak{E}_e(m) = (m_1 + e_1, m_2 + e_2, \ldots, m_r + e_r, m_{r+1} + e_1, \ldots, m_s + e_{s-r}),$$

and for $c = (c_1, c_2, \ldots, c_s) \in \mathcal{C}$, *let*

$$\mathfrak{D}_d(c) = (c_1 - e_1, c_2 - e_2, \ldots, c_r - e_r, c_{r+1} - e_1, \ldots, c_s - e_{s-r}),$$

where $+$ *is addition modulo* n. *This cryptosystem is called the* Vigenère Cipher *with period* s. *If* $r = s$, *then this cipher is often called a* Running-key Cipher.

The Vigenère Cipher is symmetric-key, given that knowing e is tantamount to knowing d. It is a Block Cipher with blocklength r, and it is polyalphabetic if we ensure that not all the keys e_j for $j = 1, 2, \ldots, r$ are the same.

Example 2.41 Let $\mathcal{A} = \mathbb{Z}/n\mathbb{Z} = \mathbb{Z}/26\mathbb{Z}$ and as numerical equivalents for our plaintext, we use Table 2.37. Let $r = 11 = s$ and set our key to be

STAMPEDBANK,

the numerical equivalent for which is given by,

$$e = (18, 19, 0, 12, 15, 4, 3, 1, 0, 13, 10) = (e_1, e_2, e_3, e_4, e_5, e_6, e_7, e_8, e_9, e_{10}, e_{11}).$$

Suppose that we wish to encipher the text

BE PREPARED
TOMORROW THE BANKS ALL FAIL

using the Vigenère Cipher. First, we convert to the numerical equivalents:

$$1, 4, 15, 17, 4, 15, 0, 17, 4, 3, 19, 14, 12, 14, 17, 17,$$

$$14, 22, 19, 7, 4, 1, 0, 13, 10, 18, 0, 11, 11, 5, 0, 8, 11$$

Then we apply the Vigenère Cipher:

$$e(1, 4, 15, 17, 4, 15, 0, 17, 4, 3, 19) = (19, 23, 15, 3, 19, 19, 3, 18, 4, 16, 3),$$

$$e(14, 12, 14, 17, 17, 14, 22, 19, 7, 4, 1) = (6, 5, 14, 3, 6, 18, 25, 20, 7, 17, 11),$$

and

$$e(0, 13, 10, 18, 0, 11, 11, 5, 0, 8, 11) = (18, 6, 10, 4, 15, 15, 14, 6, 0, 21, 21).$$

Translating the ciphertext into alphabetic equivalents via Table 2.37, we get

TXPDTTDSEQD
GFODGSZUHRL
SGKEPPOGAVV

This is sent as the cryptogram. The reader may now employ the inverse function $\mathfrak{D}_d(c)$ given by:

$$(c_1 - 18, c_2 - 19, c_3, c_4 - 12, c_5 - 15, c_6 - 4, c_7 - 3, c_8 - 1, c_9, , c_{10} - 13, c_{11} - 10),$$

to decipher.

The reader may now wish to have a look at Exercise 2.41 on page 100, where we define a variant of the Vigenère Cipher.

In Definition 2.39, we see that a Simple Substitution Cipher encrypts single plaintext symbols as single ciphertext symbols. When groups of one *or more* symbols are replaced by other groups of ciphertext symbols, then we call this cryptosystem a *Polygram Substitution Cipher*. For instance, in Chapter One we discussed *Digraph Ciphers*, where two letters are used as a single symbol. In particular, we explored the *Playfair Cipher* in Exercise 1.5 on page 15. These are examples of Polygram Substitution Ciphers. We now look at another classical Polygram Substitution Cipher that requires knowledge of some elementary matrix theory. The reader unfamiliar with such concepts, or requiring a brief review of the ideas and notation should consult pages 286–289 in Appendix A.

Definition 2.42 (The Hill Cipher)
Fix $r, n \in \mathbb{N}$, *let* $\mathcal{K} = \{e \in \mathcal{M}_{r \times r}(\mathbb{Z}/n\mathbb{Z}) : e \text{ is invertible}\}$, *and set* $\mathcal{M} = \mathcal{C} = (\mathbb{Z}/n\mathbb{Z})^r$. *Then for* $m \in \mathcal{M}$, $e \in \mathcal{K}$,

$$\mathfrak{C}_e(m) = me,$$

and

$$\mathfrak{D}_d(c) = ce^{-1},$$

where $c \in \mathcal{C}$. *(Note that* e *is invertible if and only if* $\gcd(\det(e), n) = 1$. *See Theorem A.40.) This cryptosystem is known as the* Hill Cipher, *created by L.S. Hill.*[2.17]

[2.17]Lester S. Hill devised this cryptosystem in 1929. His only published papers in the area of cryptography appeared in 1929 and 1931. Thereafter, he kept working on cryptographic ideas, but turned all of his work over to the Navy in which he had served as a lieutenant in World War I. He taught mathematics at Hunter College in New York from 1927 until his retirement in 1960. He died in Lawrence Hospital in Bronxville, New York after suffering through a lengthy illness. Hill's rigorous mathematical approach may be said to be one of the factors which has helped foster today's solid grounding of cryptography in mathematics.

Example 2.43 Let $r = 2$ and $n = 26$ with alphabet of definition $\mathcal{A} = \mathbb{Z}/26\mathbb{Z}$, where Table 2.37 gives the numerical equivalents of plaintext letters. Thus, $\mathcal{M} = \mathcal{C} = (\mathbb{Z}/26\mathbb{Z})^2$, and $\mathcal{K}$ consists of all invertible 2×2 matrices with entries from $\mathbb{Z}/26\mathbb{Z}$, so if $e \in \mathcal{K}$, then $\gcd(\det(e), 26) = 1$. Let us take

$$e = \begin{pmatrix} 1 & 4 \\ 2 & 3 \end{pmatrix}$$

for which $\det(e) = -5$. Suppose that we want to encipher $MATRIX$. First we get the numerical equivalents from Table 2.37: 12, 0, 19, 17, 8, 23. Thus, we may set $m_1 = (12, 0)$, $m_2 = (19, 17)$, and $m_3 = (8, 23)$. Now use the enciphering transformation defined in the Hill Cipher.

$$\mathfrak{E}_e(m_1) = (12, 0) \begin{pmatrix} 1 & 4 \\ 2 & 3 \end{pmatrix} = (12, 22),$$

$$\mathfrak{E}_e(m_2) = (19, 17) \begin{pmatrix} 1 & 4 \\ 2 & 3 \end{pmatrix} = (1, 23),$$

and

$$\mathfrak{E}_e(m_3) = (8, 23) \begin{pmatrix} 1 & 4 \\ 2 & 3 \end{pmatrix} = (2, 23).$$

Now we use Table 2.37 to get the ciphertext letter equivalents and send $MWBXCX$ as the cryptogram.

Now we show how decryption works. Once the cryptogram is received, we must calculate the inverse of e which is

$$e^{-1} = \begin{pmatrix} 15 & 6 \\ 16 & 5 \end{pmatrix}.$$

(See Example A.41 in Appendix A.) Now apply the deciphering transformation to the numerical equivalents of the ciphertext as follows. Given $c_1 = (12, 22)$, $c_2 = (7, 19)$, and $c_3 = (2, 23)$, we have,

$$\mathfrak{D}_d(c_1) = \mathfrak{D}_{e^{-1}}(12, 22) = (12, 22) \begin{pmatrix} 15 & 6 \\ 16 & 5 \end{pmatrix} = (12, 0),$$

$$\mathfrak{D}_d(c_2) = \mathfrak{D}_{e^{-1}}(1, 23) = (1, 23) \begin{pmatrix} 15 & 6 \\ 16 & 5 \end{pmatrix} = (19, 17),$$

and

$$\mathfrak{D}_d(c_3) = \mathfrak{D}_{e^{-1}}(2, 23) = (2, 23) \begin{pmatrix} 15 & 6 \\ 16 & 5 \end{pmatrix} = (8, 23).$$

The letter equivalents now give us back the original plaintext message $MATRIX$.

Example 2.43 is an example of *bad concealment* since two of the letters M and X in the ciphertext are exactly the same as in the plaintext, namely they are unconcealed. Cryptographers must guard against such instances.

Block Ciphers are classically broken into two classes, the Substitution Ciphers of Definition 2.39, and the class of ciphers described as follows.

Definition 2.44 (Transposition/Permutation Ciphers)

A Simple Transposition Cipher, *also known as a* Simple Permutation Cipher, *is a symmetric-key block cryptosystem having blocklength* $r \in \mathbb{N}$, *with keyspace* $\mathcal{K}$ *being the set of permutations on* $\{1, 2, \ldots, r\}$. *The enciphering transformation is given, for each* $m = (m_1, m_2 \ldots, m_r) \in \mathcal{M}$, *by*

$$\mathfrak{E}_e(m) = (m_{e(1)}, m_{e(2)}, \ldots, m_{e(r)}),$$

and for each $c = (c_1, c_2, \ldots, c_r) \in \mathcal{C}$,

$$\mathfrak{D}_d(c) = \mathfrak{D}_{e^{-1}}(c) = (c_{d(1)}, c_{d(2)}, \ldots, c_{d(r)}).$$

The cryptosystems in Definition 2.44 have keyspace of cardinality $|\mathcal{K}| = r!$. Permutation encryption involves grouping plaintext into blocks of r symbols and applying to each block the permutation e on the numbers $1, 2, \ldots, r$. In other words, the places where the plaintext symbols sit are permuted. This gives mathematical rigour to the discussion of transpositions given on page 4 in Chapter One. Note that the enciphering key e implicitly defines r, since it is a permutation on r symbols.

Example 2.45 Let $r = 6$, $\mathcal{M} = \mathcal{C} = \mathbb{Z}/26\mathbb{Z}$, with the English letter equivalents given by Table 2.37. Then if $e = (1, 2, 3, 6, 5, 4)^{2.18}$ is applied to $CIPHER$, we get $IPRCHE$ since $m_1 = C$, $m_2 = I$, $m_3 = P$, $m_4 = H$, $m_5 = E$, and $m_6 = R$, so

$$e(m) = (m_{e(1)}, m_{e(2)}, m_{e(3)}, m_{e(4)}, m_{e(5)}, m_{e(6)}) = (m_2, m_3, m_6, m_1, m_4, m_5).$$

Since the inverse transformation is $d = e^{-1} = (4, 1, 2, 5, 6, 3)$, then another way to visualize encryption is to write $4, 1, 2, 5, 6, 3$ in the first row, and the plaintext letter equivalents in the second row, then read the letters off in *numerical order*. For instance,

$$\begin{pmatrix} 4 & 1 & 2 & 5 & 6 & 3 \\ C & I & P & H & E & R \end{pmatrix}.$$

Thus, the first in numerical order is I, the second is P and so on.

An easy means for finding the inverse of a given key e such as in Example 2.45 is given as follows. The key in that example can be written as

$$e = \begin{pmatrix} 1 & 2 & 3 & 4 & 5 & 6 \\ 2 & 3 & 6 & 1 & 4 & 5 \end{pmatrix},$$

since $1 \mapsto 2$, $2 \mapsto 3$ and so on. To find the inverse, just read off in numeric order (determined by the second row), the terms in the first row. For instance, the term in the first row sitting above the 1 is 4 so 4 is the first term in e^{-1},

---

[2.18] Note that this is the traditional notation for a permutation, which means, in this case, that permutation e sends: $1 \mapsto 2$, $2 \mapsto 3$, $3 \mapsto 6$, $4 \mapsto 1$, $5 \mapsto 4$, and $6 \mapsto 5$.

The term in the first row sitting above the 2 is 1, so 1 is the second term in e^{-1} and so on. The use in practice of Permutation Ciphers is easier than the mathematical description would indicate.

Permutation Ciphers are subject to cryptanalysis by frequency analysis since they preserve the frequency distribution of each character.

The best-known Symmetric-key Block Cipher, recognized world-wide, is the *Data Encryption Standard* (DES) which was the first commercially available algorithm (namely for use with unclassified computer data) put into use in the 1970's.[2.19] Succinctly, the DES Algorithm is a symmetric-key, block enciphering cryptosystem for octograms of bytes, with a key based upon a permutation, then sixteen substitutions,[2.20] followed by another permutation. In other words, the DES Cipher encrypts data in 64-bit blocks of plaintext to produce 64-bit blocks of ciphertext, using the same key for encryption and decryption, the key being based upon a combination of substitution and permutation techniques. Now we set out to describe the DES cryptosystem in exact mathematical detail.

The central part of the DES cryptosystem involves a certain function that we will discuss before we give details of the algorithm itself. However, before we can describe this function, we must establish the following core concept of the DES Algorithm. The following provides the DES cryptosystem with its security, a fact that we will discuss in detail later.

◆ S-Boxes[2.21]

An S-Box (or substitution box) is a given, fixed 4×16 matrix with entries from $\mathbb{Z}/16\mathbb{Z}$ satisfying the following properties. The rows are labelled $0 \leq i \leq 3$ and the columns are labelled $0 \leq j \leq 15$. The given S-box operates on input, which is a binary string $(b_1 b_2 b_3 b_4 b_5 b_6)_2$ of length six, and outputs the element from the matrix which is in the row determined by the binary integer $r = (b_1 b_6)_2$, and the column corresponding to the binary integer $c = (b_2 b_3 b_4 b_5)_2$. Thus, the output $S(b_1 b_2 b_3 b_4 b_5 b_6)$ is the bitstring of length four, which is the binary representation of the entry that sits in the (r, c) position of the given matrix.

Example 2.46 Consider the given array

Table 2.47

S

14	4	13	1	2	15	11	8	3	10	6	12	5	9	0	7
0	15	7	4	14	2	13	1	10	6	12	11	9	5	3	8
4	1	14	8	13	6	2	11	15	12	9	7	3	10	5	0
15	12	8	2	4	9	1	7	5	11	3	14	10	0	6	13

[2.19] In fact, a complete description of DES is given in the U.S. Federal Information Processing Standards Publication number 46 (FIPS-46), Springfield, Virginia, April 1977. The U.S. Federal Register is first dated March 17, 1975, then updated August 1, 1975.

[2.20] For further data on why sixteen substitutions are required, see the discussion of "confusion" and "diffusion" techniques on page 102 and the subsequent discussion of cryptanalysis of DES in the optional Section Three of this chapter.

[2.21] The S-boxes are applied in the DES function, which we will describe immediately after the description of how the S-boxes work. For now, the reader should set out to understand the basics of this concept so that we can build the various components of the DES Algorithm in small increments in order to paint the whole picture at the end of the process.

Suppose that the S-box has binary input

$$(b_1 b_2 b_3 b_4 b_5 b_6) = (101001).$$

Since

$$(b_1 b_6) = (11)$$

corresponds to the row labelled $r = 3 = (11)_2$ and since

$$(b_2 b_3 b_4 b_5) = (0100)$$

corresponds to column labelled $c = 4 = (0100)$, then

$$S((101001)) = (0100),$$

the binary representation of the entry 4 in position $(3, 4)$ of the array. Note that we are indexing from 0, so that input (000110) corresponds to row (00) (the first row, which is labelled row 0), and column (0011) which is column 3 (the fourth column, labelled column 3). Hence, the output is the binary representation of 1, the entry in position $(0, 3)$, namely

$$S((000110)) = (0001).$$

Similarly, if

$$(b_1 b_2 b_3 b_4 b_5 b_6) = (111011),$$

then this corresponds to the row labelled $3 = (11)$ (which is the fourth row), and the column labelled $13 = (1101)$ (which is the fourteenth column), so

$$S((111011)) = (0000),$$

the binary representation of the entry 0 in position $(3, 13)$ of the array.

◆ **The DES Function**

Let

$$f : (\mathbb{Z}/2\mathbb{Z})^{32} \times (\mathbb{Z}/2\mathbb{Z})^{48} \mapsto (\mathbb{Z}/2\mathbb{Z})^{32}$$

be a function where $f(x, y)$ is defined by the following steps.

(1) **(The expansion step)** First we expand x to a 48-bit binary integer via a given table $\mathbf{E}$, called the *expansion table*, or *expansion permutation*, which is an 8×6 array with entries from $\{1, 2, \ldots, 32\}$. We use $E(j)$ to denote the entries in the permuted matrix where $1 \leq j \leq 48$ are read sequentially by row from upper left to lower right of the array, as shown below by the explicitly given permutation. Thus, for a given $x \in (\mathbb{Z}/2\mathbb{Z})^{32}$,

$$x = (r_1 \ldots r_{32}) \mapsto (r_{E(1)} \ldots r_{E(48)}) = \mathbf{E}(x).$$

In this way, the expansion table ensures that the 32 bits are permuted in a given manner, with 16 of the bits appearing twice.

(2) (**Addition modulo 2**)[2.22] Then we compute

$$\mathbf{E}(x) + y \in (\mathbb{Z}/2\mathbb{Z})^{48},$$

and we write the result as a concatenation

$$B_1 B_2 B_3 B_4 B_5 B_6 B_7 B_8,$$

where $B_j \in (\mathbb{Z}/2\mathbb{Z})^6$. In other words, the output is read as a concatenation of eight bitstrings of length six.

(3) (**Application of S-boxes**) We then use 8 S-boxes S_j for $1 \leq j \leq 8$ to compute $C_j = S_j(B_j)$.

(4) (**Final permutation**) Lastly, the bitstring

$$C = C_1 C_2 C_3 C_4 C_5 C_6 C_7 C_8$$

of length 32 is permuted according to a given, fixed permutation $\mathbf{P}$, called a **P**-box, which is an 8×4 array of distinct entries from $\{1, 2, \ldots, 32\}$. Then $\mathbf{P}(C)$ is defined to be $f(x, y) \in (\mathbb{Z}/2\mathbb{Z})^{32}$, namely

$$f(x, y) = \mathbf{P}(C).$$

Example 2.48 Suppose that the expansion permutation described above is given by the following.

1	2	3	4	5	6
7	8	9	10	11	12
13	14	15	16	17	18
19	20	21	22	23	24
25	26	27	28	29	30
31	32	33	34	35	36
37	38	39	40	41	42
43	44	45	46	47	48

$\xrightarrow{\;\mathbf{E}\;}$

32	1	2	3	4	5
4	5	6	7	8	9
8	9	10	11	12	13
12	13	14	15	16	17
16	17	18	19	20	21
20	21	22	23	24	25
24	25	26	27	28	29
28	29	30	31	32	1

The array on the right is often called the **E**-box. Assume that the following values of $(x, y) \in (\mathbb{Z}/2\mathbb{Z})^{32} \times (\mathbb{Z}/2\mathbb{Z})^{48}$ are given.

$$x = (11101111010010100110010101000100) =$$

$$(r_1 r_2 \ldots r_{32}),$$

and

$$y = (011110011010111011011001110110111100100111100101).$$

[2.22]This is often called the XOR step, since addition modulo 2 is equivalent to the use of the *exclusive or* operation.

For step (1), we use the permutation $\mathbf{E}$ from above to expand x from a 32-bit integer to one of length 48. For convenience's sake, we list x as a 2×32 array and $E(x)$ as a 2×48 array, with the first row being the position of the bit and the second row being the bits in the binary integer, so that the reader may more easily see the means by which the initial x value has been *expanded*.

										x							
j	1	2	3	4	5	6	7	8	9	10	11	12	13	14	15	16	17
r_j	1	1	1	0	1	1	1	1	0	1	0	0	1	0	1	0	0

j	18	19	20	21	22	23	24	25	26	27	28	29	30	31	32
r_j	1	1	0	0	1	0	1	0	1	0	0	0	1	0	0

							$\mathbf{E}(x)$							
j	1	2	3	4	5	6	7	8	9	10	11	12	13	14
$r_{E(j)}$	0	1	1	1	0	1	0	1	1	1	1	0	1	0

j	15	16	17	18	19	20	21	22	23	24	25	26
$r_{E(j)}$	1	0	0	1	0	1	0	1	0	0	0	0

j	27	28	29	30	31	32	33	34	35	36	37
$r_{E(j)}$	1	1	0	0	0	0	1	0	1	0	1

j	38	39	40	41	42	43	44	45	46	47	48
$r_{E(j)}$	0	1	0	0	0	0	0	1	0	0	1

For instance, the bit in position 32 of the input x is $r_{32} = 0$, which gets moved to positions 1 and 47 of $\mathbf{E}(x)$ since

$$E(1) = 32 = E(47).$$

Also, since

$$1 = r_5 = r_{E(6)} = r_{E(8)},$$

then the bit in position 5 of x gets permuted to positions 6 and 8 of $\mathbf{E}(x)$. This completes step (1).

The modulo 2 addition of $\mathbf{E}(x)$ and y is given by

$$\mathbf{E}(x) + y =$$

$$(011101011110101001010100001100001010101000001001)+$$
$$(011110011010111011011001110110111100100111100101) =$$
$$(000011000100100100011011110101101100011111101100) =$$
$$(000011)(000100)(010010)(001101)(111010)(110110)(001111)(101100)$$

where the latter is a concatenation of eight bitstrings of length six. Hence,

$$B_1 B_2 B_3 B_4 B_5 B_6 B_7 B_8 =$$

$$(000011)(000100)(010010)(001101)(111010)(110110)(001111)(101100).$$

which is step (2).

For step (3), we use the following S-boxes. S_1 will be taken to be the S-box given in Table 2.47. The remaining seven are given as follows.

S_2															
15	1	8	14	6	11	3	4	9	7	2	13	12	0	5	10
3	13	4	7	15	2	8	14	12	0	1	10	6	9	11	5
0	14	7	11	10	4	13	1	5	8	12	6	9	3	2	15
13	8	10	1	3	15	4	2	11	6	7	12	0	5	14	9

S_3															
10	0	9	14	6	13	15	5	1	13	12	7	11	4	2	8
13	7	0	9	3	4	6	10	2	8	5	14	12	11	15	1
13	6	4	9	8	15	3	0	11	1	2	12	5	10	14	7
1	10	13	0	6	9	8	7	4	15	14	3	11	5	2	12

S_4															
7	13	14	3	0	6	9	10	1	2	8	5	11	12	4	15
13	8	11	5	6	15	0	3	4	7	2	12	1	10	14	9
10	6	9	0	12	11	7	13	15	1	3	14	5	2	8	4
3	15	0	6	10	1	13	8	9	4	5	11	12	7	2	14

S_5															
2	12	4	1	7	10	11	6	8	5	3	15	13	0	14	9
14	11	2	12	4	7	13	1	5	0	15	10	3	9	8	6
4	2	1	11	10	13	7	8	15	9	12	5	6	3	0	14
11	8	12	7	1	14	2	13	6	15	0	9	10	4	5	3

S_6															
12	1	10	15	9	2	6	8	0	13	3	4	14	7	5	11
10	15	4	2	7	12	9	5	6	1	13	14	0	11	3	8
9	14	15	5	2	8	12	3	7	0	4	10	1	13	11	6
4	3	2	12	9	5	15	10	11	14	1	7	6	0	8	13

S_7															
4	11	2	14	15	0	8	13	3	12	9	7	5	10	6	1
13	0	11	7	4	9	1	10	14	3	5	12	2	15	8	6
1	4	11	13	12	3	7	14	10	15	6	8	0	5	9	2
6	11	13	8	1	4	10	7	9	5	0	15	14	2	3	12

S_8															
13	2	8	4	6	15	11	1	10	9	3	14	5	0	12	7
1	15	13	8	10	3	7	4	12	5	6	11	0	14	9	2
7	11	4	1	9	12	14	2	0	6	10	13	15	3	5	8
2	1	14	7	4	10	8	13	15	12	9	0	3	5	6	11

Thus,

$$S_1(B_1) = S_1((000011)) = (1111) = C_1,$$

$$S_2(B_2) = S_2((000100)) = (1000) = C_2,$$

$$S_3(B_3) = S_3((010010)) = (1101) = C_3,$$

$$S_4(B_4) = S_4((001101)) = (0000) = C_4,$$

$$S_5(B_5) = S_5((111010)) = (0011) = C_5,$$

$$S_6(B_6) = S_6((110110)) = (1010) = C_6,$$

$$S_7(B_7) = S_7((001111)) = (1010) = C_7,$$

and

$$S_8(B_8) = S_8((101100)) = (1110) = C_8,$$

which completes step (3).

Lastly, to complete step (4), we use the following permutation $\mathbf{P}$ on

$$C = C_1 C_2 C_3 C_4 C_5 C_6 C_7 C_8.$$

1	2	3	4
5	6	7	8
9	10	11	12
13	14	15	16
17	18	19	20
21	22	23	24
25	26	27	28
29	30	31	32

$\xrightarrow{\mathbf{P}}$

16	7	20	21
29	12	28	17
1	15	23	26
5	18	31	10
2	8	24	14
32	27	3	9
19	13	30	6
22	11	4	25

Again, for clarity we present C and $\mathbf{P}(C)$ in 2×32 tabular form.

C																
j	1	2	3	4	5	6	7	8	9	10	11	12	13	14	15	16	17
r_j	1	1	1	1	1	0	0	0	1	1	0	1	0	0	0	0	0

j	18	19	20	21	22	23	24	25	26	27	28	29	30	31	32
r_j	0	1	1	1	0	1	0	1	0	1	0	1	1	1	0

								$P(C)$								
j	1	2	3	4	5	6	7	8	9	10	11	12	13	14	15	16	17
r_j	0	0	1	1	1	1	0	0	1	0	1	0	1	0	1	1	1

j	18	19	20	21	22	23	24	25	26	27	28	29	30	31	32
r_j	0	0	0	0	1	1	1	1	0	1	0	0	0	1	1

Hence, applying $\mathbf{P}$ to C, we get

$$\mathbf{P}(C) = f(x,y) = (00111100101010111000011110100011).$$

There is one last notion that we need to define before describing the DES mechanism, namely the means for computing keys.

◆ **The DES Key Schedule**
The keys for DES are computed as follows.

(1) We are given as input a 56-bit key[2.23]

$$k = (e_1 e_2 \ldots e_{56}).$$

Permute the bits according to a given, fixed permutation **PC1**, which is written as an 8×7 array with entries from the natural numbers less than 64, excluding $8j$ for $1 \le j \le 7$. The permuted key is denoted by

$$\mathbf{PC1}(k) = c_0 d_0,$$

which is a concatenation of the initial bitstring c_0 of length 28 with the final bitstring d_0 of length 28. **PC1** is often called the (first) *permuted choice*.

(2) For $1 \le j \le 16$, compute keys k_j as follows. Define c_j to be a shift left[2.24] of c_{j-1} by one position if $j = 1, 2, 9$ or 16, and a shift of two positions left otherwise. Similarly, define d_j to be such a shift of d_{j-1}. We denote the outcome as the concatenation $c_j d_j$ of two bitstrings of length 28.

(3) Apply a given, fixed permutation **PC2**, which is written as an 8×6 array of *certain* of the numbers

$$\{1, 2, \ldots, 56\},$$

to $c_j d_j$ for $1 \le j \le 16$, which chooses a *subset* of 48 of the 56 bits in $c_j d_j$, as well as permuting them. For each j, the result is written

$$k_j = \mathbf{PC2}(c_j d_j),$$

which is a bitstring of length 48, so we have sixteen such keys. **PC2** is often called the *compression permutation*, or the *permuted choice two*.

[2.23] Often a 64-bit key is input, where every eighth bit is a parity-check bit for error detection. However, even in this case, the parity check bits are discarded and the remaining 56-bit key is processed. One way to get the input key is to obtain an 8-character ASCII string.
[2.24] This is also called a *left circular shift* or a *cyclic shift to the left* since the initial bit gets shifted to the end.

Example 2.49 We will construct only the first key here in detail. Suppose that the permutation **PC1** is given by the following.

1	2	3	4	5	6	7
9	10	11	12	13	14	15
17	18	19	20	21	22	23
25	26	27	28	29	30	31
33	34	35	36	37	38	39
41	42	43	44	45	46	47
49	50	51	52	53	54	55
57	58	59	60	61	62	63

$$\downarrow \text{PC1}$$

57	49	41	33	25	17	9
1	58	50	42	34	26	18
10	2	59	51	43	35	27
19	11	3	60	52	44	36
63	55	47	39	31	23	15
7	62	54	46	38	30	22
14	6	61	53	45	37	29
21	13	5	28	20	12	4

Let the input key be $k = (e_1 e_2 \ldots e_7, e_9, \ldots, e_{62}, e_{63}) =$

$(11000111100001000101111101010101110001101010101111111100011110)$.

The following illustrates the effect of **PC1** on k, where we list the key and the output as a 2×56 array with the first row being the position for the sake of increased clarity.

								k								
j	1	2	3	4	5	6	7	9	10	11	12	13	14	15	17	18
e_j	1	1	0	0	0	1	1	1	1	0	0	0	0	1	0	0

j	19	20	21	22	23	25	26	27	28	29	30	31	33	34
e_j	0	1	0	1	1	1	1	1	0	1	0	1	0	1

j	35	36	37	38	39	41	42	43	44	45	46	47	49	50
e_j	1	1	0	0	0	1	1	0	1	0	1	0	1	1

j	51	52	53	54	55	57	58	59	60	61	62	63
e_j	1	1	1	1	0	0	0	1	1	1	1	0

$PC1(k)$															
j	1	2	3	4	5	6	7	9	10	11	12	13	14	15	17	18
$e_{PC1(j)}$	0	1	1	0	1	0	1	1	0	1	1	1	1	0	1	1

j	19	20	21	22	23	25	26	27	28	29	30	31	33	34
$e_{PC1(j)}$	1	1	0	1	1	0	0	0	1	1	1	1	0	0

j	35	36	37	38	39	41	42	43	44	45	46	47	49	50
$e_{PC1(j)}$	0	0	1	1	1	1	1	1	1	0	0	1	0	1

j	51	52	53	54	55	57	58	59	60	61	62	63
$e_{PC1(j)}$	1	1	0	0	1	0	0	0	0	1	0	0

For instance, $PC1(1) = 57$, so

$$e_{PC1(1)} = e_{57} = 0$$

under the action of the permutation $PC1$, and this is the first bit in the permuted key. Similarly, $PC1(20) = 51$, so

$$e_{PC1(20)} = e_{51} = 1$$

is the bit in position 20 of the permuted key. Hence,

$$PC1(k) =$$

$$(0110101101111011110110001111)(00001111111001011100010000100) = c_0 d_0.$$

We have completed step (1) in the DES key schedule. Note that the first four rows of $PC1$ acted upon c_0 and the last four rows acted upon d_0.

To complete step (2), for just one keyround, we get,

$$c_1 d_1 = (1101011010111101111010110001110)(000111111100101110010000100) =$$

$$(b_1 b_2 \ldots b_{48}).$$

Lastly, to complete step (3), we assume that $PC2$ is given as follows.

1	2	3	4	5	6
7	8	9	10	11	12
13	14	15	16	17	18
19	20	21	22	23	24
25	26	27	28	29	30
31	32	33	34	35	36
37	38	39	40	41	42
43	44	45	46	47	48

$\xrightarrow{PC2}$

14	17	11	24	1	5
3	28	15	6	21	10
23	19	12	4	26	8
16	7	27	20	13	2
41	52	31	37	47	55
30	40	51	45	33	48
44	49	39	56	34	53
46	42	50	36	29	32

Now to display the output resulting from the application of $PC2$ to $c_1 d_1$, we list $c_1 d_1$ as a 2×56 array and $PC2(c_1 d_1)$ as a 2×56 array for clarity.

$c_1 d_1$															
j	1	2	3	4	5	6	7	8	9	10	11	12	13	14	15	16
b_j	1	1	0	1	0	1	1	0	1	1	1	1	0	1	1	1

j	17	18	19	20	21	22	23	24	25	26	27	28	29	30
b_j	1	0	1	1	0	0	0	1	1	1	1	0	0	0

j	31	32	33	34	35	36	37	38	39	40	41	42	43	44
b_j	0	1	1	1	1	1	1	1	0	0	1	0	1	1

j	45	46	47	48	49	50	51	52	53	54	55	56
b_j	1	0	0	1	0	0	0	0	1	0	0	0

$\mathbf{PC2}(c_1 d_1)$													
j	1	2	3	4	5	6	7	8	9	10	11	12	13	14
$b_{\mathbf{PC2}(j)}$	1	1	1	1	1	0	0	0	1	1	0	1	0	1

j	15	16	17	18	19	20	21	22	23	24	25	26
$b_{\mathbf{PC2}(j)}$	1	1	1	0	1	1	1	1	0	1	1	0

j	27	28	29	30	31	32	33	34	35	36	37
$b_{\mathbf{PC2}(j)}$	0	1	0	0	0	0	0	1	1	1	1

j	38	39	40	41	42	43	44	45	46	47	48
$b_{\mathbf{PC2}(j)}$	0	0	0	1	1	0	0	0	1	0	1

Hence,

$$k_1 = (111110001101011110111101100100000111100011000101) = \mathbf{PC2}(c_1 d_1)$$

Based upon the above, we are now in a position to describe the DES Cipher.

◆ The DES Cryptosystem[2.25]

The DES cryptosystem is defined by the following three steps.

(1) **(The initial permutation step)** Given a plaintext bitstring x of length 64, we construct a bitstring x_0 by permuting the bits of x according to a given fixed *initial permutation* denoted by **IP**, which we will take to be the following.

[2.25] For detailed background information on DES, see [190]. Also, for a discussion of other Block Ciphers that we do not need to cover in this text, such as *SAFER*, *RC2*, *RC5*, and *FEAL*, see [134].

1	2	3	4	5	6	7	8
9	10	11	12	13	14	15	16
17	18	19	20	21	22	23	24
25	26	27	28	29	30	31	32
33	34	35	36	37	38	39	40
41	42	43	44	45	46	47	48
49	50	51	52	53	54	55	56
57	58	59	60	61	62	63	64

$$\downarrow \text{IP}$$

58	50	42	34	26	18	10	2
60	52	44	36	28	20	12	4
62	54	46	38	30	22	14	6
64	56	48	40	32	24	16	8
57	49	41	33	25	17	9	1
59	51	43	35	27	19	11	3
61	53	45	37	29	21	13	5
63	55	47	39	31	23	15	7

The bitstring x_0 is written as $L_0 R_0$ where L_0 is the initial bitstring of length 32 and R_0 is the final bitstring of length 32. Thus, we write

$$\text{IP}(x) = x_0 = L_0 R_0.$$

(2) (**Rounds**) For $1 \leq j \leq 16$, construct keys k_j via the above key schedule algorithm, and recursively define:

$$L_j = R_{j-1},$$

and

$$R_j = L_{j-1} + f(R_{j-1}, k_j) \in (\mathbb{Z}/2\mathbb{Z})^{32},$$

where f is the DES Function determined in the above algorithm.

(3) (**Final permutation**) Apply the inverse permutation IP^{-1} to the bitstring (in inverted order), $R_{16}L_{16}$, which yields the ciphertext y, namely

$$y = \text{IP}^{-1}(R_{16}L_{16}).$$

Here the inverse is given by:

1	2	3	4	5	6	7	8
9	10	11	12	13	14	15	16
17	18	19	20	21	22	23	24
25	26	27	28	29	30	31	32
33	34	35	36	37	38	39	40
41	42	43	44	45	46	47	48
49	50	51	52	53	54	55	56
57	58	59	60	61	62	63	64

$\downarrow \text{IP}^{-1}$

40	8	48	16	56	24	64	32
39	7	47	15	55	23	63	31
38	6	46	14	54	22	62	30
37	5	45	13	53	21	61	29
36	4	44	12	52	20	60	28
35	3	43	11	51	19	59	27
34	2	42	10	50	18	58	26
33	1	41	9	49	17	57	25

The first round of the DES Cipher is illustrated as follows.

The following diagrams give a visual summary of the above discussion of the DES Cipher.

Diagram 2.50 The First Round of DES

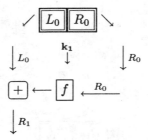

The following illustrates the entire DES Cipher with input followed by the complete sixteen rounds and the final permutation that yields the output.

Diagram 2.51 The DES Flow Chart

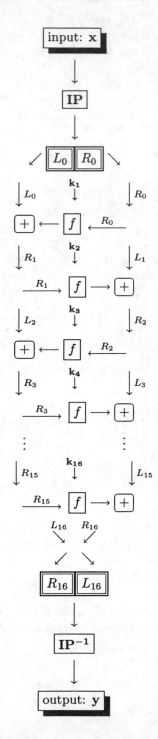

We now exhibit a concrete example using the DES Cipher, but for the sake of conservation of space, we do not exhibit each iteration as a matrix depicting the position as well as the bits in each bitstring as we did in Examples 2.48–2.49. However, the reader is encouraged to work through the following example by setting up such tabular information in order to carefully follow each iteration and verify the steps in the DES Algorithm, by solving Exercise 2.44.

Example 2.52 Suppose that the following is the bitstring of length 64 that is the input x,

(0110101110100101010000010001110000000110001000110100010110000000).

Then we apply the initial permutation given in step (1) of the DES Algorithm to get the following. $\mathbf{IP}(x) = x_0$ is given by:

(0100010100001000010110100110011110000010001000110000100100110001),

so $L_0 R_0$ is given by:

(01000101000010000101101001100111)(10000010001000110000100100110001),

and we note that $L_1 = R_0$.

Now assume that the input key k is that in Example 2.49. Thus, we have already calculated k_1 in that example, namely

$$k_1 = (111110001101011110111101100100000111100011000101).$$

We now calculate

$$\mathbf{E}(R_0) = (110000000100000100000110100001010010100110100011),$$

so addition modulo 2 yields:

$$\mathbf{E}(R_0) + k_1 = (001110001001011010111011000101010101000101100110).$$

The **S-box** output on $\mathbf{E}(R_0) + k_1$ via step (3) of the DES Function Algorithm is:

$$C = (10001111010001110010110110110001).$$

Thus, the final permutation that defines the action of the DES function f, as given in step (4) of its algorithm, is the following.

$$P(C) = f(R_0, k_1) = (11010010110010010111110010011001).$$

The next step is the following modulo 2 addition.

$$L_0 + f(R_0, k_1) = R_1 = L_2 = (10010111110000010010011011111110).$$

This completes round one of the DES Algorithm.

Exercises

2.37. Show that the inverse, if it exists, of an affine function as defined on page 79 is given by $f^{-1}(c) = a^{-1}(c-b) \pmod{n}$.

2.38. Prove that the set of all permutations on $\{1, 2, \ldots, n\}$ is a group S_n under composition with order $|S_n| = n!$, called the *permutation group on n symbols*.

2.39. A subset $H \neq \varnothing$ of a group G is called a *subgroup* of G if H is a group under the binary operation inherited from G. Prove that a subset H of a group G is a subgroup of G if and only if each of the following holds.

   (a) $h_1 h_2 \in H$ for each $h_1, h_2 \in H$.

   (b) $1_G \in H$ where 1_G is the multiplicative identity of G.

   (c) For each $h \in H$, $h^{-1} \in H$.

2.40. With reference to Exercise 2.38, a *transposition* is a permutation on $\mathcal{S} = \{1, 2, \ldots, n\}$ that interchanges two elements and leaves the balance unchanged. For $a, b \in \mathcal{S}$, we denote a transposition by (a, b), which means that $a \mapsto b$ and $b \mapsto a$. If $\sigma \in S_n$ and σ is a product of an odd number of transpositions, then σ is called *odd*. If σ is the product of an even number of transpositions, then σ is called *even*. Prove that if $n \geq 2$, then the set of all even transpositions forms a subgroup A_n of S_n, called the *alternating group on n symbols*, of order $n!/2$. (*Hint: Use Exercise 2.39.*)

2.41. A variant of the Vigenère Cipher given in Definition 2.40 is the *Beaufort Cipher*, which is defined as follows.[2.26]

   Fix $r, n \in \mathbb{N}$. Both the encryption and decryption functions are given by

   $$x \mapsto (e_1 - x_1, e_2 - x_2, \ldots, e_r - x_r),$$

   for

   $$e = (e_1, \ldots, e_r) \in \mathcal{K} \text{ and } x = (x_1, \ldots, x_r) \in (\mathbb{Z}/n\mathbb{Z})^r.$$

[2.26]The self-decrypting Beaufort Cipher was used in a rotor-based cipher machine called the Hagelin M-209, invented in the early 1940's by Boris Caesar Wilhelm Hagelin. Hagelin was born on July 2, 1892 in the Caucasus, the area of Russia between the Black and Caspian seas, sometimes regarded as part of the border between southern Europe and western Asia. In 1922, Emanuel Nobel, nephew of the famed Alfred Nobel, put Hagelin to work in the firm *Aktiebolaget Cryptograph* or *Cryptograph Incorporated*. This was a company owned by Avid Gerhard Damn, who invented cipher machines of his own. Hagelin simplified and improved one of Damn's machines, much to the liking of the Swedish army, who placed a large order with the Damn firm. After Damn's death in 1927, Hagelin ran the firm. Later he developed the M-209, which became so successful that in the early 1940's more than $140,000$ were manufactured. The royalties from this alone made Hagelin the first to become a millionaire from cryptography. The reader will recall another cipher mechanism, the Jefferson *wheel cypher*, which we discussed in Chapter One (see page 11). This is an example of a rotary device using a Polyalphabetic Substitution Cipher, without the use of complex machinery. It may be said that the class of rotor machines, especially those used during World War II, were inspired by Jefferson's device.

In other words, the encryption and decryption functions are the same, namely they are their own inverses.[2.27]

Assuming that $r = 4$, $n = 26$, and the following was enciphered using the Beaufort Cipher, with the key $EASY$, find the plaintext. (Use Table 2.37.)

CJUJ   LAFY   TCTU   LTKG

2.42. Assume that $XOQSMMDGRXZLHIUCLW$ has been encrypted with with $n = 26$, and $r = 3$ using the Hill Cipher with key

$$e = \begin{pmatrix} 1 & 2 & 3 \\ 0 & 5 & 1 \\ 2 & 0 & 1 \end{pmatrix}.$$

Find the plaintext.

2.43. Assuming a Hill enciphering of $MGSDYGPPHLH$ with $n = 26$, $r = 2$ and key $e = \begin{pmatrix} 3 & 2 \\ 1 & 1 \end{pmatrix}$, find the plaintext.

☆ 2.44. Complete the task begun in Example 2.52 by computing the remaining fifteen tables with k_j, $\mathbf{E}(R_{j-1})$, $\mathbf{E}(R_{j-1}) + k_j$, S-box outputs, $f(R_{j-1}, k_j)$, and

$$L_{j-1} + f(R_{j-1}, k_j) = R_j$$

for each $j = 2, 3, \ldots, 16$. Conclude with the output y generated by the permutation $\mathbf{IP}^{-1}$ acting on $R_{16}L_{16}$.

2.45. Decipher y from Exercise 2.44 to get the value of x given in Example 2.52 using the decryption algorithm for DES (described on page 102). Assuming that the binary value of x represents English letters in five bit increments reading from left to right and the numerical decimal value of the letters is given by Table 2.37, find the English plaintext. Ignore any "left over" zeros on the right.

2.46. Let $c(k)$ denote the bitwise complementation of an input key k in DES. This means the replacement of all 0's with 1's and all 1's with 0's. Also, let $\mathfrak{E}_k(x)$ represent the DES enciphering transformation on plaintext x with key k. Prove that

$$\mathfrak{E}_{c(k)}(c(x)) = c(\mathfrak{E}_k(x)).$$

(*Hence, complementation of the plaintext results in complementation of the ciphertext. Consequently, a chosen-plaintext attack against DES only has to test half of the 2^{56} keys, namely 2^{55} of them. This is not a serious concern in practice.*)

[2.27]This cryptosystem was invented by Admiral Sir Francis Beaufort, Royal Navy, who was also the creator of the *Beaufort scale*, which is an instrument that meteorologists use to indicate wind velocities on a scale from 0 to 12, where 0 is calm and 12 is a hurricane.

2.3   ☞DES Cryptanalysis & Successor AES

In this optional section, we look at cryptanalysis of DES, studied in the preceding section, and describe its successor, the *Advanced Encryption Standard* (AES) — *Rijndael*.

◆ DES Decryption

With y as input, we apply **IP** to cancel the effect of the final step in the DES Cipher. This yields $R_{16}L_{16}$ as output. Then we apply the DES Algorithm with the keys in reverse order $k_{16}, k_{15} \ldots k_1$. Thus, since $R_{16} = L_{15} + f(R_{15}, k_{16})$ and $L_{16} = R_{15}$, then for the first round of decryption, we have $R_{16} + f(L_{16}, k_{16}) = L_{15} + f(R_{15}, k_{16}) + f(R_{15}, k_{16}) = L_{15}$, since addition modulo 2 means that the addition of $f(R_{15}, k_{16})$ to itself is the additive identity. Thus, the output is (R_{15}, L_{15}). Each step in the decryption algorithm follows in a similar fashion. Observe that these deciphering operations are independent of the definitions of f and k_j in each round. The reader may now go to Exercise 2.45 on the page before for an interesting interpretation of the input in Example 2.52. Hence, DES decryption is essentially a use of the DES Algorithm with the keys in reverse order. Since enciphering and deciphering are essentially identical, this allows for the use of the same hardware device or software code for both. (Compare this with enciphering and deciphering of AES at the end of this section.)

DES, at its most basic level is nothing more than a combination of two basic techniques of cryptography: *confusion* and *diffusion*. *Confusion* obscures the relationship between the plaintext and the ciphertext, whereas diffusion dissipates the redundancy of the plaintext by spreading it over the ciphertext. Confusion, therefore, gets in the way of a cryptanalyst's attempts to study the ciphertext by looking for redundancies and statistical patterns, whereas diffusion frustrates a cryptanalyst's attempts to look for redundancies in the plaintext through observation of the ciphertext. The best way to cause confusion is through the use of substitutions and the simplest way to cause diffusion is through the use of permutations. Since DES involves an initial permutation, followed by sixteen rounds of a substitution, then a final permutation, DES essentially employs a sequence of confusion and diffusion techniques.

Essentially, IBM is responsible for the inception of DES (see Footnote 2.33 on page 107) since the algorithm resulted from IBM's need for enciphering algorithms to protect computer data. In 1974, IBM made a submission to the American National Bureau of Standards (NBS)[2.28] for such algorithms. The result was the description of DES given in the publication cited in Footnote 2.19. This has since been updated twice with the latest being in 1993, called FIPS 46-2, which succeeded FIPS 46-1 from 1988. The former is what we now call the DES Algorithm. The American National Standards Institute (ANSI) approved DES as a private-sector standard in 1981.[2.29]

[2.28]The NBS is now called NIST for the *National Institute of Standards and Technology*. See the NIST homepage: *http://www.nist.gov/*.

[2.29]ANSI X3.92, *American National Standard for Data Encryption Algorithm (DEA)* — American National Standards Institute, 1983.

◆ Modes of Operation

The following *modes of operation* to encipher messages are defined in FIPS 81, December 2, 1980 as well as in ANSI X3.106-1983. They may be used with *any* Block Cipher, not just DES, but since we have expended so much effort on an understanding of DES, then we apply them only to this cipher here.

(a) Electronic Code Book (ECB),

(b) Cipher Block Chaining (CBC),

(c) Cipher Feedback Mode (CFB),

(d) Output Feedback Mode (OFB).

In what follows $\mathfrak{E}_k$ is the enciphering function for DES using the key k. In ECB mode, we input a sequence of x_j for $j \geq 1$ of 64-bit plaintext blocks, each of which is enciphered with the same key, producing a string of ciphertext blocks c_j. In other words, enciphering is $\mathfrak{E}_k(x_j) = c_j$ and deciphering is $\mathfrak{E}_k^{-1}(c_j) = x_j$.

In CBC mode, we first let IV be an initialization vector, meaning a 64-bit input bitstring, set $c_0 = IV$, and let k be the 64-bit input key. Given a sequence x_j of 64-bit plaintext blocks, for $j \geq 1$, we recursively define encryption by $c_j = \mathfrak{E}_k(c_{j-1} + x_j)$, and decryption by $x_j = \mathfrak{E}_k^{-1}(c_j) + c_{j-1}$, where $+$ is addition modulo 2.

In CFB mode, again we input IV, x_j as above, and set $y_0 = IV$. Then we produce subkeys by enciphering the previous ciphertext block. In other words, $\mathfrak{E}_k(y_{j-1}) = k_j$ for $j \geq 1$. Then we produce ciphertext by addition modulo 2 as $y_j = x_j + k_j$.

In OFB mode, we input IV, k, x_j for $j \geq 1$ as above, and set $k_0 = IV$. Then subkeys are computed by repeatedly encrypting the initialization vector, $k_j = \mathfrak{E}_k(k_{j-1})$. Then the plaintext sequence is enciphered by addition modulo 2 as $y_j = x_j + k_j$ for $j \geq 1$.

In ECB and OFB modes, changing one input block x_j causes *exactly one* ciphertext block y_j to be changed. This is valuable in such applications as the encryption of satellite transmissions. In CBC and CFB modes, a change to input block x_j changes $y_j, y_{j+1}, \ldots$ This turns out to be useful in applications involving message authentication. In other words, these latter two modes can be used to produce *message authentication code* (MAC). What this means is that the MAC can be used as an *electronic signature* (or *digital signature*, see page 147) that will convince the receiving party of the authenticity of the message (See Exercise 2.49, and for a more detailed analysis of DES modes see [138].)

As we have seen above, DES is quite a lengthy process but, in practice, the implementations of DES are efficiently done, since only arithmetic modulo 2 is performed. The **E**-, **S**-, and **P**-boxes, as well as computation of the sixteen keys can all be done in constant time by hard-wiring them into a circuit, or by software implementations of tables. Applications of DES (in the public domain) included enciphering of personal and corporate identification numbers (PINs) of individual banking customers, as well as their account transactions on automated teller machines (ATMs). The U.S. government maintains export controls over such encryption algorithms as DES.

The S-boxes provide the DES Cipher with its security since they provide the nonlinear component of the algorithm. It was long suspected that the NSA had somehow embedded a so-called *trap-door*, which is a secret means of inverting the enciphering transformation (and about which we will learn much more later when we study public-key cryptography). Such a trap-door would mean that the NSA would have an easy method for deciphering messages. This suspicion was even investigated in 1978 by the U.S. Senate Select Committee on Intelligence, the findings of which are classified. However, an unclassified summary of the investigation stated that the NSA had not had such an improper involvement in the design of DES. Nevertheless, many remain skeptical since no details were made public. Thus, no trapdoors are publicly known, but we cannot prove that they do not exist. See Coppersmith's paper [50] for a discussion of the design principles of DES.

◆ **Security of DES**

Although the stated primary aim of this text is to study cryptography proper, with cryptanalysis relegated to mostly parenthetical comments, it is worth further discussion since we will be looking at the security of DES and AES.

First, we discuss what *attacks* on a cryptosystem mean. Basically, an attack on a cryptosystem is any method that starts with some information about the plaintext and the corresponding ciphertext, enciphered under a key, which is yet unknown to the cryptanalyst. Determining the key, and thus the plaintext in its entirety is the end-goal. There are two major classes of attacks. One is *passive*, where the cryptanalyst monitors the channels of communications, thereby only *threatening* confidentiality of data. The other is *active*, where the cryptanalyst attempts to add, delete, or somehow alter the message, so threatens not only confidentiality, but also integrity and authentication of data. Passive attacks may be further subdivided into six classes. They are *chosen-plaintext, chosen-ciphertext, known-plaintext, ciphertext-only, adaptive chosen-plaintext*, and *adaptive chosen-ciphertext*. In *chosen plaintext*, the cryptanalyst chooses plaintext, is then given corresponding ciphertext, and analyzes the data to determine the enciphering key to obtain plaintext from previously unseen ciphertext. In a *chosen-ciphertext* attack the cryptanalyst chooses the ciphertext and is given the corresponding plaintext. Of course, given the required data, by their very nature such attacks are difficult to mount. For example, one way to mount a chosen-ciphertext attack is to gain access to the equipment used to encipher, as for instance was the case with the Americans gaining access to the Japanese cipher machines prior to World War II (see page 13). Then the cryptanalyst can use such equipment to deduce plaintext from other intercepted ciphertext, since the equipment does not give them the decryption key.

A *known-plaintext* attack is more practical than chosen-plaintext since the cryptanalyst has some amount of both plaintext and corresponding ciphertext. More practical still is the *ciphertext-only* attack, wherein the cryptanalyst has only the ciphertext as data to deduce the key and subsequent plaintext. Any cryptosystem that is vulnerable to this type of attack is completely insecure. An *adaptive chosen-plaintext* attack is a chosen-plaintext attack where the choice of plaintext depends upon the previously received plaintexts. An *adaptive chosen-ciphertext* attack is a chosen-ciphertext attack with chosen ciphertext depending upon previously received ciphertexts.

One of the most significant cryptanalytic techniques developed in the early 1990's, which works well against numerous Block Ciphers, including DES, is *differential cryptanalysis* (DC), a chosen-plaintext attack developed by Biham and Shamir [30] in 1993. DC involves the comparisons of pairs of plaintext with pairs of ciphertext, wherein the cryptanalyst looks at ciphertext pairs whose plaintexts have certain "differences." Some of these differences have a high probability of reappearing in the ciphertext pairs. Those which do are called *characteristics*. Differential cryptanalysis uses these characteristics to assign probabilities to the possible keys, with an end-goal being the location of the most probable key. It turns out that this attack is heavily dependent upon the structure of the S-boxes. However, DES has its S-boxes optimized against DC. The reason for this, despite the fact that DC appeared (publicly) much later than DES, is best explained by Don Coppersmith. He is known to have stated in an internal IBM newsgroup that the IBM group knew about DC in 1974, so designed the S- and P-boxes to deflect it. However, they kept quiet about it since they wanted to keep discovery of DC out of the public domain as long as possible, given their knowledge of the potency of this form of cryptanalytic attack.

The number of rounds has an impact upon the DC attack. If only eight rounds of DES are performed, then DC can break DES in a couple of minutes on a personal computer. At the full sixteen rounds, the DC attack is only slightly more efficient than an exhaustive search of the keyspace.[2.30] However, if one increases the number of DES rounds to seventeen or eighteen, then the DC attack takes about the same time as an exhaustive search of the keyspace. If DES is improved to nineteen rounds, then exhaustive search of the keyspace is easier than DC. Thus, although DC is a theoretical breakthrough, it is impractical because of both the enormous amounts of time and data requirements to mount such an attack. What the DC attack showed, nevertheless, is that DES with any number of rounds less than the full sixteen can be broken with a known-plaintext attack more efficiently than by exhaustive search.

[2.30] An *exhaustive search of the keyspace*, also called *brute force attack*, means that all possible keys are tried to see which one is being used by the communicating parties. For an n-bit block with an m-bit key k, given a small number (say $\lceil m/n \rceil$) of plaintext-ciphertext pairs enciphered with k, then k can be obtained after 2^{m-1} operations (*expected time*, which means the *expectation* (in the probability sense) of the runtimes over all the possible inputs, expressed as a function of the input size. This means that one needs to estimate the probability that a given input occurs – not an easy task in general.) For DES, $m = 56$ and $n = 64$, so expected time is 2^{55} decipherments on a single plaintext-ciphertext pair to recover the key. This is based on the fact that one expects to find the correct key after searching half the keyspace. (Note that the *ceiling function* is defined by $\lceil x \rceil = \lfloor x \rfloor + 1$ if $x \notin \mathbf{Z}$ and $\lceil x \rceil = x$ if $x \in \mathbf{Z}$ (see Exercises 1.12–1.14).)

In 1994, Matsui [127] introduced an attack called *linear cryptanalysis* (LC), which is is a known-plaintext attack, using linear approximations to describe the behavior of the Block Cipher. LC proved to be more successful than DC against DES. In fact, in [128] Matsui was able to recover a DES key under experimental conditions. LC can break 8-round DES with 2^{21} known plaintexts, and a full 16-round DES with 2^{43} known plaintexts. In 1994, Langford and Hellman [114] introduced an attack called *differential-linear cryptanalysis*, which combines aspects of both DC and LC. There is also the *truncated differential analysis* of Knudsen [103], introduced in 1995.

◆ AES — Successor for DES

With all of the above being said, the DES Cipher has reached the end of its credibility. There are several reasons for this, the most germane being that the 56-bit keylength is too small to be secure today. Also, see Exercises 2.47–2.48 on page 112 for a look at the notion of *weak* and *semi-weak* keys in DES. An oft-cited fact is that at the CRYPTO conference in 1993 (see Footnote 1.4), M.J. Wiener presented an efficient key-search design for a machine that would take (at that time) 3.5 hours, on a machine costing one million dollars U.S., to do an exhaustive search of the keyspace.[2.31]

On January 2, 1997, NIST announced the initiation of an effort to develop the *Advanced Encryption Standard* (AES) as an unclassified, publicly disclosed encryption algorithm for protecting sensitive data. As an interim measure before the AES was finalized and adopted, NIST proposed *DRAFT FIPS 46-3* in 1999, which announced the new DES *interim* standard — Triple DES, which is described as follows. Let $\mathfrak{E}_k$ and $\mathfrak{D}_k$ denote the DES enciphering and deciphering transformations, and k denote a DES key. We employ three keys k_1, k_2 and k_3, which may range from all being different to the possibility that $k_1 = k_3$. (However, either of $k_1 = k_2$ or $k_2 = k_3$ is useless.) Then encryption of plaintext block x is achieved via: $y = \mathfrak{E}_{k_3}(\mathfrak{D}_{k_2}(\mathfrak{E}_{k_1}(x)))$, and deciphering is achieved via: $x = \mathfrak{D}_{k_1}(\mathfrak{E}_{k_2}(\mathfrak{D}_{k_3}(y)))$.

On August 9, 1999, NIST announced five finalists for the AES (in round two of their competition): *MARS, RC6, Rijndael, Serpent*, and *Twofish*. On October 2, 2000, NIST announced that *Rijndael* was selected as the proposed AES. Before we describe Rijndael in detail, we must define a cryptosystem of

---

[2.31]For the reader with some interest in the group theoretic aspects, the following will be of interest. For a fixed DES key k, DES essentially defines a permutation from $(\mathbb{Z}/2\mathbb{Z})^{64}$ to $(\mathbb{Z}/2\mathbb{Z})^{64}$. The set of DES keys has, at least potentially, 2^{56} different permutations. For many years cryptanalysts could not determine whether DES satisfies the property that the set of permutations is closed as a group under composition. Closure under composition here would mean that given keys k_1 and k_2, there would exist a key k_3 such that $\mathfrak{E}_{k_3}(x) = \mathfrak{E}_{k_2}(\mathfrak{E}_{k_1}(x))$ for all inputs x. This is not the case for DES, since it was finally proved in 1992 that DES is not a group by Campbell and Wiener [43]. In fact, it is known that a lower bound on the size of the group generated by composing this set of permutations is 10^{2499}. This is a good thing for DES since, if DES were a group, then cryptanalysis would be easier. The reason is that multiple encryption would be rendered useless, where multiple encryption means enciphering the same plaintext block multiple times. Thus, if DES were a group, then multiple encryption would be equivalent to single encryption (see Exercise 2.50).

which DES is an example, and the reign of which Rijndael brought to an end.[2.32] It is essentially due to a man who is known as the *father of the Data Encryption Standard*, Horst Feistel.[2.33]

◆ Feistel Ciphers

A Feistel Cipher is a Block Cipher that inputs a plaintext pair (L_0, R_0), where both L_0 and R_0 have bitlength $b \in \mathbb{N}$ each, and outputs a ciphertext pair (R_r, L_r) of bitlengths b each, for $r \in \mathbb{N}$, according to the following iterative process. A key k is input and *subkeys* k_j are derived from it for $j = 1, 2, \ldots, r$. A function f called a *round function* (iterated over r rounds) acts on plaintext pairs (R_{j-1}, k_j) for $j = 1, 2, \ldots, r$ in a prescribed fashion.[2.34] The ciphertext output pair is (L_j, R_j) where $L_j = R_{j-1}$ and $R_j = L_{j-1} + f(R_{j-1}, k_j)$, with $+$ denoting modulo 2 addition. More succinctly, we may take the input plaintext pair to be (R_{-1}, R_0), where the jth round for $j = 1, 2, \ldots, r + 1$ has input (R_{j-2}, R_{j-1}), and output $(R_{j-1}, R_{j-2} + f(R_{j-1}, k_j))$.

Example 2.53 The DES Cipher is an example of a Feistel Cipher with round function given as the DES function above and subkeys generated by the DES key schedule also given above.

[2.32]See the Rijndael fan club home page: *http://www.rijndael.com/*, where it is stated: "This page is dedicated to the fans of the Rijndael Block Cipher, whose selection, in an upset of Karelinean proportions, as the National Institute of Standards and Technology's proposed Advanced Encryption Standard has brought down the Feistel cipher dynasty."

[2.33]Horst Feistel was born in Germany. In 1934, when he was only twenty years old, he emigrated to the United States. In 1941, he was placed under *house arrest*, since he was not yet an American citizen, and Germany had just declared war on the United States. This meant that he could move around Boston, where he resided, but had to report any movements outside of Boston. On January 31, 1944, the house arrest was lifted and he became an American citizen. The very next day, he was given a security clearance and began a job at the Air Force Cambridge Research Center (AFCRC) where he established a research group in cryptography. His group made great strides in developing cryptographic algorithms, especially in the development of the first practical Block Ciphers. A cryptographic system they developed was the *MARK XII*, which is widely used in U.S. aircraft. However, it is known that the NSA was instrumental in shutting down the work by Feistel's group, which dissolved in the late 1950's. At this point Feistel moved to MIT's Lincoln Laboratory, then to Mitre Corporation, which was a spinoff of the MIT lab. Again, it is known that NSA put pressure on Mitre to abandon Feistel in his desire to set up a research group in cryptography there. As a result, he moved to IBM's Watson Laboratory in Yorktown Heights in New York. Feistel's work at IBM resulted in a cryptosystem used in the IBM2984 banking system, which today is called the *Alternative Encryption Technique*. At that time, however, it was called *Lucifer*, which today refers to two very different algorithms. One is a cryptosystem designed by Feistel in 1973, which uses 4×4 invertible S-boxes. The second is essentially a weak version of DES, which was described in 1971 by J.L. Smith in [191], wherein the structure of the Feistel Cipher was first introduced. This is viewed as the direct predecessor of DES. In the early 1970's patent secrecy orders were placed on some of Feistel's inventions by the U.S. government.

[2.34]This iterative procedure tells us that a Fiestel Cipher is an example of an *Iterated (Block) Cipher*. One must ensure that the iteration of any cryptosystem actually results in increased security. For instance, see the discussion at the end of Footnote 2.31, and the above discussion of Triple DES.

Most Block Ciphers employ the Fiestel structure in the round function. However, the round function used by *Rijndael* does *not* have the Fiestel structure. Instead, the round function in Rijndael is comprised of three distinct invertible functions, the details of which we will learn in what follows.

◆ The Advanced Encryption Standard (AES) — Rijndael

The name "Rijndael", pronounced as any of: "Rhine Dahl", "Rain Doll", or "Reign Dahl", was derived from the names of Rijndael's Belgian designers, Vincent **Rijmen** and Joan **Daemen**. The Rijndael Cipher is based upon the 128-bit Block Cipher, called *Square*, which Rijmen and Daemen originally designed with a concentration on resistance against linear cryptanalysis. Later Lars Knudsen engaged in more cryptanalysis of the Square Cipher. A paper by these three authors, describing the details of Square, was presented at the workshop for *Fast Software Encryption* in the Spring of 1997 in Haifa, Israel.[2.35] In that Spring of 1997, Daemen and Rijmen began working on a variant of the Square Cipher that would allow for key and block lengths of 128-, 192-, and 256-bits. They called their new cipher design "Rijndael" and submitted it to NIST by the June 1998 deadline. The rest, as noted above, is history.

In order to give even a brief description of Rijndael, we need to describe the essential components of it.

◆ The State

The *State*, is the intermediate cipher resulting from application of the round function. The State can be depicted as a $4 \times \mathbf{Nb}$ matrix, with bytes as entries, where $\mathbf{Nb}$ is the block length divided by 32. For instance, if the input block has 256 bits, then $\mathbf{Nb} = 8 = 256/32$, and the State would appear as a matrix $(a_{i,j}) \in \mathcal{M}_{4 \times 8}((\mathbb{Z}/2\mathbb{Z})^8)$ of bytes. In this case, the State has 32 bytes. For an input block of 192 bits, the State would have 24 bytes as a $4 \times \mathbf{Nb} = 4 \times 6$ matrix, and for a block of length 128, it would have 16 bytes as a $4 \times \mathbf{Nb} = 4 \times 4$ matrix. Thus, we have variable State size.

Note that the input block (or *plaintext* if the mode of operation is ECB) is put into the State (matrix) by column: $a_{0,0}, a_{1,0}, a_{2,0}, a_{3,0}, a_{0,1}, a_{1,1} \ldots$ and at the end of the execution of the cipher the bytes are taken from the State in the same order.

◆ The Cipher Key

As with the State, the *Cipher Key* is portrayed as a $4 \times \mathbf{Nk}$ matrix of bytes, where $\mathbf{Nk}$ is the key length divided by 32. For instance, if the key length is 128 bits, then the Cipher Key is $(k_{i,j}) \in \mathcal{M}_{4 \times 4}((\mathbb{Z}/2\mathbb{Z})^8)$. Hence, we have variable key size 16, 24, or 32 bytes, depending on key length 128, 192, or 256 bits.

[2.35] Rijndael has been called *Son of Square* and alternatively Square has been called *Mother of Rijndael* by their creators. See: *http://www.esat.kuleuven.ac.be/~rijmen/square/index.html* for a description of and implementation details for *Square*.

◆ **Key Schedule and Round Keys**

The *Round keys* can be derived from the Cipher key by means of the following *Key Schedule*. There are two parts.

(1) The total number of round key bits equals $B(\mathbf{Nr}+1)$, where B is the block length and $\mathbf{Nr}$ is the number of rounds defined for each case in Table 2.54 below. For instance, if the block length is 128 bits and $\mathbf{Nr} = 12$, then 1664 round key bits are required.

(2) The Cipher Key is expanded into the *Expanded Key* in the following fashion. The Expanded Key is a linear array of 4-byte *words* (i.e. columns of the Key matrix), where the first $\mathbf{Nk}$ words contain the Cipher Key. All other words are defined recursively in terms of previously defined words.[2.36]

Then Round Keys are extracted from the Expanded Key as follows. The first Round Key consists of the first $\mathbf{Nb}$ words, the second Round Key consists of the following $\mathbf{Nb}$ words, and so on.

◆ **Round Function**

First, we note that the *number of rounds*, denoted by $\mathbf{Nr}$, is defined via the following table.

Table 2.54

Nr	Nb = 4	Nb = 6	Nb = 8
Nk = 4	10	12	14
Nk = 6	12	12	14
Nk = 8	14	14	14

In this table, we are including the final round, which we will describe below, which slightly differs from the other rounds in that step (3) below is eliminated.

The round function consists of four steps, each with its own name and its own particular function.

(1) **Bytesub (BSB)**: In this step, bytes are mapped by an invertible S-box, and there is only one single S-box for the complete cipher. Thus, for instance, the State (position) matrix $(a_{i,j}) = (8i + j - 9)$ (for $1 \le i \le 32$, $1 \le j \le 8$) would be mapped, elementwise, by the S-box to State matrix $(b_{i,j})$ via: $a_{i,j} \longrightarrow \boxed{\text{S-box}} \longrightarrow b_{i,j}$. This guarantees a high degree of non-linearity by operating on each of the State bytes $a_{i,j}$ independently. To view the S-box explicitly, together with a description of how it was constructed, see Appendix $\mathbf{A_0}$ on pages 269–270.

(2) **Shift Row (SR)**: In this step, depending upon the value of Nb, row j for $j = 2, 3, 4$ of The State matrix are shifted x_j units to the right, where x_j is defined by Table 2.55 below.

[2.36] For a detailed account of how this is done, see the original description of Rijndael at: *http://www.esat.kuleuven.ac.be/~rijmen/rijndael/*.

Table 2.55

Nb	x_2	x_3	x_4
4	1	2	3
6	1	2	3
8	1	3	4

For instance, if **Nb** = 4, then

$$
\begin{pmatrix}
a_{0,0} & a_{0,1} & a_{0,2} & a_{0,3} \\
a_{1,0} & a_{1,1} & a_{1,2} & a_{1,3} \\
a_{2,0} & a_{2,1} & a_{2,2} & a_{2,3} \\
a_{3,0} & a_{3,1} & a_{3,2} & a_{3,3}
\end{pmatrix}
\xrightarrow{\boxed{SR}}
\begin{pmatrix}
a_{0,0} & a_{0,1} & a_{0,2} & a_{0,3} \\
a_{1,3} & a_{1,0} & a_{1,1} & a_{1,2} \\
a_{2,2} & a_{2,3} & a_{2,0} & a_{2,1} \\
a_{3,1} & a_{3,2} & a_{3,3} & a_{3,0}
\end{pmatrix}
$$

The **SR** step introduces high diffusion over multiple rounds and interacts with the next step.

(3) **Mix Column (MC)**: In this step, the columns in the State matrix are treated as polynomials $a(x)$ over $\mathbb{F}_{2^8} \cong \mathbb{F}_2[x]/(m(x))$, where $m(x) = x^8 + x^4 + x^3 + x + 1$ is the irreducible Rijndael polynomial (see Appendix A_0 on pages 269–270 and pages 289–292 in Appendix A). Then $a(x)$ is multiplied modulo $M(x) = x^4 + 1$ with a fixed invertible polynomial $c(x) = 3x^3 + x^2 + x + 2$, denoted by $c(x) \otimes a(x)$. Here multiplying modulo $x^4 + 1$ means that $x^i \pmod{x^4 + 1} = x^{i \pmod 4}$. It can be shown that if $a_j(x) = a_{3,j}x^3 + a_{2,j}x^2 + a_{1,j}x + a_{0,j}$ represents column j in the State matrix, then $c(x) \otimes a(x)$ can be represented by the matrix product:

$$
CA_j =
\begin{pmatrix}
2 & 3 & 1 & 1 \\
1 & 2 & 3 & 1 \\
1 & 1 & 2 & 3 \\
3 & 1 & 1 & 2
\end{pmatrix}
\begin{pmatrix}
a_{0,j} \\
a_{1,j} \\
a_{2,j} \\
a_{3,j}
\end{pmatrix}
=
\begin{pmatrix}
b_0 \\
b_1 \\
b_2 \\
b_3
\end{pmatrix}
= B,
$$

where the matrix A_j is column j of the State matrix and C is the *circulant* matrix representing $c(x)$. Hence, each column A_j of the State matrix is multiplied in this fashion by C. For instance, if $a(x) = x^3 + 1$, then $c(x) \otimes a(x) = 5x^3 + 4x^2 + 2x + 3$, which is given by the matrix product:

$$
\begin{pmatrix}
2 & 3 & 1 & 1 \\
1 & 2 & 3 & 1 \\
1 & 1 & 2 & 3 \\
3 & 1 & 1 & 2
\end{pmatrix}
\begin{pmatrix}
1 \\
0 \\
0 \\
1
\end{pmatrix}
=
\begin{pmatrix}
3 \\
2 \\
4 \\
5
\end{pmatrix}
= B,
$$

This step linearly combines bytes in the columns, and creates high intra-column diffusion. This technique is based upon the theory of error correcting codes, such as in cyclic redundancy checks (for example see [134, Example 9.80, p. 363]).

(4) **Round Key Addition (RKA)**: In this step, a Round Key is added modulo 2 to the State. For example, $(a_{i,j}) \oplus (k_{i,j}) = (b_{i,j})$, where $\oplus$ is addition

modulo 2, $(a_{i,j})$ is the State matrix, $(k_{i,j})$ is the Round Key matrix, and $(b_{i,j})$ is the resulting State matrix. Thus, this step makes the Round function key dependent.

There is significant parallelism in the Round Function. All four steps of a given round operate in a parallel manner on bytes, rows, or columns of the State.

◆ **Stepwise Description of the Rijndael Cipher**

Step 1 (Initial Addition Round) There is an initial RKA step.

Step 2 (Rounds) There are **Nr** − 1 Rounds executed.

Step 3 (Final Round) A final Round is executed, where the **MC** step is omitted.

Hence, the detailed sequence of steps for Rijndael is an initial round key addition, then **Nr** − 1 Rounds of **BSB, SR, MC, RKA** each, followed by a final Round consisting of **BSB, SR, RKA**. Unlike DES, Rijndael does not require a "swapping step" in its rounds since the **MC** step causes every byte in a column to alter every other byte in the column.

Deciphering Rijndael is executed by reversing the steps using inverses and a modified key schedule.

◆ **Security of Rijndael**

The design of Rijndael practically eliminates the possibility of weak or semi-weak keys, which exist for DES (see Exercises 2.47–2.48). Moreover, the design of the Key Schedule virtually eliminates the possibility of equivalent keys. Although the mechanisms of differential and linear cryptanalysis can be adjusted to present attacks on Rijndael, it appears that Rijndael's design is sufficient to withstand these cryptanalytic onslaughts, since its S-box is nearly perfect for resistance to differential cryptanalysis and the $\mathbb{F}_{2^8}$ equivalent of linear cryptanalysis.

A chosen plaintext attack, called the *Square attack*, which is a dedicated attack on the Square Cipher, can be used as well, since Rijndael inherited many features from Square. However, for seven or more rounds in Rijndael, no such attack, faster than exhaustive key search, has been found. Other attacks such as Biham's *related-key attack*, or the *interpolation attacks* introduced by Jakobsen and Knudsen have little chance of success against Rijndael due to the diffusion and non-linearity of Rijndael's Key Schedule and the complicated construction of the S-box.

◆ **Concluding Comments**

Unlike the Feistel structure of the round function, such as in DES, where some of the bits of the intermediate State or simply put into a different position unchanged, the Rijndael Round function is comprised of three different invertible transformations, called *layers*, through which every bit of the State is treated in a similar fashion, called *uniformity*. The BSB step in each round is a *non-linear mixing layer* (confusion). SR is a *linear mixing layer* (inter-column diffusion), and **MC** is also a linear mixing layer (inter-byte diffusion within columns). Then there is the *Key addition layer*. These layers ensure that Rijndael Round does *not* have a Fiestel structure. The layers are predominantly based upon the application of the *Wide Trail Strategy*, which is a devised system for providing resistance against Linear and Differential Cryptanalysis, discussed in Daemen's Doctoral Dissertation of March 1995.

Rijndael is well-tailored to modern processors (Pentium, RISC, and parallel processors). It is also ideally suited for ATM, HDTV, Voice, and Satellite. Uses for Rijndael include MAC by employing it in a CBC-MAC algorithm. It is also possible to use it as a *Synchronous Stream Cipher*, a *Pseudorandom Number Generator*, or a *Self-Synchronizing Stream Cipher* (the latter, by using it in CFB mode), and we will learn about all of these concepts in the next section.

Exercises

2.47. For a Block Cipher, a *weak key*, also called a *palindromic key*,[2.37] is a secret key with a certain value for which the Block Cipher will exhibit certain irregularities in encryption including the possibility of a very poor level of encryption. For example, in DES a weak key is a key such that $\mathfrak{E}_k(\mathfrak{E}_k(x)) = x$ for all $x \in \mathcal{M}$. Find the four weak DES keys.

2.48. In DES, a pair of keys (k_1, k_2) is called *semi-weak* if $\mathfrak{E}_{k_1}(\mathfrak{E}_{k_2}(x)) = x$ for all $x \in \mathcal{M}$. Find the six pairs of semi-weak keys for DES. (*These keys will encipher plaintext to the identical ciphertext. This means that one key in the pair can decipher messages enciphered with the other key in the pair. Hence, given the description of the DES Cipher, these key pairs generate only two different subkeys, each of which are used eight times in the DES Algorithm. These keys should be avoided, and since they are so easy to recognize under a simple check, they may be discarded from consideration.*)

2.49. Construct a MAC using DES's CBC mode as follows. Produce ciphertext $y_1, \ldots, y_n$ with key k from plaintext input blocks $x_1, \ldots, x_n$. Define the MAC to be y_n. If A sends $x_1, \ldots, x_n$ and y_n, how can receiver B, who has the secret key k, verify that y_n is indeed an authentic electronic signature?

2.50. In the closing of this section, we discussed Triple DES. Consider the case of Double DES where we have two keys k_1, k_2 so that encryption is given by $\mathfrak{E}_{k_2}(\mathfrak{E}_{k_1}(x)) = y \in \mathcal{C}$ for any $x \in \mathcal{M}$. Find a decryption transformation such that $\mathfrak{D}_{k_1}(\mathfrak{D}_{k_2}(y)) = y$ for all $y \in \mathcal{C}$. *Note that this system would be useless as a cryptosystem since* $\mathfrak{E}_{k_2}(\mathfrak{E}_{k_1}(x)) = x$ *for all* $x \in \mathcal{M}$.

[2.37]The definition of a *palindrome* is a sequence of symbols that reads the same forwards or backwards. For example: *never odd or even* and *a toyota* are palindromes. The classical palindrome is *able was I ere I saw elba*.

2.4   Stream Ciphers

In Section Two, we studied the first kind of symmetric-key cryptosystem, the Block Cipher. This section is devoted to the second kind, Stream Ciphers. In order to place ourselves in the position of being able to define such cryptosystems, we need the following.

Definition 2.56 (Keystreams, Seeds, and Generators)

If $\mathcal{K}$ is the keyspace for a set of enciphering transformations, then a sequence $k_1 k_2 \cdots \in \mathcal{K}$ is called a keystream. *A keystream is either randomly chosen, or is generated by an algorithm, called a* keystream generator, *which generates the keystream from an initial small input keystream called a* seed. *Keystream generators that eventually repeat their output are called* periodic.[2.38]

Remark 2.57 (Randomness)

A naive definition of randomly chosen *would be any method that generates a sequence of digits in a way that nobody could predict or duplicate. One method would be to toss a coin, with heads as a 1 and tails as a 0, and assuming that the probability of tossing a head with this coin is exactly one-half, then by repeatedly tossing the coin, we can generate a random bitstring, which will correspond to our randomly generated key. Of course, this primitive method is nothing more than illustrative. To generate random numbers in this fashion is time-consuming, so is of little value when random sequences are required to be generated quickly and often. Therefore, we require a computer to "randomly generate" bitstrings. However, the notion of "randomness" is not easy to define precisely. In [104, p. 149], Knuth admits that with respect to randomness: " The mathematical theory of probability and statistics scrupulously avoids the issue." He goes on for forty pages [104, pp. 149–189] devoted to answering "What is a random sequence?," and the conclusion is not exactly satisfactory. If we state our mathematical case too strongly, then we find that there are no sequences that are truly random! In other words, if we demand too much, we define randomness out of existence. If we weaken our demands on what randomness means, then we do arrive at a notion of randomness that is (perhaps) satisfactory, but at the expense of some mathematical discourse that is beyond the scope of this book. In practice, we use a computer program that generates a sequence of digits in a fashion that appears to be random, called a* pseudorandom number generator. *Here we say "appears to be random" since computers are* finite-state *devices, so any random-number generator on a computer must be periodic, which means it is predictable, so it cannot be truly random. The most that one can expect therefore from a computer is* pseudorandomness. *Of course, these pseudorandom number generators are periodic, but if the periods are large enough, then they can be used for cryptographic applications. Actually designing them is not so easy*

[2.38] All keystream generators are periodic except for *one-time pads*. See the discussion of the *Vernam* Cipher on pages 115–116.

as it sounds, and there is a vast literature concerning how to obtain a secure means of generating a sequence of digits that has the statistical properties of a truly random sequence, at least in appearance. Even more than that, these keystreams must be cryptographically secure. This means they must satisfy the additional property that, for a given output bit, the next output bit must be computationally infeasible (see Footnote 2.14) to predict, even given knowledge of all previous bits, knowledge of the algorithm being used, and knowledge of the hardware. These issues, however, are not our concern here. We will assume that we have such a cryptographically secure pseudorandom number generator for our keystream, or a truly randomly chosen keystream, and proceed with learning cryptographic techniques.[2.39]

Definition 2.58 (Stream Ciphers)

Let $\mathcal{K}$ be a keyspace for a cryptosystem and let $k_1 k_2 \cdots \in \mathcal{K}$ be a keystream. This cryptosystem is called a Stream Cipher if encryption upon plaintext strings $m_1 m_2 \cdots$ is achieved by repeated application of the enciphering transformation on plaintext message units, $\mathfrak{E}_{k_j}(m_j) = c_j$, and if d_j is the inverse of k_j, then deciphering occurs as $\mathfrak{D}_{d_j}(c_j) = m_j$ for $j \geq 1$. If there exists an $\ell \in \mathbb{N}$ such that $k_{j+\ell} = k_j$ for all $j \in \mathbb{N}$, then we say that the Stream Cipher is periodic with period ℓ.

The following is the simplest flow-chart for a Stream Cipher.

Diagram 2.59 A Stream Cipher

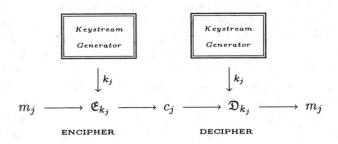

Generally speaking, Stream Ciphers are faster than Block Ciphers from the perspective of hardware. The reason is that Stream Ciphers encrypt individual plaintext message units, usually one binary digit at a time.[2.40] However, in view of our general Definition 2.58, we may view a Block Cipher as a special

---

[2.39]For the reader interested in the details and methods for ensuring cryptographic security of pseudorandom number generators and the design of them, see [173, pp. 356–375].

[2.40]In fact, some authors choose to define Stream Ciphers in this more restricted fashion, namely as those ciphers that convert plaintext to ciphertext one bit at a time. See [173, p. 168]. In practice, the Stream Ciphers used are most often those which do indeed encipher one bit at a time.

case of a Stream Cipher having constant keystream $k_j = k$ for all $j \geq 1$. At least historically, the distinction between Block Ciphers and Stream Ciphers is not clear-cut. Modern distinctions deem that Stream Ciphers can encrypt a single bit of plaintext at a time, whereas Block Ciphers take a number of bits (typically 64 in modern cryptosystems such as the DES Cipher just studied in Section Two), and encipher them as a single block. However, even with our more general definition, we can "stream" a Block Cipher as we did with the CBC mode of operation for DES (described on page 103) which results in a Stream Cipher with large blocks.

Certain modes of operation for a Block Cipher can transform it into a Stream Cipher. Hence, in this way, a Block Cipher can be used as a Stream Cipher. For instance, DES in CFB or OFB modes may be considered to be the transforming of DES into a Stream Cipher. This is done by using CFB mode, for instance, to encipher a character stream using the DES as a keystream generator of plaintext-dependent type. With all this being said, it is still the case that Stream Ciphers with a dedicated design are usually much faster.

Example 2.60 This is an example of a periodic Stream Cipher that we actually discussed in Chapter One and formally defined in Definition 2.40. The Vigenére Cipher with key e of length r may be considered to be a Periodic Stream Cipher with period r. The key $e = (e_1, e_2, \ldots, e_r)$ provides the first r elements of the keystream $k_j = e_j$ for $1 \leq j \leq r$, after which the keystream repeats itself.

One of the simplest illustrations of the Stream Cipher is the following, which was discovered by Gilbert Vernam[2.41] in 1917.

◆ **The Vernam Cipher**

The Vernam Cipher is a Stream Cipher with alphabet of definition $\mathcal{A} = \{0, 1\}$ that enciphers in the following fashion. Given a bitstring $m_1 m_2 \cdots m_n \in \mathcal{M}$, and a keystream $k_1 k_2 \cdots k_n \in \mathcal{K}$, the enciphering transformation is given by

$$\mathfrak{E}_{k_j}(m_j) = m_j + k_j = c_j \in \mathcal{C},$$

and the deciphering transformation is given by

$$\mathfrak{D}_{k_j}(c_j) = c_j + k_j = m_j,$$

[2.41]Gilbert S. Vernam was born in Brooklyn, New York, and was a graduate of Massachusetts College, where he was president of the Wireless Association. He worked for the American Telegraph and Telephone Company (AT&T), where his rare ability with electrical circuitry earned him an assignment to a special secrecy project. On December 17, 1917, he wrote down the cipher for which he has earned the title of the *Father of Automated Cryptography*. A patent was filed in September of 1918, and issued July 1919. This was the first Polyalphabetic Cipher automated using electrical impulses. However, his device was a commercial failure. The stock market crash caused Vernam to lose his job at AT&T. He then went to work for a company that eventually merged with Western Union. From that time he was granted sixty-five patents, including the fully automated telegraph switching system. He even invented one of the first versions of a binary digital enciphering of pictures. However, for all his accomplishments, he died in relative obscurity on February 7, 1960 in Hackensack, New Jersey after years of battling Parkinson's disease.

where $+$ is addition modulo 2. The keystream is randomly chosen and never used again. For this reason, the Vernam Cipher is also called the *one-time pad*.

Current interest in Stream Ciphers is probably due to the palatable theoretical properties of the one-time pad. A one-time pad can be shown to be theoretically unbreakable.[2.42] What this means is that since the key is used only once then discarded, a cryptanalyst with access to the ciphertext $c_1 c_2 \cdots c_n$ can only guess at the plaintext $m_1 m_2 \cdots m_n$, since both are equally likely. Conversely, it has been shown that to have a theoretically unbreakable system means that the keylength must be at least that of the length of the plaintext. This vastly reduces the practicality of the system. The reason of course is that since the secret key (which can only be used once) is as long as the message, then there are serious key-management problems. Nevertheless, it is part of the folklore that Soviet spies used one-time pads to send messages, and that they were also used in German diplomatic systems starting in the late 1920's. Even the much-ballyhooed *hot-line* between Washington and Moscow, inspired by the Cuban missile crisis of the 1960's, used what they called the *one-time tape*, which was a physical manifestation of the Vernam Cipher. At the American end, this took the form of the ETCRRM II or *Electronic Teleprinter Cryptographic Regenerative Repeater Mixer II*. The manner in which the one-time tape worked was that there existed two magnetic tapes, one at the enciphering source, and one at the deciphering end, both having the same keystream on them. To encipher, one performs addition modulo 2 with the plaintext and the bits on the tape. To decipher, the receiver performs addition modulo 2 with the ciphertext and the bits on the (identical) tape at the other end. Thus, they had instant deciphering and perfect secrecy if they used truly random keystreams, each used only once, and the tapes were burned after each use. The same keystream cannot be used twice since the one-time pad would then be open to a known-plaintext attack, given that the key k can be computed by addition modulo 2 of the plaintext with the ciphertext, as seen above. Today, one-time pads are in use for military and diplomatic purposes when unconditional security is of the utmost importance.

Typically, Stream Ciphers are classified as follows.

Definition 2.61 (Synchronous and Self-Synchronizing Ciphers)

A Stream Cipher is said to be synchronous *if the keystream is generated without use of the plaintext and of the ciphertext, called keystream generation independent of the plaintext and ciphertext. A Stream Cipher is called* self-synchronizing (*or* asynchronous) *if the keystream is generated as a function of the key and a fixed number of previous ciphertext units. If the Stream Cipher utilizes plaintext in the keystream generation, then it is called* nonsynchronous.

An example of a synchronous Stream Cipher is DES operating in OFB mode (see page 103). The following gives a general flow chart for a synchronous Stream Cipher.

[2.42]It was not until 1949 with Shannon's development [183] of the concept of *perfect secrecy* that the one-time pad was proved to be unbreakable, something assumed to be true for many years prior to the proof.

Diagram 2.62 A Synchronous Stream Cipher

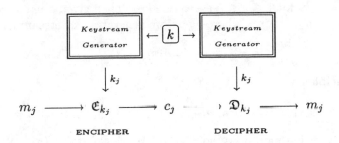

An example of an asynchronous Stream Cipher is DES in CFB mode. The following is the general flow-chart for the self-synchronizing Stream Cipher.[2.43]

Diagram 2.63 A Self-Synchronizing Stream Cipher

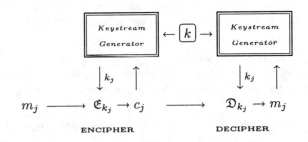

The distinctions between Block and Stream Ciphers are more readily seen in practice than in theory. Stream Ciphers encrypting one bit at a time are not suitable for software implementation since bit manipulation is time-consuming. Where Stream Ciphers win out is in the arena of error propagation. Obviously, with a Block Cipher, a single error will corrupt at least a block's worth of data, whereas implementation of a synchronous Stream Cipher can guarantee that a single bit error will result in only a single bit of corrupted plaintext. Thus, synchronous Stream Ciphers would be useful where lack of error propagation is critical. However, use of self-synchronizing Stream Ciphers can result in error propagation. If the keystream is acting on the nth ciphertext digit and an error occurs, then the deciphering of up to n subsequent ciphertext digits may be incorrect.

In Chapter One, we discussed autokeys, and in Example 2.60 we showed how the Vigenére Cipher may be viewed as a synchronous Stream Cipher. Another idea due to Vigenére provides us with a nonsynchronous Stream Cipher.

[2.43]On August 6, 1946, G. Guanella was granted a patent for the basic idea of a Self-synchronizing Cipher.

Example 2.64 This is a description of the *Autokey Vigenére Cipher* informally discussed in Chapter One. Let $n = |\mathcal{A}|$ where $\mathcal{A}$ is the alphabet of definition. We call $k_1 k_2 \cdots k_r$ for $1 \leq r \leq n$ a *priming key*. Then given a plaintext message unit $m = (m_1, m_2, \ldots, m_s)$ where $s \geq r$, we generate a keystream as follows.

$$k = k_1 k_2 \cdots k_r m_1 m_2 \cdots m_{s-r}.$$

Then we encipher via:

$$\mathfrak{E}_k(m) = (m_1 + k_1, \ldots, m_r + k_r, m_{r+1} + m_1, m_{r+2} + m_2, \cdots, m_s + m_{s-r}) = c,$$

where $+$ is addition modulo n, so $m_j + k_j, m_{r+j} + m_j \in \mathbb{Z}/n\mathbb{Z}$. We decipher via:

$$\mathfrak{D}_k(c) = (c_1 - k_1, \ldots, c_r - k_r, c_{r+1} - m_1, \ldots, c_s - m_{s-r}) = m.$$

This cryptosystem is nonsynchronous since the plaintext serves as the key, from the $(r+1)^{st}$ position onwards, with the simplest case being $r = 1$.

The reader should now solve Exercises 2.53–2.54 on the Vigenére Autokey Cipher. Example 2.64 provides us with an illustration of the following concept.

Definition 2.65 (Autokey Ciphers)
   An Autokey Cipher *is a cryptosystem wherein the plaintext itself (in whole or in part) serves as the key (usually after the use of an initial priming key).*

As is the case with the Autokey Vigenére Cipher, the plaintext is introduced into the key generation after the priming key has been exhausted.
   There is also a means of interpreting the Vernam Cipher as a Vignére Cipher.

Example 2.66 The definition of the Vignére Cipher given in Definition 2.40 is a Running-key Cipher. In other words, the keystream is as long as the plaintext. The Vignére Cipher becomes a Vernam Cipher if we assume that the keystream is truly randomly generated and never repeated.

Perhaps the most common of the Stream Ciphers is the following type.

Definition 2.67 (Binary Additive Stream Ciphers)
   A Binary Additive Stream Cipher *is a synchronous Stream Cipher for which all of the digits in the ciphertext, keystream, and plaintext are binary, and output is achieved by addition modulo 2.*

An illustration of a Binary Additive Stream Cipher is easily accomplished via Diagram 2.62 by replacing both $\mathfrak{E}_{k_j}$ and $\mathfrak{D}_{k_j}$ by $\boxed{+}$ signifying addition modulo 2.

Many, if not most, keystream generators have the following as their basic component. We will discuss the uses for the following after we have described and illustrated the concept.

◆ **Linear Feedback Shift Registers**

A *linear feedback shift register* (LFSR) is comprised of two parts. The first part is a *shift register* of length $\ell \in \mathbb{N}$, consisting of a sequence of ℓ registers (memory cells) labelled $0, 1, 2, \ldots, \ell - 1$, each capable of holding one bit and each having one input and one output. The second part is a *tap sequence* (also called a *clock* or *clock pulse*). In the initialization step, an ℓ-tuple of bits $k = (k_{\ell-1}k_{\ell-2}\ldots k_0)$ is input to the shift register with k_j going to register j for $j = \ell - 1, \ldots, 0$. This is called the *seed* or *initial state*. The clock controls the movement of data as follows. For each unit of time (clock pulse) the following actions are carried out.

(a) The bit in register 0 is *output* as the next bit in the keystream. In other words, the bit in register 0 is *tapped* as the next keystream bit, and becomes (is stored as) part of the output sequence.

(b) The bit in register j is shifted one register to the right for $1 \le j \le \ell - 1$.

(c) Register $\ell - 1$ is given as input the bit achieved by addition modulo 2 of a certain fixed subset of the previous registers 0 through $\ell - 1$. In other words, if the bit in register j is b_j for $j = 0, 1, \ldots, \ell - 1$, then the bit input to register $\ell - 1$ is

$$\sum_{j=1}^{\ell} c_j b_{j-1} \text{ where } c_j \in \{0, 1\}.$$

The values of j for which $c_j = 1$ (namely those bits in register j that are tapped) determine the subset that is added modulo 2 to get the new bit for register $\ell - 1$. This step is called the *linear feedback*, and the sequence $c_1, c_2, \ldots, c_\ell$ is the *tap sequence*.

We will call the completion of (a)–(c) an *LFSR-bit-iteration* (or simply a *bit-iteration*). An *LFSR-bit round* (or simply a *round*) will refer to the completion of ℓ bit-iterations. Succinctly put, the above description of an LFSR says that if the internal state (contents of the registers) after $n \ge 0$ bit-iterations is $(b_{\ell-1}, \ldots, b_0)$, the the internal state after the $(n+1)^{th}$ bit-iteration is

$$\left(\sum_{j=1}^{\ell} c_j b_{j-1}, b_{\ell-1}, \ldots, b_1 \right),$$

where the initial state is bit-iteration $n = -1$. When no confusion can arise we will suppress the commas between the register's bit contents.

We label the initial (given) state as $s_0 = (k_{\ell-1}k_{\ell-2}\ldots k_0)$. Thus, the internal state after round $n + 1$ for any integer $n \geq 0$ is given by

$$s_{n+1} = (k_{(n+2)\ell-1}k_{(n+2)\ell-2}\cdots k_{(n+1)\ell}),$$

where

$$k_{(n+2)\ell-1} = \sum_{i=1}^{\ell} c_i k_{(n+1)\ell-i}, \tag{2.68}$$

the summation being addition modulo 2, and

$$k_{(n+2)\ell-r} = k_{(n+1)\ell-r+1} \qquad (r = 2, \ldots, \ell). \tag{2.69}$$

The following shows the result of the first bit-iteration in round 1.

Diagram 2.70 A Linear Feedback Shift Register

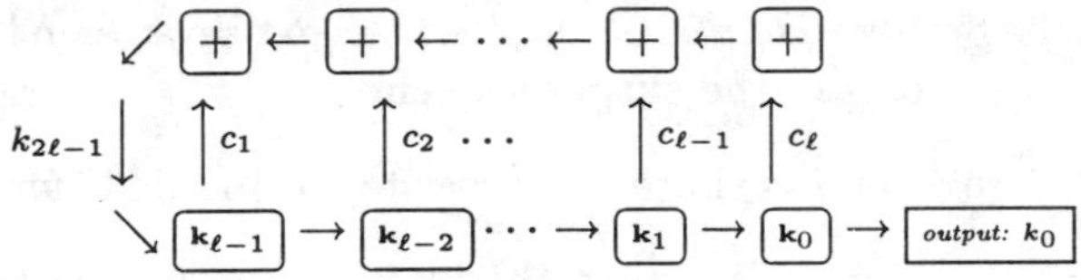

Thus, the internal state after the completion of the first bit-iteration given in Diagram 2.70 is

$$s_1 = (k_{2\ell-1}, k_{2\ell-2}, k_{2\ell-3}, \ldots, k_\ell) = \left(\sum_{j=1}^{\ell} c_j k_{\ell-j}, k_{\ell-1}, k_{\ell-2}, \ldots, k_1\right).$$

The first iteration of bitround $n \geq 0$ may be illustrated in Diagram 2.70 by replacing the left value $k_{2\ell-1}$ by $k_{(n+2)\ell-1}$, replacing the subscripts on the $k_{\ell-r}$ by $(n+2)\ell - r$ for $r = 1, 2, \ldots, \ell$, and replacing the output bit k_0 by $k_{(n+1)\ell}$.

The polynomial

$$t(x) = 1 + c_1 x + c_2 x^2 + \cdots + c_\ell x^\ell \in (\mathbb{Z}/2\mathbb{Z})[x]$$

is called the *connection polynomial* or *tap polynomial*. It can be shown that an LFSR is *periodic* (meaning that $s_{n+1} = s_0$ for some $n \geq 0$) if and only if $t(x)$ has degree ℓ (namely $c_\ell = 1$) in which case the LFSR is said to be *nonsingular*. Hence, based upon the above, the final (distinct) internal state for a nonsingular LFSR is s_{m-1}, where $m \in \mathbb{N}$, the *period*, is the least nonnegative integer such that $k_{(m+1)\ell-r} = k_{\ell-r}$ for $r = 1, 2, \ldots, \ell$, since $s_{n+1} = s_m = s_0$. Therefore, once the mth round is completed, the *output bitstring* is uniquely determined as

$$b = (k_{(m-1)\ell}k_{(m-2)\ell}\ldots, k_\ell k_0).$$

Another way of viewing b is that bitstring consisting of the least significant digits output by each internal state, namely the right-most bit that is output at each round after the first bit-iteration.

Example 2.71 Suppose that we have an LFSR with $\ell = 4$, $c_4 = c_2 = 1$, $c_1 = c_3 = 0$ and initial state

$$(k_3 k_2 k_1 k_0) = (1101).$$

Then we calculate from (2.68)–(2.69) the following.

n	k_{4n+7}	k_{4n+6}	k_{4n+5}	k_{4n+4}
-1	1	1	0	1
0	0	1	1	0
1	1	0	1	1
2	1	1	0	1

For instance, for $n = 0$, the bitstring after bit-round 1 is given by,

$$s_{n+1} = s_1 = (k_7 k_6 k_5 k_4) = (0110),$$

since

$$k_7 = \sum_{i=1}^{4} c_i k_{4-i} = 0 \cdot 1 + 1 \cdot 1 + 0 \cdot 0 + 1 \cdot 1 = 2 \equiv 0 \,(\mathrm{mod}\,2),$$

and

$$k_6 = k_3 = 1, \ k_5 = k_2 = 1, \ \text{and} \ k_4 = k_1 = 0.$$

Thus, the LFSR has period $m = 3$, where

$$k_{(m+1)\ell-r} = k_{16-r} = k_{4-r}$$

for $r = 1, 2, 3, 4$. Thus,

$$s_{m-1} = s_2 = (1011)$$

is the last distinct internal state bitstring output for this LFSR. The output bitstring consists of the right-most entry in each of the above table's rows for each round $j = -1, 0, \ldots m - 2$, namely

$$b = (101) = (k_{(m-1)\ell} k_{(m-2)\ell} \cdots, k_\ell k_0) = (k_8 k_4 k_0).$$

It is possible for an LFSR to cycle through all $2^\ell - 1$ internal states, unlike Example 2.71, which only cycled through three of the $2^4 - 1 = 15$ possible internal states. (Observe that we cannot consider the full 2^ℓ states since a shift register with only zeros cannot exit from that state, namely only zeros will continue to be output.) It can be shown that an LFSR has period $2^\ell - 1$ (for any nonzero initial state) provided that $t(x)$ is *primitive*, which means that (in $\mathbb{F}_2[x]$),

$$t(x) \mid (x^{2^\ell - 1} - 1)$$

but

$$t(x) \nmid (x^k - 1)$$

for any proper positive divisor number k of $2^\ell - 1$. (Another (equivalent) way of defining $t(x)$ as a primitive polynomial is to say that any of its roots generate the group of nonzero elements of the field of 2^ℓ elements. For the reader needing a reminder of the basic concepts involving polynomials and finite fields, see pages 289–294 in Appendix A.) In this case, the LFSR is called a *maximum-length LFSR*. The bitstring (or sequence) produced by a maximum-length LFSR is called a *pn-sequence*[2.44] or (*pseudo-noise sequence*). (Some sources call this an *m-sequence*.)

Example 2.72 Let $\ell = 4$, $c_1 = c_4 = 1$, $c_2 = c_3 = 0$, with initial state

$$k = (0001) = (k_3 k_2 k_1 k_0).$$

Then the output of the LFSR is given by the following.

n	k_{4n+7}	k_{4n+6}	k_{4n+5}	k_{4n+4}	n	k_{4n+7}	k_{4n+6}	k_{4n+5}	k_{4n+4}
-1	0	0	0	1	7	1	0	1	0
0	1	0	0	0	8	1	1	0	1
1	1	1	0	0	9	0	1	1	0
2	1	1	1	0	10	0	0	1	1
3	1	1	1	1	11	1	0	0	1
4	0	1	1	1	12	0	1	0	0
5	1	0	1	1	13	0	0	1	0
6	0	1	0	1	14	0	0	0	1

Here, $m = 15$ since

$$s_{15} = s_0 = (0001)$$

and

$$k_{(m+1)\ell - r} = k_{64-r} = k_{4-r} \text{ for } r = 1, 2, 3, 4.$$

Hence, this is a maximum-length LFSR. The pn-sequence output by this LFSR is

$$b = (001101011110001) = (k_{(m-1)\ell} k_{(m-2)\ell} \dots, k_\ell k_0) = (k_{56} k_{52} \dots k_4 k_0).$$

Here the tap polynomial is

$$t(x) = x^4 + x + 1,$$

which is primitive since

$$(x^{15} - 1)/t(x) = (x - 1)(x^{10} + x^5 + 1)$$

and $t(x)$ does not divide $x^d - 1$ for any proper divisor d of 15 in $\mathbb{F}_2[x]$.

[2.44]The reason for the name *pn-sequence* has to do with its satisfying certain good statistical properties called *Golomb's Randomness Postulates*. See [92, pp. 43–48].

LFSRs are amenable to hardware implementations, the costs of which are low. The choice of the bits to be tapped can ensure a statistically random appearance of output bits. However, bitstrings output by a single LFSR are not secure since sequential bits are linear, so it only takes 2ℓ output bits of the LFSR to determine it, even if the feedback scheme is unknown to the cryptanalyst. Nevertheless, LFSRs are useful as building blocks for more secure systems. One such means is to use several of them in parallel. This means that $n \in \mathbb{N}$ LFSRs are input to a function f, called the *combining function*, which outputs a keystream. Such a system is called a *nonlinear combination generator*.

Diagram 2.73 (Nonlinear Combination Generator)

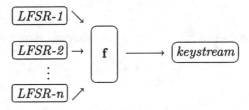

The LFSR-based keystream should exhibit good statistical properties and have large period. Moreover, they should satisfy certain complexity requirements involving the following notion. The reader requiring a reminder of the properties and notation surrounding sequences should refer to page 298.

Definition 2.74 (Linear Complexity and LFSRs)
An LFSR is said to generate a binary sequence if there exists an initial state for which the sequence is the output of the LFSR. The linear complexity of a binary sequence $k = \{k_j\}$ is defined to be the length of the shortest LFSR that generates k. If there is no such LFSR that generates k, the linear complexity is said to be infinite. If k is the zero sequence, then the linear complexity is defined to be 0.

For instance, if $t(x)$ is an irreducible polynomial over $\mathbb{Z}/2\mathbb{Z}$ with degree ℓ, and is the tap polynomial of an LFSR, then that LFSR has linear complexity ℓ for each of its $2^\ell - 1$ nonzero initial states.

A necessary (although possibly not sufficient) property for the security of an LFSR is that it should have large linear complexity.[2.45]

◆ Geffe Generator
This nonlinear combination generator uses three maximum-length LFSRs with period lengths L_j for $j = 1, 2, 3$ that are pairwise relatively prime, and having combining function f defined by

$$f(k_1, k_2, k_3) = k_1 k_2 + (1 + k_2) k_3 = k_1 k_2 + k_2 k_3 + k_3,$$

[2.45]There is an efficient test for determining the linear complexity of a finite binary sequence, called the *Berlekamp-Massey Algorithm*. This test shows that if $k = \{k_j\}$ is a binary sequence of length $n \in \mathbb{N}$ and having linear complexity L, then there is a unique LFSR of length L which generates k if and only if $L \leq n/2$. See [134, pp. 200–202].

where the summation is addition modulo 2, and k_j for $j = 1, 2, 3$ is the bit output by the jth LFSR for a given bit-iteration, as follows, where the downarrow $\Downarrow$ signifies the binary addition of 1 to k_2.

Diagram 2.75

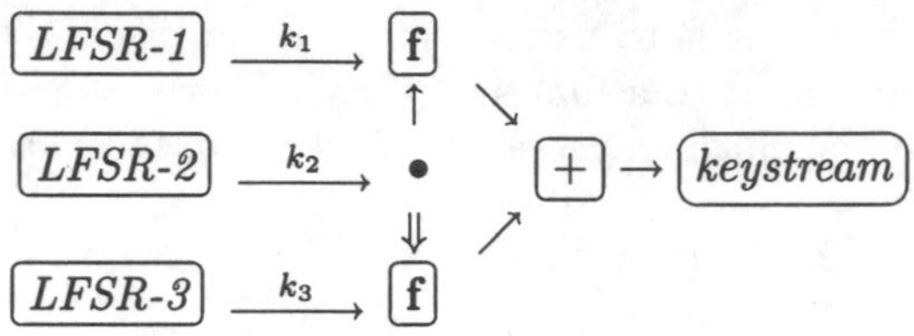

This keystream output by the Geffe generator has period

$$(2^{L_1} - 1)(2^{L_2} - 1)(2^{L_3} - 1)$$

and linear complexity

$$L = L_1 L_2 + L_2 L_3 + L_3.$$

Although the Geffe generator looks good given the rather large period and linear complexity, it is cryptographically weak. The reason is that the output of the Geffe generator is the output of LFSR-3 about seventy-five percent of the time. With this correlation, the keystream can be cracked fairly easily.

A *correlation attack* on nonlinear combination generators, first developed in 1985 by T. Siegenthaler in [186], is described as follows. If $n \in \mathbb{N}$ maximum-length LFSRs R_j of length L_j for $j = 1, 2, \ldots, n$, respectively, are used in a nonlinear combination generator, then the number of different initial states (namely, secret keys) is given by

$$\prod_{j=1}^{n} (2^{L_j} - 1) = K.$$

If there is a known correlation between the keystream and the output sequences of each R_j, for instance by a known-plaintext attack on a Binary Additive Stream Cipher, then the secret keys of each R_j can be determined in a total of

$$T = \sum_{j=1}^{n} (2^{L_j} - 1)$$

trials. Notice that T is significantly smaller than K. Hence, the combining function must be carefully designed to avoid statistical dependence between the LFSR sequences (or any subset thereof) and the keystream.

There exists a method of nonlinear combination generation called *clock-controlled generation* that attempts to foil attacks based upon the regular action of LFSRs such as that described above. Clock control means that nonlinearity is introduced into LFSR-based keystream generators by having the output of one LFSR control the clocking (or stepping) of the LFSR. There is such a keystream generator which is secure if properly set up.

◆ The Shrinking Generator

The shrinking generator, proposed by Coppersmith, Krawczyk, and Mansour in 1993 (see [51]) is described as follows. Let R_1 and R_2 be two LFSRs. If $k_0^{(j)}, k_1^{(j)}, \ldots$ is the output bitstring of R_j for $j = 1, 2$, then the keystream output by the shrinking generator is $s_i = k_{i(1)}^{(2)}$, where $i(1)$ is the position of the ith 1 in $k_0^{(1)}, k_1^{(1)}, \ldots$ for $i \geq 0$. In other words, if a 1 occurs in the ith position of R_1, then $k_i^{(2)}$ becomes the next bit in the output keystream. Otherwise, it is discarded. Hence, the output keystream is a "shrunken" version of the output of R_2. This is illustrated as follows.

Diagram 2.76

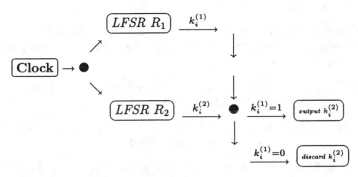

If $\gcd(L_1, L_2) = 1$, then the keystream has period $(2^{L_2} - 1)2^{L_1-1}$, and the linear complexity $L(k)$ of the output keystream k satisfies,

$$L_2 \cdot 2^{L_1-2} < L(k) < L_2 \cdot 2^{L_1-1}.$$

The most efficient attack on the shrinking generator takes

$$O(2^{L_1} \cdot L_2^2)$$

steps, but this attack requires

$$2^{L_1} \cdot L_2$$

consecutive bits from the output sequence. Thus, when R_1 and R_2 are chosen as maximum-length LFSRs with $\gcd(L_1, L_2) = 1$, then for $\max(L_1, L_2) \geq 64$, the shrinking generator appears to be secure against presently known attacks.

There are numerous other LFSRs and LFSR-based Stream Ciphers. However, for the purposes of this text, we have more than covered what is necessary. The reader interested in further information in the realm is referred to [134].

Later we will exploit the notion of modular exponentiation introduced in Section One of this chapter, in order to study Exponentiation Ciphers. These ciphers are a precursor to the important notion of public-key cryptography to be addressed in Chapter Three.

Exercises

2.51. Assume that the following bitstring is a randomly chosen key for the one-time pad.

$$k = (11001010001100111100010101110001011111110101010001).$$

Also, assume that the following was enciphered using k.

$$c = (10111001011110101111000101000001111100000101010010).$$

Find the plaintext string.

2.52. Interpret the plaintext solution to Exercise 2.51 as a bitstring that is the concatenation of bitstrings of length 5, each corresponding to an English letter equivalent given in decimal form by Table 2.37. Find the English text equivalent of the bitstring.

2.53. Given a priming key $k = k_1 k_2 = 7, 2$, if $n = 26$ in the autokey Vignére Cipher, decrypt **LPXEHGM**. (For the numerical equivalents, use Table 2.37.)

2.54. Given a priming key $k = k_1 = 3$ and $n = 26$ in the autokey Vignére Cipher, decrypt **DFYXVDMTA**.

2.55. Find the pn-sequence output by the LFSR given by $\ell = 4$, $c_1 = c_4 = 1$, and $c_2 = c_3 = 0$ with initial state (1000).

2.56. Let an LFSR be given by $\ell = 5$, and $(c_1, c_2, c_3, c_4, c_5) = (1, 0, 0, 1, 0)$ having input state (11101). By calculating the pn-sequence output by it, verify that this is a maximum-length LFSR.

Exercises 2.57–2.58 are devoted to the following notion of a generator. Let $a, b \in \mathbb{N}$ with $n \geq 2$ and $a, b \leq n - 1$. Suppose that an integer s_0 is given with $0 \leq s_0 \leq n - 1$, called a seed. Define

$$s_j \equiv as_{j-1} + b \pmod{n}$$

for $1 \leq j \leq \ell$ where $\ell \in \mathbb{N}$ is the least value such that $s_{\ell+1} = s_j$ for some $j \in \mathbb{N}$ such that $j \leq \ell$. Then $f(s_0) = (s_1, s_2, \ldots, s_\ell)$ is called a linear congruential pseudorandom number generator — or simply linear congruential generator where a is called the multiplier, *ℓ is called the* period, *and b is called the* increment. *It can be shown (see [104]) that the maximum period $\ell = n$ is achieved if and only if $\gcd(b, n) = 1$, $a \equiv 1 \pmod{p}$ for all primes $p \mid n$, and $a \equiv 1 \pmod{4}$ if $4 \mid n$.*

2.57. Find the maximum period of f, the above-described linear congruential generator, if $b = 0$.

2.58. Find the linear congruential generator f when $a = 23$, $s_0 = 1$, $b = 0$, and $n = 1806$.

Chapter 3

Public-Key Cryptosystems

3.1 Exponentiation, Discrete Logs, & Protocols

The notion of modular exponentiation introduced and explored in Section One of Chapter Two is one of the most important arithmetic operations for the type of cryptography — public-key — that we will study in Section Two of this chapter. In preparation, this section is devoted to a description of some Exponentiation Ciphers and related phenomena which, as we shall see, are of interest in their own right. We begin with a private key cipher in preparation for public key cryptosystems to come.

◆ **The Pohlig-Hellman Symmetric Key Exponentiation Cipher**[3.1]

Let p be an odd prime and $e \in \mathbb{N}$ a secret key with $1 \leq e \leq p-2$ and $\gcd(p-1, e) = 1$. Compute a second key $d \equiv e^{-1} \pmod{p-1}$ with $1 \leq d \leq p-2$. Then enciphering is accomplished by

$$\mathfrak{E}_e(m) = m^e \equiv c \pmod{p}$$

for any plaintext message unit $m \in \mathcal{M}$ and deciphering is achieved via

$$\mathfrak{D}_d(c) = c^d \equiv m \pmod{p}$$

for any ciphertext message unit $c \in \mathcal{C}$. The communicating entities must agree in advance to share the symmetric keys e and d. If a cryptanalyst had e and p, then they could calculate d, but without knowledge of e or d, a cryptanalyst, given m, c, and p, would be forced to solve $c \equiv m^e \pmod{p}$ for e. In other words, they would have to calculate,

$$e \equiv \log_m(c) \pmod{p}. \tag{3.1}$$

[3.1]A patent for the Pohlig-Hellman Cryptosystem was filed on May 1, 1978, about four and one-half months after the RSA patent (see Section Two).

Solving (3.1) for the unique integer e with $0 \leq e \leq p - 2$, given m, c, and $p > 2$, is called the *discrete logarithm problem*, or *discrete log*. Another way to think of the discrete log problem based upon the notion of a finite field given in Theorem 2.13, is the following (also see the discussion on page 292). Given a prime p, a generator α of $\mathbb{F}_p^*$ and an element $\beta \in \mathbb{F}_p^*$, find the unique nonnegative integer $x \leq p - 2$ such that $\alpha^x \equiv \beta \pmod{p}$. If p is "properly chosen," then this is a very difficult problem, for which there is no known polynomial-time algorithm to solve it. It is generally agreed that if p has more than one-hundred and fifty digits, and $p-1$ has at least one "large" prime factor, then p is properly chosen. Later in this section, we will learn about a fast method of Pohlig and Hellman for computing discrete logs in $\mathbb{Z}/p\mathbb{Z}$ when $p-1$ has only "small" prime factors.

The complexity of finding e in (3.1), when p has n digits, is virtually the same as factoring an n-digit integer. In other words, computing discrete logs is roughly as hard as factoring. No tractable factorization algorithms are known, so factoring is intrinsically difficult, but no nontrivial lower bounds for the complexity of factorization have yet been established. Thus, when used as the basis for a cryptosystem, computation of discrete logs is assumed to be intractable (see page 51).

Given that there is no efficient algorithm (publicly) known for computing discrete logs, whereas the inverse operation of exponentiation can be efficiently computed using either the extended Euclidean Algorithm (see Theorem A.22), or the repeated squaring method given on page 71 (see Exercises 3.3–3.4), then exponentiation modulo p, as used in the Pohlig-Hellman Cipher for example, is an instance of a special kind of function, which is defined as follows.

Definition 3.2 (One-Way Functions)

A one-to-one function f from a set $\mathcal{M}$ to a set $\mathcal{C}$ is called one-way *if $f(m)$ is "easy" to compute for all $m \in \mathcal{M}$, but for a randomly selected $c \in \text{img}(f)$, finding an $m \in \mathcal{M}$ such that $c = f(m)$ is computationally infeasible. In other words, we can easily compute f, but it is computationally infeasible to compute f^{-1}.*

Definition 3.2 is not a mathematically rigorous definition since there is no rigorous *definition* of "computational infeasibility." To obtain such a rigorous result, we would need to establish nontrivial lower bounds for the number of bit operations that would be required to find the deciphering algorithm without the deciphering key. In other words, we would require the minimum number of bit operations required in order to cryptanalyze the algorithm. Nevertheless, the description of *computational infeasibility* given in Footnote 2.14 is relatively clear in *practical terms*. Also, *computationally easy to compute* is made clear in pragmatic terms, for instance for exponentiation, by the method of repeated squaring. Thus, from a strictly *pragmatic* viewpoint, we can proceed with at least an empirical mathematical foundation. This means that we are dealing with *computational number theory*, which is the newest branch of number theory. Computational number theory, unlike many areas of mathematics, typically employs a significant amount of experimental method in practice. The need for computers to execute algorithms and the need for complexity theory to measure their performance renders computer science an essential tool in this experimental science.

Certain special one-way functions will be introduced in Section Two as a primary component of public-key cryptosystems. For now, we are going to look at some other uses for exponentiation in cryptography as a warm-up for the next section. The following is a description of a means for *electronic coin flipping*.

◆ Coin Flipping via Exponentiation

Suppose that two entities A and B want to make a decision based upon a random coin flip, but they are not in the same physical space. We now describe how they can do this over a channel remotely.[3.2]

The following is the cryptographic protocol[3.3] for electronic coin flipping using exponentiation modulo a prime $p > 2$.

(1) A and B agree upon a prime p such that the factorization of $p-1$ is known.

(2) A selects two generators $\alpha, \beta \in \mathbb{F}_p^*$ and sends them to B.[3.4]

(3) B randomly chooses an integer x, relatively prime to $p - 1$. Then B computes *one* of $y \equiv \alpha^x \pmod{p}$ or $y \equiv \beta^x \pmod{p}$, and sends y to A.

(4) A *guesses* whether y is a function of α or β and sends the guess to B.

(5) If A's guess is correct, then the result of the coin flip is deemed to be heads, and if incorrect, it is tails. B sends the result of the coin flip to A.

In order for B to cheat on the coin toss, two integers x_1 and x_2 must be known where, $\alpha^{x_1} \equiv \beta^{x_2} \pmod{p}$. To compute x_2 given x_1, B must compute $\log_\alpha \beta^{x_2} \pmod{p}$, which is possible if B knows $\log_\alpha \beta$. However, A chooses α and β in step 2. Hence, B is in the position of having to compute a discrete log. Also, B could cheat by choosing an x such that $\gcd(x, p - 1) > 1$. That can be avoided by the following verification step added on the end, called the *verification protocol.*

[3.2]Henceforth, a *channel* will mean any means of communicating information from one entity to another, where an *entity* is any person or thing, such as a computer terminal, which sends, receives, or manipulates information. A *secure channel* is one that is not physically accessible to an adversary, whereas an *unsecured channel* is one from which entities, other than those for whom the information was intended, can delete, insert, read, or reorder data. A cryptosystem is said to be *secure* (against eavesdropping) if an adversary, who eavesdrops on a channel which is sending enciphered messages, gains nothing over an entity which does not tap into the channel (see Section Two of Chapter Six).

[3.3]A *cryptographic protocol* or simply a *protocol* is an algorithm defined by a sequence of steps specifying the actions of two or more entities to achieve a specific security objective. The definition of a cryptographic protocol is sometimes given more loosely in the literature as *any orderly method for entities to accomplish a task using cryptography* or even simply as *prearranged etiquette*. However, we will maintain the first characterization as our definition herein.

[3.4]In Chapter Four, we will learn how generators of such groups may be chosen. For instance, see Example 4.6 on page 162.

(6) B reveals x to A. Then A computes $\alpha^x \pmod{p}$ and $\beta^x \pmod{p}$ to verify both the outcome of the coin toss, and that B has not cheated.

Step (6) ensures that B did not cheat at step (3) since A can then check $\gcd(x, p-1)$.

Characteristics implicit in the above protocol are that neither entity learns about the coin flip at the same time. Also, at some point in the protocol, one entity knows the result of the coin flip, but cannot alter it. This type of coin flipping protocol is called *flipping coins into a well*. This is a metaphor for the following situation. Suppose that B is next to a well, and that A is physically removed from this well. B throws a coin into the well, and can see (but not reach) the coin at the bottom of it. A cannot see the result until B allows A to come to the well to have a look.

The electronic coin flipping protocol relies on the difficulty of the discrete log problem, which is tantamount to the difficulty of factoring. We now describe a general protocol for electronic coin flipping involving one-way functions.

◆ Coin Flipping Using One-Way Functions

(1) A and B both know a one-way function f but not the inverse f^{-1}.

(2) B selects an integer x at random and sends the value $f(x) = y$ to A.

(3) A makes a guess concerning a property of the number x that is valid fifty percent of the time, such as parity (whether x is odd or even) and sends the guess to B.

(4) B tells A whether or not the guess is correct.

(5) B sends A the value x.

(6) A confirms that $f(x) = y$ (verification step).

The security of this protocol relies on the choice of $f(x)$, which must reliably produce, for instance, even and odd numbers with equal probability. If it does not, and were to produce say even numbers sixty percent of the time, then A could guess even every time in step (3) and win sixty percent of the coin tosses. Also, of course, f must be one-to-one. If not then there could exist an x_1 even and an x_2 odd such that $f(x_1) = f(x_2) = y$, and B can cheat A every time.

The above protocol is another example of tossing coins into a well. This is intimately related to another protocol described as follows. The problem is that B wants to commit to a prediction (of either a 0 or a 1), and does not want to reveal this prediction to A until sometime later (where *commit* means that B cannot change the choice). A, on the other hand, wants to ensure that B cannot change the bit prediction once made. To do this using symmetric-key cryptography, we follow the steps given below.

◆ Bit Commitment Protocol Using Symmetric-Key

(1) A generates a random bitstring R and sends it to B.

(2) B creates a message consisting of A's random bitstring R and the bit b to which B wants to commit, producing (R, b). B uses a random (secret symmetric) key e to encipher and sends the cryptogram $\mathfrak{E}_e(R, b)$ to A. This concludes the commitment portion of the protocol. Later, when B is ready to reveal the committed bit, we continue as follows.

(3) B sends the key e to A.

(4) A deciphers the message to reveal the bit and the random bitstring to verify the bit's validity.

The bitstrings that B sends to A to commit to a bit are typically called *blobs* or *envelopes* in the cryptographic community. Blobs have these properties: (1) B, by committing to a blob, commits to a bit. (2) B can open any committed blob, and once opened can convince A of the value of the committed bit (called *binding*), but A cannot open the blob (called *concealing*). (3) Blobs do not carry any information other than B's committed bit, and the blobs themselves. The means by which B commits to and opens them are not related to any other data that B might want to keep secret from A.

The method for enciphering a blob is called a *bit commitment cryptosystem* or *bit commitment scheme* (see Footnote 2.13). There are also various means of bit commitment using one-way functions (see Exercises 6.29–6.30 on page 261), as well as bit commitment protocols using pseudorandom number generators, among others, but we will not cover them since the above protocol is sufficient for our purposes. Historically, the term "bit commitment" developed as follows. In 1982, Manuel Blum introduced the problem of fair electronic coin flipping in [32]. He solved his problem using a bit commitment protocol in the fashion described below.

(1) B commits to a random bit using, say, the above bit commitment protocol.

(2) A makes the guess.

(3) B reveals the committed bit to A, and A wins if the predicted bit is correct.

To be "fair," this protocol must ensure that a guess is correct fifty percent of the time, that B cannot change the bit once it is committed, and A cannot know the predicted bit in advance of the guess. The above protocols for coin flipping via exponentiation are one way of ensuring that the fairness properties are satisfied. See Exercises 3.7–3.8.

We are now ready to describe the aforementioned method of Pohlig and Hellman for computing discrete logs.

◆ **The Pohlig-Hellman Algorithm for Computing Discrete Logs**

Let α be a generator of $\mathbb{F}_p^*$ and let $\beta \in \mathbb{F}_p^*$, and assume that we have a factorization

$$p - 1 = \prod_{j=1}^{r} p_j^{a_j} \qquad a_j \in \mathbb{N},$$

where the p_j are distinct primes. The general overall technique for computing $a = \log_\alpha \beta$ is to compute a modulo $p_j^{a_j}$ for $j = 1, 2, \ldots, r$, then use the Chinese Remainder Theorem to piece together the information. To compute a modulo $p_j^{a_j}$ we need to determine a in its base p_j representation:

$$a = \sum_{i=0}^{a_j - 1} b_i^{(j)} p_j^i \qquad \text{where } 0 \le b_i^{(j)} \le p_j - 1 \text{ for } 0 \le i \le a_j - 1.$$

To find these $b_i^{(j)}$, we proceed as follows.

(1) Compute $b_0^{(j)}$.

By Exercise 3.9, we have $\beta^{(p-1)/p_j} \equiv \alpha^{(p-1)b_0^{(j)}/p_j} \pmod{p}$. Thus, we compute $\alpha^{(p-1)i/p_j}$ modulo p for each $i = 0, 1, 2 \ldots$ until we get that $\alpha^{(p-1)i/p_j} \equiv \beta^{(p-1)/p_j} \pmod{p}$, in which case this i must be $b_0^{(j)}$.

(2) Compute $b_k^{(j)}$ for $k = 1, 2, \ldots, a_j - 1$ as follows. For each such k, recursively define

$$\beta_k = \beta \alpha^{-\sum_{i=0}^{k-1} b_i^{(j)} p_j^i}, \text{ and } x_k = \log_\alpha \beta_k.$$

Then

$$x_k = \sum_{i=k}^{a_j - 1} b_i^{(j)} p_j^i,$$

since

$$\log_\alpha \beta_k = \log_\alpha(\beta \alpha^{-\sum_{i=0}^{k-1} b_i^{(j)} p_j^i}) = \log_\alpha \beta - \sum_{i=0}^{k-1} b_i^{(j)} p_j^i =$$

$$\sum_{i=0}^{a_j - 1} b_i^{(j)} p_j^i - \sum_{i=0}^{k-1} b_i^{(j)} p_j^i = \sum_{i=k}^{a_j - 1} b_i^{(j)} p_j^i.$$

By Exercise 3.9,

$$\beta_k^{(p-1)/p_j^{k+1}} \equiv \alpha^{(p-1)b_k^{(j)}/p_j} \pmod{p}, \tag{3.3}$$

so we compute $\alpha^{(p-1)i/p_j}$ modulo p for $i \ge 0$ until (3.3) occurs in which case i is $b_k^{(j)}$.

If $n = p - 1$, then given a factorization of n, the running time of the Pohlig-Hellman Discrete Log Algorithm is

$$O\left(\sum_{j=1}^{r} a_j \left(\ln n + \sqrt{p_j} \right) \right)$$

group multiplications. This implies that the Pohlig-Hellman Algorithm is only efficient if the prime divisors of $p-1$ are small. This is the reason why we talked about a proper choice of p in the beginning of this section for the intractability of the discrete log problem. We now illustrate with a small example.

Example 3.4 Let $p = 37$. Then $\alpha = 2$ generates $\mathbb{F}_{37}^*$. Select $\beta = 19$. We want to compute $a = \log_2(19)$. We have $\beta^{(p-1)/p_1} = 19^{18} \equiv 36 \,(\text{mod } 37)$, and $\alpha^{(p-1)i/p_1} = 2^{18i} \equiv 36 \,(\text{mod } 37)$ for $i = 1$. Thus, $b_0^{(1)} = 1$. Next, $\beta_1 = \beta\alpha^{-1} = 19 \cdot 2^{-1} \equiv 19^2 \equiv 28 \,(\text{mod } 37)$, $\beta_1^{(p-1)/p_1^{k+1}} = 28^9 \equiv 36 \,(\text{mod } 37)$, and $\alpha^{(p-1)i/p_1} = 2^{18i} \equiv 36 \,(\text{mod } 37)$ for $i = 1$. Therefore, $b_1^{(1)} = 1$. Hence,

$$a \equiv 3 \,(\text{mod } 4). \tag{3.5}$$

We now move to $p_2 = 3$, where $\beta^{(\nu-1)/p_2} = 19^{12} = 10 \,(\text{mod } 37)$, and $\alpha^{(p-1)i/p_2} = 2^{12i} \equiv 10 \,(\text{mod } 37)$ for $i = 2$. Thus, $b_0^{(2)} = 2$. Next, $\beta_1 = \beta\alpha^{-2} = 19 \cdot 2^{-2} \equiv 19^3 \equiv 14 \,(\text{mod } 37)$, $\beta_1^{(p-1)/p_2^{k+1}} = 14^4 \equiv 10 \,(\text{mod } 37)$, and $\alpha^{(p-1)i/p_2} \equiv 2^{12i} \equiv 10 \,(\text{mod } 37)$ for $i = 2$. Therefore, $b_1^{(2)} = 2$. Hence,

$$a \equiv 8 \,(\text{mod } 9). \tag{3.6}$$

Solving (3.5)–(3.6) by the Chinese Remainder Theorem, we get that $a = \log_2 19 = 35$ in $\mathbb{F}_{37}^*$.

The following is one of the components used as building blocks for certain protocols. Their purpose is largely to ensure that the original message has not been subject to tampering and they are also typically used for digital signatures (see Definition 3.17).

Definition 3.7 (Hash Functions)

A hash function is a computationally efficient function that maps bitstrings of arbitrary length to bitstrings of fixed length, called hash values. A one-way hash function is a hash function that satisfies Definition 3.2. The process of using a hash function on a message is called hashing the message.

In order for a hash function to be cryptographically useful, it must be a one-way hash function to prevent easy unauthorized retrieval of the original bitstring. Thus, one-way hash functions are sometimes called *cryptographic hash functions*. We will mean a cryptographic hash function when we use the term *hash function* henceforth. One class of hash function is the MAC (message authentication code) discussed on page 103. MACs are essentially hash functions taking two inputs, a plaintext message and a key, and outputting a fixed-size bitstring, with the underlying built-in assumption that it is infeasible to produce the same output without knowledge of the key. For instance, consider the following.

Suppose that entities A and B share a secret key k. If entity A desires to authenticate a message M and send it to B, then a means of accomplishing this MAC is to concatenate M and k and compute a one-way hash function of that concatenation. Since B knows k, reproduction of M is possible, but not to a cryptanalyst who does not know k. For instance, in 1992 the following was introduced in [112].

◆ **Speedy Stream Cipher MAC**

Suppose that we have a cryptographically secure pseudorandom-number generator (see Remark 2.57), $PRNG$ which takes an input message stream $M = (m_1m_2 \ldots m_j \ldots)$ and outputs two substreams S_1 and S_2 as follows. If the jth output bit of $PRNG$ is 1, message bit m_j is switched into S_1, and if it is 0, m_j is switched into S_2. Then S_j is fed into a linear feedback shift register, LFSR-j for $j = 1, 2$. The output of the two LFSRs is the hashed message output by the Stream Cipher MAC. This is illustrated as follows.

Diagram 3.8 (Stream Cipher MAC Hashing)

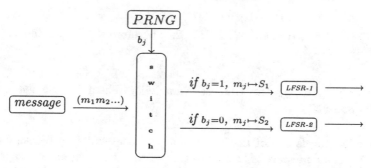

The discrete log problem was the basis for an important idea first given in [63] in 1976.

◆ **The Diffie-Hellman Key-Exchange Protocol**

Suppose that entities A and B, who have never met in advance or exchanged keys, want to establish a shared secret key k by exchanging messages over an open (unsecured) channel. First A and B agree on a large[3.5] prime p and a generator α of $\mathbb{F}_p^*$ ($2 \le \alpha \le p - 2$). These need not be kept secret, so A and B can agree over an unsecured channel. Then the protocol proceeds as follows.

(1) A chooses a random (large) $x \in \mathbb{N}$ and computes the least positive residue X of α^x modulo p, then sends X to B (and keeps x secret).

(2) B chooses a random (large) $y \in \mathbb{N}$ and computes the least positive residue Y of α^y modulo p, then sends Y to A (and keeps y secret).

(3) A computes the least positive residue k of Y^x modulo p.
     (*Note that $Y^x \equiv \alpha^{yx} \pmod{p}$.*)

(4) B computes the least positive residue k of X^y modulo p.
     (Note that $X^y \equiv \alpha^{xy} \pmod{p}$.)

[3.5]Today a prime of 1024 bits would be secure. Furthermore, given the relative ease of computing discrete logs using the Pohlig-Hellman Algorithm studied and illustrated above, one should also ensure that $(p-1)/2$ is prime. Notice that the Diffie-Hellman Protocol differs from the Pohlig-Hellman Cipher in that the latter requires that both p and e be kept secret since d could be deduced from them, whereas in the former p and α are made public.

Hence, k is the shared secret key independently generated by both A and B. The key exchange is complete, since A and B are in agreement on k. A cryptanalyst C listening to the channel would know p, α, X, and Y, but neither x nor y. Thus, C faces what is called the *Diffie-Hellman Problem*: find α^{xy} given α, α^x and α^y (but not x or y). If C can solve the Discrete Log Problem, then C can clearly solve the Diffie-Hellman Problem. Whether the converse is true or not is unknown. In other words, it is not known if it is possible for a cryptanalyst to solve the Diffie-Hellman Problem without solving the Discrete Log Problem. Nevertheless, the consensus is that the two problems are equivalent. Thus, for practical purposes, one may assume that the Diffie-Hellman Key-exchange Protocol is secure as long as discrete log is intractable.[3.6]

This provides the first step in public-key cryptography, which is the topic of the next section.

Exercises

3.1. Let $p = 3049$, and $e = 5$ in the Pohlig-Hellman Cryptosystem. Decipher the cryptogram

$$(76, 1200, 2278, 2503, 1024, 243, 0, 1678, 0, 2278, 2364).$$

Once the numerical plaintext equivalents have been found, use Table 2.37 to get the plaintext message in English.

3.2. Given $p = 3371$ and $e = 3$, use the same instructions as given in Exercise 3.1 to decipher: $(0, 117, 117, 0, 8, 1000, 1728, 512, 27, 535, 0, 340)$.

3.3. Given $m, e, n \in \mathbb{N}$ with $m < n$, show that the least positive residue of m^e modulo n can be computed in $O((\log_2 n)^2 \log_2 e)$ bit operations.

3.4. Given $e, n \in \mathbb{N}$ with $\gcd(e, n) = 1$ and $e < n$, prove that the multiplicative inverse of e modulo n can be found in $O(\log_2^3 n)$ bit operations.

3.5. Prove that the difficulty of the discrete log problem, as described on pages 127–128, is independent of the generator of $(\mathbb{Z}/p\mathbb{Z})^*$.

3.6. The *generalized discrete log problem* is described as follows. Given a finite cyclic group G of order $n \in \mathbb{N}$, a generator α of G, and an element $\beta \in G$, find that unique nonnegative integer $x \leq n - 1$ such that $\alpha^x = \beta$. Given the fact that such a group G is isomorphic to $\mathbb{Z}/n\mathbb{Z}$ (see page 295), one would expect that an efficient algorithm for computing discrete logs in one group would imply an efficient algorithm for the other group. Explain why this is *not* the case.

3.7. Suppose that entities A and B each have computers that input bitstrings of length $m \in \mathbb{N}$ and output bitstrings of length $n \in \mathbb{N}$. Also, assume that they send each other copies of the circuits of their respective computers. The following coin tossing protocol is followed.

[3.6]The equivalence of the two problems has been proved for certain special classes of groups. See [129] and [130].

(1) B chooses a random bitstring R of length m, which is then run through the circuitry of both computers. The two outputs are added modulo 2 and the result is sent to A.

(2) A guesses the parity of B's input (the number of 1's in it) and sends the guess to B.

(3) B sends R to A.

(4) A can verify that the modulo 2 addition of the outputs from the two circuits is indeed what B sent earlier, and determine the correctness of the guess.

What two properties must hold for the above coin flipping algorithm to be fair? (*Hint: See the comments after the description of the coin flipping algorithm using one-way functions on page 130.*)

☆   3.8.   Let $n \equiv 1 \, (\text{mod } 4)$ be a product of two primes. A *Blum Integer* is a natural number $n = pq$ where p and q are primes such that $p \equiv q \equiv 3 \, (\text{mod } 4)$. The following is a coin flipping protocol using Blum Integers.

(1) B generates a Blum Integer n, a random $x \in \mathbb{N}$ such that $\gcd(x, n) = 1$, $x_0 \equiv x^2 \, (\text{mod } n)$, and $x_1 \equiv x_0^2 \, (\text{mod } n)$. Then B sends n and x_1 to A.

(2) A guesses the parity of x and sends the guess to B.

(3) B sends x and x_0 to A.

(4) A checks that both $x_0 \equiv x^2 \, (\text{mod } n)$ and $x_1 \equiv x_0^2 \, (\text{mod } n)$. Thus, A can determine if the guess is correct.

Explain why this protocol would fail to be fair to A if n were not a Blum Integer, namely if $p \equiv q \equiv 1 \, (\text{mod } 4)$.

3.9.   Given $j \in \{1, 2, \ldots, r\}$, establish that for each $k = 0, 1, \ldots, a_j - 1$,

$$\beta_k^{(p-1)/p_j^{k+1}} \equiv \alpha^{(p-1)b_k^{(j)}/p_j} \, (\text{mod } p),$$

in the Pohlig-Hellman Algorithm for computing discrete logs, described on page 132.

3.10.   Let $p = 1481$, $\beta = 78$ and $\alpha = 3$, where α generates $\mathbb{F}_p^*$. Compute $\log_\alpha \beta$ using the Pohlig-Hellman Algorithm for computing discrete logs.

3.2 Public-Key Cryptography

With the advent of the Diffie-Hellman Key-exchange Protocol discussed at the end of the preceding section, cryptography was in for a radical facelift. Previous to their algorithm, *both* the enciphering key and the deciphering key had to be kept secret since one could be recovered from the other in a symmetric-key cryptosystem. With the Diffie-Hellman idea, the enciphering key could be made *public* since it is computationally infeasible to obtain the deciphering key from it. Hence, the central idea of public-key cryptography was born — a public enciphering key.

Definition 3.9 (Public-Key Cryptosystems)

A cryptosystem consisting of a set of enciphering transformations $\{\mathfrak{E}_e\}$ and a set of deciphering transformations $\{\mathfrak{D}_d\}$ is called a Public-key Cryptosystem *or an* Asymmetric Cryptosystem *if, for each key pair (e, d), the enciphering key e, called the* public key, *is made publicly available, while the deciphering key d, called the* private key,[3.7] *is kept secret. The cryptosystem must satisfy the property that it is computationally infeasible to compute d from e.*

An analogy for public-key cryptography is the following. Suppose that entity A has a wall safe with a secret combination lock known only to A, and the safe is left open and made available to passers-by. Then anyone, including B, can put messages in the safe and lock it. However, only A can retrieve the message, since even those, such as B, who left a message in the box have no way of retrieving the message.

What the infeasibility of inverting the enciphering function in a secure public-key cryptosystem tacitly assumes is that the receiver of the message must have some additional information for deciphering. This is the content of the next idea that is central to public-key cryptography.

Definition 3.10 (Trapdoor One-Way Functions)

A trapdoor one-way function *or* public-key enciphering function *is a one-way function*

$$f : \mathcal{M} \mapsto \mathcal{C}$$

satisfying the additional property that there exists information, called trapdoor *information, or simply* trapdoor, *the knowledge of which makes it feasible to find $m \in \mathcal{M}$ for a given $c \in \text{img}(f)$ such that $f(m) = c$, but without the trapdoor this task becomes infeasible.*

The very heart of the Diffie-Hellman idea is the use of trapdoor one-way functions. As observed in the discussion of the Diffie-Hellman Protocol given in

---

[3.7]We use the convention that the term *private key* is reserved for use in association with public-key cryptography, whereas the term *secret key* is reserved for use in association with symmetric-key cryptosystems. This convention is used in the cryptographic community since it is justified by the fact that it takes two or more entities to share a secret, but a key is truly private when only one entity knows about it.

Section One, entities who have never met or exchanged information could now establish a shared secret by exchanging messages over an unsecured channel. This introduction of public-key cryptosystems meant that number theory would play an increasingly vital role in cryptography. One reason is that number theory is a treasure trove of one-way trapdoor functions. This is quite evident in the most widely used of public-key cryptosystems named after its inventors R. Rivest, A. Shamir, and L. Adleman, who introduced it in 1978 (see [170]).

◆ The RSA Public-Key Cryptosystem

We break the algorithm into two parts.

(I) RSA Key Generation

Each communicating entity should perform the following steps.

(1) Generate two large, random primes $p \neq q$ of roughly the same size.

(2) Compute both $n = pq$ and $\phi(n) = (p-1)(q-1)$.[3.8] The integer n is called the (*RSA*) *modulus*.

(3) Select a random $e \in \mathbb{N}$ such that $1 < e < \phi(n)$ and $\gcd(e, \phi(n)) = 1$. The integer e is called the (*RSA*) *enciphering exponent*.[3.9]

(4) Use the extended Euclidean Algorithm, Theorem A.22, to compute the unique $d \in \mathbb{N}$ with $1 < d < \phi(n)$ such that $ed \equiv 1 \,(\mathrm{mod}\ \phi(n))$.

(5) The (*RSA*) *public-key* is (n, e) and the (*RSA*) *private key* is d. The integer d is called the (*RSA*) *deciphering exponent*.

(II) RSA Public-Key Cipher

If entity A wishes to send a message to entity B, then A must perform the following steps.

enciphering stage:

(1) Obtain B's public-key (n, e).

(2) Translate the plaintext message into base-N numerical equivalents for a suitable $N > 1$. These numerical equivalents are then subdivided into blocks of equal size $\ell \in \mathbb{N}$.

(For instance, ℓ may be chosen such that $N^\ell < n < N^{\ell+1}$.)

[3.8]Instead of $\phi(n)$, one may use the Carmichael Function $\lambda(n)$, which is defined in the solution to Exercise 2.57 on page 314. If p and q are chosen such that $\gcd(p-1, q-1)$ is small, then $\phi(n)$ and $\lambda(n)$ are about the same size.

[3.9]If one needs to execute RSA key generation *often*, say for numerous communications between banks, then we must ensure that finding such an e over a large number of trials is fast. In other words, we need to know that calculating $1 - \phi(n)/n$ over N trials can be made arbitrarily small for large N. For instance, this will happen if $(1 - \phi(n)/n)^N < (1/2)^N$. In particular, if $n = pq$ is an RSA modulus, then $\phi(n)/n = (1 - 1/p)(1 - 1/q) > (4/5)^2 = .64 > 1/2$. It can be shown by more sophisticated techniques that for arbitrary n, $\phi(n)/n > (1 + o(1))e^{-\gamma}/\ln\ln n$, where $\gamma = \lim_{n \to \infty}(1 + 1/2 + \cdots + 1/n - \ln n)$ is *Euler's Constant*.

(3) Encipher each block $m \in \mathcal{M}$ separately by computing $c \equiv m^e \pmod{n}$ using the repeated squaring method given on page 71.

(4) Send each block $c \in \mathcal{C}$ to B.

Once B receives c, then the following is performed.

deciphering stage:

(1) Use d to compute $m \equiv c^d \pmod{n}$.

By Exercise 3.11, m is indeed deciphered using this technique. Since $\mathfrak{E}_e(m)$, given by $m^e \equiv c \pmod{n}$, is a one-way function with trapdoor d (or (p,q)), the RSA Cryptosystem is an example of a Public-key Cryptosystem. A cryptanalyst faces the problem of recovering c knowing only (n,e), the public key. This is called the *RSA Problem*. (In other words, the RSA Problem is that of finding eth roots in $(\mathbb{Z}/n\mathbb{Z})^*$ for a composite integer n (see Definition 3.13 on page 141).) The consensus is that the RSA Problem is equivalent to the Factoring Problem, although there is no known proof.[3.10] If a cryptanalyst can factor the modulus, then the RSA Cryptosystem can be broken. This is made clearer by Exercises 3.19–3.20, where we see that factoring n is equivalent to computing $\phi(n)$ in terms of complexity, and this is in turn "computationally equivalent" to computing d from the public key (n, e). In other words, knowing one can be converted into an algorithm for determining the other.

Public-key Ciphers are usually much slower than those using symmetric keys. For instance, at its fastest RSA is roughly a thousand times slower than DES. Thus, in practice, public-key cryptography is used for the sending of keys that are later used for the bulk of the data enciphering by Symmetric Ciphers. Public-key cryptosystems are also used for other applications including enciphering small data items such as PINs and credit card numbers.

We now present a small example, which is not realistic in terms of the size of p and q, since these primes would have to be of 154 digits (512 bits) each, ensuring a modulus of 308 digits (1024 bits) for long-term security. Thus, the following is strictly for illustrative and pedagogical purposes.

Example 3.11 Let $p = 29$, $q = 67$, so $n = 1943$ and $\phi(n) = 1848$. Choose base $N = 26$. Then since $26^2 < n < 26^3$, we will choose blocks of length $\ell = 2$. Select $e = 701$. Then using the extended Euclidean Algorithm, we may calculate d from $1 = ed + \phi(n)x = 701d + 1848x$, the solution of which yields $d = 29$ (and $x = -11$). We use the following table to convert plaintext into numerical equivalents. The only difference between this table and Table 2.37 is that we have two digits for all numerical equivalents so there are extra leading zeros.

Table 3.12

A	B	C	D	E	F	G	H	I	J	K	L	M
00	01	02	03	04	05	06	07	08	09	10	11	12
N	O	P	Q	R	S	T	U	V	W	X	Y	Z
13	14	15	16	17	18	19	20	21	22	23	24	25

[3.10]However, in 1998, Boneh and Venkatesan [35] provided some evidence that they may *not* be equivalent, so caution should be exercised.

Suppose that we want to encipher **EXAMS ARE HARD**. Then by using Table 3.12, we obtain the following numerical equivalents of the plaintext in base-26 blocks of 2 each.

$$\mathcal{M} = \{04, \quad 23, \quad 00, \quad 12, \quad 18, \quad 00, \quad 17, \quad 04, \quad 07, \quad 00, \quad 17, \quad 03\}.$$

Then we encipher each $m \in \mathcal{M}$ via $m^{701} \equiv c \,(\mathrm{mod}\ 1943)$. For instance, for the block $m = 04$, we have $4^{701} \equiv 613 \,(\mathrm{mod}\ 1943)$. Similarly, we get the remaining ciphertext block messages:

$$\mathcal{C} = \{613, \quad 458, \quad 0, \quad 1926, \quad 1439, \quad 0, \quad 1119, \quad 613, \quad 616, \quad 0, \quad 1119, \quad 206\}.$$

The reader may verify that we decipher each ciphertext block via

$$c^{29} \equiv m \,(\mathrm{mod}\ 1943).$$

For instance, $613^{29} \equiv 4 \,(\mathrm{mod}\ 1943)$.

For further examples, see Exercises 3.12–3.13.

It is worthy of discussion to look at the possible choices for the various parameters in the RSA Algorithm (or simply RSA henceforth for convenience). For instance, in the choice of the primes p and q, one should not choose them too close together. The reason is hidden in a discussion that we had earlier on page 24, where we talked about Fermat's "Difference of Squares Method". Suppose that $p > q$ and p is "close" to q, in the sense that $(p + q)/2$ is only slightly bigger than $\sqrt{n} = \sqrt{pq}$. Given that

$$\left(\frac{p+q}{2}\right)^2 - n = \left(\frac{p-q}{2}\right)^2,$$

then we have a solution $(x, y) \in \mathbb{N} \times \mathbb{N}$ to $x^2 - n = y^2$. To factor n we need only test those integer values $x > \sqrt{n}$ until $x^2 - n = y^2$ for some $y \in \mathbb{N}$, since $n = (x-y)(x+y)$. Thus, choosing p and q too close together will make this task much easier. Take, for instance, our modulus in Example 3.11, $n = 1943$ where $\lfloor \sqrt{n} \rfloor = 44$. Testing for $x = 45, 46, 47$ does not yield a square, but for $x = 48$ we get $48^2 - 1943 = 19^2$, so we factor n as $n = 1943 = (48 - 19)(48 + 19) = 29 \cdot 67$. Of course, our example is too small to be realistic, for reasons discussed above, but even with much larger primes the idea works, so caution must be exercised. Certain precautions should also be taken in the choice of p and q to ensure that $p - 1$ and $q - 1$ do not have any large common factor. If they have such a factor, then it becomes easier to obtain the deciphering exponent d since the inverse of the enciphering exponent e modulo $\mathrm{lcm}(p - 1, q - 1)$ suffices for d. Take an example of the extreme case where $p = 23$, $q = 67$, and $e = 5$. Here $22 = (p - 1) \mid (q - 1) = 66$, so we need only compute the inverse of e modulo $q - 1$, which is $d = 53$. Another caution is that $\phi(n)$ should have a large prime factor, or more precisely not be a product of only small prime factors. The reasons for this have been discussed when we talked about the Diffie-Hellman

Algorithm where we provided a solution to this problem, namely choose $(p-1)/2$ (and $(q-1)/2$) to be prime, which solves the aforementioned lcm problem as well (see Footnote 3.5). Such primes are called *safe primes*. It is suspected, but there is no proof, that there are infinitely many safe primes. Lastly, on these cautionary issues, although one may select an enciphering exponent e in the range $1 < e < \phi(n)$, some of these choices are very bad. For instance, if $e = \phi(n)/2 + 1$, then by Exercise 3.14,

$$m^e \equiv m \,(\text{mod } n)$$

for all $m \in \mathfrak{M}$, clearly not a desirable outcome.

In order to consider the next cryptosystem we need the following notion.

Definition 3.13 (Modular Roots and Power Residues)

If $m, n \in \mathbb{N}$, $c \in \mathbb{Z}$, $\gcd(c, n) = 1$, then c is called an mth power residue modulo n if $x^m \equiv c \,(\text{mod } n)$ for some $x \in \mathbb{Z}$, and x is called an mth root modulo n.

For instance, if $m = 2$, then x is called a *square root modulo* n, and c is called a *quadratic residue modulo* n; if $m = 3$, then c is called a *cubic residue modulo* n, and x is called a *cube root modulo* n, and so on.

If we know a factorization of n into prime powers, then we may find m^{th} roots modulo each prime power and use the Chinese Remainder Theorem to obtain an mth root modulo n.

The reader may now solve Exercise 3.16 in preparation for the following. The RSA cryptosystem is conjectured to have security equivalent to the intractability of factoring, but as noted earlier, nobody has been able to prove this. The following, on the other hand, is an example of a Public-key Cipher that is *provably* secure.

◆ The Rabin Public-Key Cryptosystem[3.11]

As with RSA, we break the algorithm into two parts.

(I) Rabin Key Generation

Each communicating entity should perform the following steps.

(1) Generate two large, random primes $p \neq q$ of roughly the same size.

(2) Compute $n = pq$.

(3) The public key is n and the private key is (p, q).

[3.11]Rabin's original system was not meant to be a cryptosystem, but rather was intended to be a signature scheme (see page 149). The problem with this scheme is the ambiguity in plaintext, which makes it unsuitable for a cryptosystem unless certain modifications such as those made by H. C. Williams are put in place (see [134]).

(II) Rabin Public-Key Cipher

If entity A wishes to send a message to entity B, then A must perform the following steps.

enciphering stage:

(1) Obtain B's public-key n.

(2) Convert the plaintext message into numerical equivalents in the range $0, 1, \ldots, n - 1$.

(3) Compute $c \equiv m^2 \pmod{n}$.

(4) Send c to B.

Once B receives c, then the following is performed.

deciphering stage:

(1) Find the four square roots m_j for $j = 1, 2, 3, 4$ of c modulo n (see Exercise 3.16).

(2) The plaintext is one of the m_j.

The clear problem is that B has the (possibly difficult) task of determining which of the four m_j is the plaintext message. If it is deciphered into English text, then the task should be easy. However, if they are random bitstrings, which is more likely, then there may be no easy way to determine the correct m_j. (One can see that this is particularly bad if the "plaintext" is a key for a single-key system.) One way to overcome this problem is to build redundancy into the original plaintext. For instance, by repeating a fixed length bitstring at the end of the message, it is highly unlikely that the other square roots will have this redundancy so B will choose this one as the intended plaintext.

Example 3.14 Let $p = 31$, $q = 167$ and $n = 5177$, so n is the public key, and $(31, 167)$ is the private key. Let the plaintext be $m = 101110$. Build in redundancy by repeating the last three digits to get $m' = 101110110$, which is 374 in decimal. Then compute

$$m'^2 = 374^2 \equiv 97 = c \pmod{n}.$$

Using Exercise 3.16 we compute the four square roots of 97 modulo 5177, $m_1 = 3133$, $m_2 = 2044$, $m_3 = 374$, and $m_4 = 4803$. Their binary representations are $m_1 = 110000111101$, $m_2 = 11111111100$, $m_3 = 101110110$, and $m_4 = 1001011000011$, only m_3 of which has the redundancy, so B deciphers m_3 to recover m.

Another public-key cryptosystem, the security of which depends upon the intractability of discrete log and the Diffie-Hellman Problem, is the following.

◆ The ElGamal Cryptosystem

(I) ElGamal Key Generation

Each communicating entity should perform the following steps.

(1) Generate a large random prime p and a generator α of the multiplicative group $(\mathbb{Z}/p\mathbb{Z})^*$.

(2) Select a random natural number $a < p - 1$ and compute $\alpha^a \pmod{p}$.

(3) The public key is (p, α, α^a) and the private key is a.[3.12]

(II) ElGamal Public-Key Cipher

If entity A wishes to send a message to entity B, then A must perform the following steps.

enciphering stage:

(1) Obtain B's public-key (p, α, α^a).

(2) Convert the plaintext message into an integer m in the range $0, 1, \ldots, p-1$.

(3) Choose a random natural number $b < p - 1$.

(4) Compute $\beta \equiv \alpha^b \pmod{p}$ and $\gamma \equiv m(\alpha^a)^b \pmod{p}$.

(5) Send the ciphertext $c = (\beta, \gamma)$ to B.

Once B receives c, then the following is performed.

deciphering stage:

(1) Use the private key to compute $\beta^{p-1-a} \pmod{p}$.

(2) Decipher m by computing $\beta^{-a}\gamma \pmod{p}$.

The deciphering actually is accomplished due to the following fact.

$$\beta^{-a}\gamma \equiv \alpha^{-ab}m\alpha^{ab} \equiv m \pmod{p}.$$

Example 3.15 Let $p = 1777$ and choose a generator $\alpha = 5$ of $\mathbb{F}_{1777}^*$. Entity B chooses a private key $a = 146$, and computes

$$\alpha^a = 5^{146} \equiv 1729 \pmod{p}.$$

Thus, B's public key is $(p, \alpha, \alpha^a) = (1777, 5, 1729)$, which A gets from a public file. To encipher $m = 1483$, say, A selects $b = 1066$ randomly and computes

$$\beta = 5^{1066} \equiv 1664 \pmod{1777},$$

[3.12]In 1985, a modification of ElGamal enciphering using the unit group of $(\mathbb{Z}/n\mathbb{Z})^*$ for $n \in \mathbb{N}$ was proposed by McCurley [132]. Cryptanalyzing his key-exchange system is provably *at least* as difficult as factoring the modulus n. Moreover, if a cryptanalyst C factors n, then C is still left with the task of solving the Diffie-Hellman Problem modulo the factors of n.

and
$$\gamma \equiv 1483 \cdot 1729^{1066} \equiv 625 \,(\mathrm{mod}\ 1777).$$

Then A sends $(\beta, \gamma) = (1664, 625)$ to B. To decipher, B computes

$$\beta^{p-1-a} = 1664^{1777-1-146} \equiv 1768 \,(\mathrm{mod}\ 1777),$$

and recovers m by computing

$$m = \beta^{-a}\gamma \equiv 1768 \cdot 625 \equiv 1483 \,(\mathrm{mod}\ 1777).$$

The reader should now solve Exercises 3.17–3.18.

To see that the ElGamal Cryptosystem is based on the discrete log problem in $\mathbb{F}_p^*$, we observe that, in order to recover

$$m = \beta^{-a}\gamma,$$

a cryptanalyst C knows only $p, \alpha, \alpha^a, \beta$, and γ. Thus, ElGamal is essentially a Diffie-Hellman Key Exchange on $k = \alpha^{ab}$, which is used to encipher m via km (step 4). However, as noted earlier, there is no proof of the equivalence of discrete log and the security of ElGamal. As with RSA, a modulus of 1024 bits is recommended for long-term security. Later we will see that the *Elliptic Curve Cryptosystem* is essentially a generalization of the ElGamal Cryptosystem.

Exercises

3.11. Prove that decryption in the RSA Public-key Cipher actually recovers m. In other words, prove that computing c^d yields m as the least positive residue modulo n.

3.12. Assume that using the RSA Cipher with $p = 167$, $q = 1009$, $N = 26$, $e = 9547$, and $\ell = 3$, the following ciphertext message units were obtained: 20908, 111777, 2828, 17029, 68641, 130441, 39468, 34112, 153750, 70491, 47669. Use RSA to decipher the numerical equivalents of the plaintext. Then use Table 3.12 to give the English plaintext.

3.13. Given $p = 97$, $q = 167$, $N = 26$, $\ell = 2$, and $e = 797$, in the RSA Cryptosystem, assume that the following are the ciphertext message units obtained: 14238, 7052, 4454, 2444, 2684, 14560, 0, 3314, 11500, 14238, 2684, 10952. Decipher and use Table 3.12 to give the English plaintext.

3.14. Let
$$e = \phi(n)/2 + 1$$

be an RSA enciphering exponent. Prove that, for any $m \in \mathfrak{M}$, $m^e \equiv m$ (mod n).

3.15. Suppose that a cryptanalyst C wants to decipher

$$c \equiv m^e \pmod{n}$$

to recover ciphertext c enciphered using RSA, and sent by entity A to entity B. Furthermore, suppose that C can intercept and disguise c by selecting a random $x \in (\mathbb{Z}/n\mathbb{Z})^*$ and computing

$$\bar{c} \equiv cx^e \pmod{n}.$$

Not knowing this, B computes

$$\bar{m} \equiv \bar{c}^d \pmod{n},$$

and sends it to A. How can m be recovered if C intercepts $\bar{m}$? (*This is an example of an adaptive chosen-ciphertext attack on RSA (see page 104). Such attacks on RSA can be thwarted by ensuring that the plaintext messages have a certain structure, which is unlikely to be maintained if disguised by C. Then if B receives a ciphertext that decrypts to a plaintext without this structure, c is rejected by B as being fraudulent. For numerous other attacks on RSA, see [134].*)

3.16. Suppose that $a \in \mathbb{N}$ and

$$a \equiv z^2 \pmod{pq} \text{ where } p \equiv q \equiv 3 \pmod{4} \text{ are primes.}$$

Prove that there are only four possible square roots of a modulo pq, and they are given as follows. For $x, y \in \mathbb{Z}$ given by the extended Euclidean Algorithm, Theorem A.22, such that

$$xp + yq = 1$$

we have:

$$z = \pm(xpa^{(q+1)/4} + yqa^{(p+1)/4}), \text{ and } z = \pm(xpa^{(q+1)/4} - yqa^{(p+1)/4}).$$

(*Hint: Use Lagrange's Theorem on page 294.*)

3.17. As in Example 3.15, let $p = 1777$, $\alpha = 6$, $\mathbb{F}_{1777}^*$, and $a = 1009$. Use the ElGamal Cryptosystem to encipher $m = 1341$ assuming that A selects $b = 701$.

3.18. Using the same values of p, α, a, b as in Exercise 3.17, decipher $(\beta, \gamma) = (1664, 1031)$ using ElGamal.

3.19. Assume that $n = pq$ where $p \neq q$ are primes. Prove that the computational complexity of computing $\phi(n)$ is the same as that for factoring n.

3.20. Prove that computing the deciphering exponent d in the RSA algorithm is computationally equivalent to factoring $n = pq$, namely they have the same complexity. In other words, since the above discussion already shows that knowing how to factor n allows us to compute d, prove that knowing how to compute d can be converted into an algorithm for factoring n.

3.3   Authentication

What has not been addressed in the discussion of public-key cryptography is the need for *authentication*, meaning that both the origin of data and the entity who sent it must somehow be verified. To see why this is necessary in public-key cryptography, one may witness one type of interception by a cryptanalyst as described in Exercise 3.15. The problem, in general, is described and illustrated as follows. Suppose that entities A and B are communicating via some public-key cryptosystem, and a cryptanalyst C is listening to the channel. C can impersonate B by sending A a public key e' (which A will assume to be the public key $e \neq e'$ of B). Then A sends

$$\mathfrak{E}_{e'} = m^{e'} = c'$$

to B. C intercepts the enciphered message and deciphers using private key d' via

$$(c')^{d'} = m^{e'd'} = m.$$

Then C enciphers m with public key e and sends $c = m^e$ to B, impersonating A, and neither A nor B is the wiser. This process is illustrated in the following.

Diagram 3.16 (Impersonation Attack on Public-Key Cryptosystems)

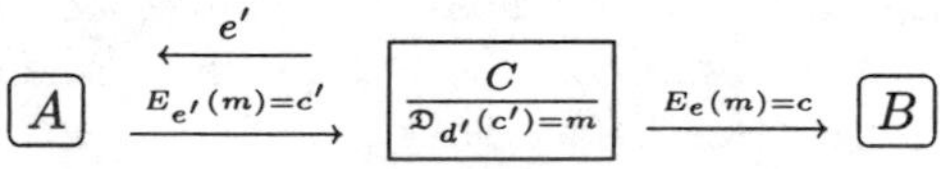

Before formalizing the notion of a digital signature, we should compare and contrast the features of a digital signature vs. a conventional (handwritten) signature. For instance, once a conventional signature is put upon a document, then the signature becomes a physical part of that document. With a digital signature algorithm, there must be some means incorporated for "binding" the signature to the message. Also, there is the issue of verification. A conventional signature, such as that on a credit card, is verified by comparing the signature on the credit card with that on the sales slip. This is, of course, very insecure since a criminal can forge a signature. Digital signatures, however, are secure since they can be verified with a (publicly known) verification algorithm, which must be part of the signature algorithm. Since people cannot disavow signatures, the verification algorithm must take care of this in some fashion as well. Finally, copies of documents with handwritten signatures can usually be identified as distinct from the original, whereas "copies" of digital signatures are identical to each other. Hence, care must be exercised in order to prevent unauthorized reuse, for instance by binding the date to the signature.

We have already discussed some methods for authentication, called MACs (see page 103). We now look at some more in greater detail starting with a formalization of the notions surrounding authentication.

Definition 3.17 (Digital Signatures)

Let $\mathcal{M}$ be a message space, and $\mathcal{K}$ be a key space.

(1) A digital signature *is a digital data string that associates a given* $m \in \mathcal{M}$ *with its sender. In other words,* $\mathcal{M}$ *is the set of elements to which a signer can affix a digital signature.*

(2) *Let* S *be a set of elements, called the* signature space, *(usually bitstrings) of fixed length used to bind the signer to the message.* A redundancy function[3.13] *is an injective function*

$$R : \mathcal{M} \mapsto \mathcal{M}_S,$$

where $\mathcal{M}_S$ *is a set of elements called the* signing space. *For each* $k \in \mathcal{K}$, *there is* digital signature transformation, *or* signature transformation, *which is an injective mapping*

$$sig_k : \mathcal{M}_S \mapsto S.$$

(3) *A method for producing a digital signature,* $sig_k \circ R : \mathcal{M} \mapsto S$, *is called a* digital signature generation algorithm, *or* signature generation algorithm.

(4) *A* digital verification algorithm *or* verification algorithm *is a method for verifying that a digital signature is authentic.*

(5) *A* digital signature scheme, *or* signature scheme *is comprised of two parts: a signature generation algorithm, and a signature verification algorithm.*

(6) *A* signature scheme with *message recovery is a signature scheme for which the message being sent is not required as input to the validation algorithm. In this case, the original message is recovered from the signature itself. A* signature scheme with appendix *is a signature scheme where the message is required as input for the verification algorithm.*

We now describe signature schemes related to the public-key cryptosystems just studied. Note that any public-key system with $\mathcal{M} = \mathcal{C}$ can be used for signatures. The sender can use the private key to encipher the message m, thereby obtaining the signature s and the receiver can decipher the signature s using the sender's public key to compare it with the message m.

◆ **The RSA Signature Scheme**

Let $\mathcal{M} = \mathcal{C} = S = \mathcal{M}_S = \mathbb{Z}/n\mathbb{Z}$ where $n = pq$ with $p \neq q$ randomly chosen primes. A publicly known redundancy function $R : \mathcal{M} \mapsto \mathcal{M}_s$ is chosen.

[3.13]Selection of the redundancy function is critical in signature schemes. For example, such a selection should not be made independently of sig_k since this could compromise the security of the signature scheme. The first international standard for digital signatures published in 1991 by the International Standards Organization is called ISO/IEC 9796 (see [134, pp. 442–444]) This provides a secure redundancy function, which can be used with both the RSA and Rabin Signature Schemes described below.

The first step in the process is that each entity generates public and private keys using the RSA key generation algorithm given on page 138. Then the following is executed.

RSA Signature Generation and Verification
Entity A signs $m \in \mathcal{M}$ and entity B verifies A's signature as follows.

signature generation stage:
A performs the following steps.

(1) Compute $R(m) = m'$.

(2) Compute

$$s = \mathrm{sig}_d(m) \equiv (m')^d \pmod{n},$$

and send s to B.

verification stage:
B performs the following steps.

(1) Obtain A's public-key (n, e).

(2) Compute $m' \equiv s^e \pmod{n}$.

(3) Verify that $m' \in \mathrm{img}(R)$, and reject if it is not.

(4) Recover $m = R^{-1}(m')$.

Thus, we see that the RSA Signature Scheme is a signature scheme with message recovery. We now illustrate, where as usual we must use unrealistically small values for pedagogical purposes.

Example 3.18 Let $p = 907$ and $q = 557$ be the primes chosen by A, and let $R(m) = m$ for all $m \in \mathcal{M}$. Thus, $n = 505199 = pq$ and $\phi(n) = 503736$. Assuming that A chooses $e = 769$, then using the extended Euclidean Algorithm, A computes $d = 49129$. Therefore, A's public key is $(e, n) = (769, 505199)$ and the private key is $d = 49129$. Suppose that the message to be signed is $m = 10411$. Then A computes

$$R(m) = m' = m = 10411,$$

and

$$\mathrm{sig}_d(m') = (m')^d = 10411^{49129} \equiv 495549 \equiv s \pmod{505199}$$

and sends s to B, which completes the signature generation stage. Now B completes the verification stage by computing

$$s^e \equiv 495549^{769} \equiv 10411 \equiv m \pmod{505199}.$$

The signature and message are accepted since $m \in \mathrm{img}(R)$.

The reader should now try Exercises 3.21–3.22.

A signature scheme similar to *RSA* is the following.

◆ **The Rabin Public-Key Signature Scheme**

The first step is for the communicating entities to generate public and private keys using the Rabin Key Generation Algorithm given on page 141. Then for entity A to sign a message for delivery to B, the following is executed.

Rabin Signature Generation and Verification

Assume that for $n \in \mathbb{N}$, $\mathcal{M}_S$ consists of the set of distinct least nonnegative quadratic residues modulo n (see Definition 3.13). Also, assume that $\mathcal{M} = \mathbb{Z}/n\mathbb{Z}$ and $\mathrm{sig}_n : \mathcal{M}_S \mapsto \mathcal{M}_S$ is defined as the taking of square roots modulo n and that $R : \mathcal{M} \mapsto \mathcal{M}_S$ is chosen and made publicly available.

<u>signature generation stage:</u>

A performs the following steps.

(1) Compute $R(m) = m'$.

(2) Compute a square root s of m' modulo n using Exercise 3.16, and send s to B.[3.14]

<u>verification stage:</u>

B performs the following steps.

(1) Obtain A's public-key n.

(2) Compute $m' \equiv s^2 \pmod{n}$.

(3) Verify that $m' \in \mathrm{img}(R)$, and reject if it is not.

(4) Compute $R^{-1}(m') = m$ to recover the message.

Example 3.19 Let $(p, q) = (647, 919)$ be A's private key, so $n = 594593$ is A's public key. The signing space $\mathcal{M}_S$ consists of all of the quadratic residues modulo n. Let $m = 298081$,

$$R(m) = m - 1^{3.15}$$

and $\mathrm{img}(R) = \{0, 1, \ldots, 300000\}$. Then A computes a square root of $m' = 298080$ which is $s = 220578$ and sends it to B. B computes

$$m' = s^2 \equiv 298080 \in \mathrm{img}(R),$$

so B accepts it and recovers

$$m = R^{-1}(m') = m' + 1 = 298081.$$

[3.14]Choosing the redundancy function here is crucial since there is no guarantee that $R(m)$ is a quadratic residue modulo n, so computing a square root could be impossible. To overcome this problem, there is a method similar to ISO/IEC 9796, discussed in Footnote 3.13, for associating the message with elements of $\mathcal{M}_S$ such that computing square roots is virtually always possible (see [134, pp. 439–440]).

[3.15]In this case, we will assume that m is chosen such that $m - 1$ is a nonnegative quadratic residue modulo n.

The last signature scheme that we will study here is our first example of a signature scheme with appendix as follows.

◆ **The ElGamal Signature Scheme**[3.16]

This scheme requires a hash function $h : \mathcal{B} \mapsto \mathbb{Z}/p\mathbb{Z}$ where $\mathcal{B}$ denotes the set of bitstrings of arbitrary length.

ElGamal Signature Generation and Verification

First, each entity must generate a public and a private key according to the ElGamal Key Generation Algorithm given on page 143. Then A has public key (p, α, α^a) and private key a. Now, entity A signs $m \in \mathcal{M}$ and entity B verifies A's signature as follows.

signature generation stage:
A performs the following steps.

(1) Randomly choose $r \in \mathbb{N}$ such that $r < p - 1$ and $\gcd(r, p - 1) = 1$.

(2) Compute $\beta \equiv \alpha^r \pmod{p}$.

(3) Compute r^{-1} modulo $p - 1$.

(4) Compute $s = r^{-1}(h(m) - a\beta)$ modulo $p - 1$.

(5) The signature for m is (β, s) which is sent to B.

verification stage:
B performs the following steps.

(1) Obtain A's public key (p, α, α^a).

(2) Test that $\beta \in \mathbb{N}$ with $\beta < p$, and reject if it is not.

(3) Compute $x \equiv \alpha^{a\beta} \beta^s \pmod{p}$.

(4) Compute $h(m)$ and $z \equiv \alpha^{h(m)} \pmod{p}$.

(5) Accept the signature only if $x = z$, in which case m is a valid message.

Observe that the verification stage works for the following reasons. Since A's signature satisfies

$$s \equiv r^{-1}(h(m) - a\beta) \pmod{p - 1},$$

then by rearranging, we have

$$h(m) \equiv sr + a\beta \pmod{p - 1}.$$

Therefore,

$$z \equiv \alpha^{h(m)} \equiv \alpha^{sr + a\beta} \equiv (\alpha^r)^s \alpha^{a\beta} \equiv \beta^s \alpha^{a\beta} \equiv x \pmod{p}.$$

[3.16]There is another signature scheme that became FIPS 186 on May 19, 1994, called the *Digital Signature Standard* (DSS), which we will not study here. The DSS is a modification of the ElGamal Signature Scheme. For more information see [134] and [193].

Example 3.20 Suppose that A selects $p = 3677$, $a = 107$, and $\alpha = 2$ as a generator of $(\mathbb{Z}/p\mathbb{Z})^*$. Then A computes

$$\alpha^a = 2^{107} \equiv 1867 \,(\text{mod } p).$$

Thus,

$$(p, \alpha, \alpha^a) = (3677, 2, 1867)$$

is A's public key and $a = 107$ is the private key. Suppose that the message to be sent is

$$m = 1964 \in (\mathbb{Z}/p\mathbb{Z})^*$$

and

$$h(m) \equiv m - 982 \,(\text{mod } p),$$

so $h(1964) = 982$. A selects $r = 69$ randomly and computes

$$\beta = \alpha^r = 2^{69} \equiv 2808 \,(\text{mod } 3677),$$

$$r^{-1} \equiv 1545 \,(\text{mod } 3676),$$

and

$$s = r^{-1}(982 - 107 \cdot 2808) \equiv 3438 \,(\text{mod } 3676).$$

Thus, A signs $m = 1964$ with

$$(\beta, s) = (2808, 3438)$$

and sends it off to B. Using A's public key, B checks that $\beta = 2808 < 3677$ and accepts it. Then B computes

$$x = \alpha^{a\beta}\beta^s = 1867^{2808} \cdot 2808^{3438} \equiv 2737 \,(\text{mod } 3677),$$

$h(m) = 982$, and

$$z = \alpha^{h(m)} = 2^{982} \equiv 2737 \,(\text{mod } 3677).$$

Since $x = z$, then B accepts m as a valid message.

Exercises 3.25–3.26 are intended to test knowledge of the ElGamal Signature Scheme.

As with both the RSA and Rabin Signature Schemes, a modulus of 1024 bits is recommended for long-term security of the ElGamal Signature Scheme. Given that a cryptanalyst must essentially solve the Discrete Log Problem in order to recover s, then the security of the ElGamal Scheme rests upon the intractability of discrete log.

There are numerous other signature schemes (covered for instance in [134]). The original concept of a digital signature scheme was introduced by Diffie and Hellman in 1976 (see [62]), which was a signature scheme with message recovery. However, it took until 1978 for the first practical signature scheme to come into being with the inception of the RSA Scheme given in [170], which was based on public-key methods. The RSA Signature Scheme remains viable today as an important, practical technique. The Rabin Signature Scheme introduced in 1979 ([168]) is a highly efficient method since it needs only one modular multiplication. The ElGamal Signature Scheme was introduced by ElGamal in 1984 (see [68]–[69]), and its variant DSS, mentioned in Footnote 3.16, is due originally to Kravitz (see [109]).

Exercises

3.21. Given $p = 157$, $q = 919$, $e = 5$, $n = 144283$,

$$R(m) = m - 1 \text{ for all } m \in \mathcal{M} = \mathbb{Z}/n\mathbb{Z},$$

and

$$\text{img}(R) = \{0, 1, 2, \ldots, 3000\}.$$

Should B accept the message $s = 49000$ sent by A under the RSA Signature Scheme? If so, what is the recovered message?

3.22. Suppose that $p = 1297$, $q = 3209$, $e = 2701$, $n = 4162073$

$$R(m) = 2m \text{ for any } m \in \mathbb{Z}/n\mathbb{Z},$$

and

$$\text{img}(R) = \{0, 1, 2, \ldots, 5001\}.$$

Should B accept signature $s = 4053313$ from A under the RSA Signature Scheme? If so, what is the recovered message?

3.23. Let p, q, n, R be given as in Example 3.19. Should $s = 12535$ be accepted by B? If so, what is m?

3.24. Given the same information as in Exercise 3.23, should B accept message $s = 131313$?

3.25. Given the values for p, a, α, and $h(m)$ in Example 3.20 should B accept the message $m = 3011$ signed with

$$(\beta, s) = (32, 3397)$$

from A?

3.26. With the same input as for Exercise 3.25, should B accept message $m = 1103$ from A with signature

$$(\beta, s) = (32, 1545)?$$

3.4 Knapsack

In this section, we study the class of problems named in the section header and some Public-key Cryptosystems based upon them. This class of problems is a generalization of the following notion.

Definition 3.21 (The Subset Sum Problem)
   Given $m, n \in \mathbb{N}$ and a set $S = \{b_j : b_j \in \mathbb{N}, \text{ for } j = 1, 2, \ldots, n\}$, called a knapsack set, determine whether or not there exists a subset S_0 of S such that the sum of the elements in S_0 equals m. In other words, given S and $m \in \mathbb{N}$, determine whether or not there exist $a_j \in \{0, 1\}$, for $j = 1, 2, \ldots, n$, such that

$$\sum_{j=1}^{n} a_j b_j = m. \tag{3.22}$$

It is known that the Subset Sum Problem is **NP**-complete (see page 53). The Subset Sum Problem is often mistakenly identified as the following more general problem.

Definition 3.23 (The Knapsack Problem)
   Given natural numbers m_1, m_2 and sets $\{b_j : b_j \in \mathbb{N}, \text{ for } j = 1, 2, \ldots, n\}$ and $\{c_j : c_j \in \mathbb{N}, \text{ for } j = 1, 2, \ldots, n\}$, determine whether or not there exists a set $S_0 \subseteq \{1, 2, \ldots, n\}$ such that

$$\sum_{j \in S_0} b_j \leq m_1 \text{ and } \sum_{j \in S_0} c_j \geq m_2.$$

(The Subset Sum Problem is the special case where $b_j = c_j$ for $j = 1, 2, \ldots, n$ and $m_1 = m_2 = m$.)

The knapsack problem derives its name from the special case of the Subset Sum Problem, which may be restated in nonmathematical terms as follows. Given a collection of items, each of different sizes, is it possible to put some of the items into a knapsack (of a given size) so that the knapsack is full? Computationally, this is equivalent to actually determining the a_js in Equation (3.22), given that such a_js in fact exist. This computational version of the Subset Sum Problem is **NP**-hard (which means that the existence of a polynomial time algorithm for its solution would imply that $\mathbf{P} = \mathbf{NP}$).
   Knapsack public-key cryptography is based on the Subset Sum Problem. The basic idea is to choose a case of the Subset Sum Problem which can be easily solved, then cryptographically disguise it to be an instance of the Subset Sum Problem that is hard to solve. In 1978, the first practical incarnation of such

a cryptosystem was introduced by Merkle and Hellman in [137].[3.17] Although this has been shown to be insecure, we present it for its historical implications in that we may see how the Knapsack Cryptosystems evolved. Later, we will present a secure Knapsack Cipher.

The Merkle-Hellman Cipher is based upon the following easily solved instance of the Subset Sum Problem.

Definition 3.24 (Superincreasing Sequences)

A superincreasing sequence is a sequence $(b_1, b_2, \ldots, b_n)$ *with* $b_j \in \mathbb{N}$ *for* $j = 1, 2, \ldots, n$ *satisfying the property that*

$$b_i > \sum_{j=1}^{i-1} b_j,$$

for each $i \in \{2, 3, \ldots, n\}$.

Solving the Subset Sum Problem for superincreasing sequences is shown to be easy as follows. To find a subset $\mathcal{S}_0$ of $\mathcal{S} = \{b_1, b_2, \ldots, b_n\}$ which sums to a given $d \in \mathbb{N}$ (assuming that such an $\mathcal{S}_0$ exists) one first sets $x_n = 1$ if $b_n \leq d$, and $x_n = 0$ otherwise. Then one merely looks at each successive b_{n-i} for $i = 1, 2, \ldots, n-1$ and one puts $x_{n-i} = 1$ if

$$b_{n-i} \leq d - \sum_{j=n-i+1}^{n} b_j$$

and put $x_{n-i} = 0$ otherwise. Then

$$d = \sum_{j=1}^{n} x_j b_j. \tag{3.25}$$

◆ **The Merkle-Hellman Knapsack Cryptosystem**

(I) Merkle-Hellman Key Generation

Given fixed $n \in \mathbb{N}$, entity B performs the following steps.

(1) Let $(b_1, b_2, \ldots, b_n)$ be a superincreasing sequence and $s \in \mathbb{N}$ such that $s > \sum_{j=1}^{n} b_j$.

(2) Randomly choose $r \in \mathbb{N}$ such that $r \leq s - 1$ and $\gcd(r, s) = 1$.

(3) Let $\sigma \in S_n$ (see Exercise 2.38 on page 100).

[3.17]The original algorithm could only be used for enciphering data, and later Shamir refined it for use with digital signatures (see [179]), but this was later broken by Odlyzko (see [147]). It was soon after the Merkle-Hellman Algorithm that the first complete cryptosystem for public-key cryptography came into being, used for both enciphering and digital signatures. This was the RSA Cryptosystem studied in the preceding section.

(4) Compute $k_j \equiv r b_{\sigma(j)} \pmod{s}$ for $j = 1, 2, \ldots, n$.

(5) B's public key is $(k_1, \ldots, k_n)$ and the private key is $(\sigma, r, s, (b_1, \ldots, b_n))$.

(II) Merkle-Hellman Knapsack Public-Key Cipher

A performs the following steps.

enciphering stage:

(1) Obtain B's public-key $(k_1, k_2, \ldots, k_n)$.

(2) Let $m = m_1 m_2 \ldots m_n$ be the representation of the message m as a bitstring of length n.

(3) Compute $c = \sum_{j=1}^{n} k_j m_j$ and send c to B.

Once B receives c, then the following is performed.

deciphering stage:

(1) Compute $d \equiv r^{-1} c \pmod{s}$.

(2) By solving the superincreasing Subset Sum Problem as described prior to Equation (3.25), obtain $x_j \in \{0, 1\}$ such that $d = \sum_{j=1}^{n} x_j b_j$.

(3) Recover $m = x_{\sigma(1)} x_{\sigma(2)} \cdots x_{\sigma(n)}$.

To see why step (3) of the deciphering stage works, we observe that

$$d \equiv r^{-1} c = r^{-1} \sum_{j=1}^{n} k_j m_j = \sum_{j=1}^{n} (r^{-1} k_j) m_j \equiv \sum_{j=1}^{n} b_{\sigma(j)} m_j \pmod{s}.$$

Since $s > d \geq 0$, then

$$d = \sum_{j=1}^{n} b_{\sigma(j)} m_j = \sum_{j=1}^{n} b_j x_j = \sum_{j=1}^{n} b_{\sigma(j)} x_{\sigma(j)},$$

so $x_{\sigma(j)} = m_j$ for each $j = 1, 2, \ldots, n$.

Example 3.26 Let $n = 7$, and assume that entity B selects superincreasing sequence $(b_1, b_2, b_3, b_4, b_5, b_6, b_7) = (1, 2, 4, 8, 16, 32, 70)$, $s = 200$, $r = 27$, and σ : $(1, 2, 3, 4, 5, 6, 7) \mapsto (7, 4, 5, 6, 3, 2, 1)$. Then B computes $k_j \equiv 27 b_{\sigma(j)} \pmod{200}$ for $1 \leq j \leq 7$ to get: $(k_1, k_2, k_3, k_4, k_5, k_6, k_7) = (90, 16, 32, 64, 108, 54, 27)$, which is B's public key, and the private key is $(\sigma, 27, 200, (1, 2, 4, 8, 16, 32, 70))$. Suppose that the message to be sent is $m = 1010011$. Using B's public key, A computes

$$c = \sum_{j=1}^{n} k_j m_j = 1 \cdot 90 + 0 \cdot 16 + 1 \cdot 32 + 0 \cdot 64 + 0 \cdot 108 + 1 \cdot 54 + 1 \cdot 27 = 203,$$

which gets sent to B, who then computes

$$d \equiv r^{-1}c = 27^{-1} \cdot 203 \equiv 163 \cdot 203 \equiv 89 \,(\mathrm{mod}\ 200).$$

Then B solves the superincreasing Subset Sum Problem as follows. Since $b_n = b_7 = 70 < d = 89$, then $x_7 = 1$. Since $b_{n-1} = b_6 = 32 > 89 - 70 = d - b_n$, then $x_6 = 0$; $b_{n-2} = b_5 = 16 < 89 - 70 = 19$ implies $x_5 = 1$; $b_{n-3} = b_4 = 8 > 89 - 70 - 16$ implies $x_4 = 0$; $b_{n-4} = b_3 = 4 > 89 - 70 - 16$ implies $x_3 = 0$; $b_{n-5} = b_2 = 2 < 89 - 70 - 16$ implies $x_2 = 1$; and $b_{n-6} = b_1 = 1$ implies $x_1 = 1$. Thus,

$$d = 89 = 70 + 16 + 2 + 1 = \sum_{j=1}^{n} x_j b_j.$$

Therefore, since $m_j = x_{\sigma(j)}$, then $m_1 = x_{\sigma(1)} = x_7 = 1$; $m_2 = x_{\sigma(2)} = x_4 = 0$; $m_3 = x_{\sigma(3)} = x_5 = 1$; $m_4 = x_{\sigma(4)} = x_6 = 0$; $m_5 = x_{\sigma(5)} = x_3 = 0$; $m_6 = x_{\sigma(6)} = x_2 = 1$; $m_7 = x_{\sigma(7)} = x_1 = 1$, and m is recovered.

Exercises 3.27–3.28 are related to Example 3.26.

In 1982, unfortunately for this historically important and elegant cipher, a polynomial-time algorithm for breaking it was produced by Shamir (see [180]–[181]). Also, in 1982 (see [47]–[48]) a new Knapsack Cryptosystem was proposed. Until recently, the following was the only known secure knapsack cipher. However, it was broken by S. Vaudenay (see J. Cryptology (2001), 87–100).

☞ The Chor-Rivest Knapsack Cryptosystem

(I) Chor-Rivest Key Generation

Entity B performs the following steps.

(1) Choose a finite field $\mathbb{F}_q$ of characteristic p, where $q = p^n$, $p > n$, and $p^n - 1$ has only small prime factors.

(2) Choose a random monic irreducible polynomial $r(x)$ with $\deg(r) = n$ over $\mathbb{F}_p$.[3.18] We may view the elements of $\mathbb{F}_q$ as those from $(\mathbb{Z}/p\mathbb{Z})[x]/(r(x))$ (see Example A.63 and the material preceding it).

(3) Let $\ell(x) \in (\mathbb{Z}/p\mathbb{Z})[x]/(r(x)) \cong \mathbb{F}_q$ be randomly chosen as a generator of $\mathbb{F}_q^*$ (see page 292).

(4) For each $\alpha \in \mathbb{Z}/p\mathbb{Z}$, find the discrete logarithm, $a_\alpha = \log_{\ell(x)}(x + \alpha)$ using the Pohlig-Hellman Algorithm on page 131.

(5) Choose a random permutation σ on $\{0, 1, \ldots, p - 1\}$.

(6) Randomly choose $d \in \mathbb{Z}$ such that $0 \leq d \leq p^n - 2$.

(7) For any nonnegative integer $j \leq p-1$ compute $c_j \equiv (a_{\sigma(j)} + d) \,(\mathrm{mod}\ p^n - 1)$.

[3.18] A method for generating irreducible polynomials is given for instance in Corollary A.50 on page 292.

(8) B has public key $((c_0, c_1, \ldots, c_{p-1}), p, n)$ and private key $(r(x), \ell(x), \sigma, d)$.

(II) Chor-Rivest Knapsack Public-Key Cipher

A performs the following steps.

enciphering stage:

(1) Obtain B's public key as given above.

(2) Let $\binom{p}{n}$ be the binomial coefficient (see Definition A.33) and represent the message m as a bitstring of length $\lfloor \log_2 \binom{p}{n} \rfloor$.

(3) Replace m by the binary p-tuple $V = (V_0, V_1, \ldots, V_{p-1})$ having exactly n values of i such that $V_i = 1$ for $0 \le i \le p-1$ determined as follows. For $i = 1, \ldots, p$, execute:

Step i. If $m \ge \binom{p-i}{n}$, then execute the following steps.

  (a) Set $V_{i-1} = 1$.
  (b) Reset $m = m - \binom{p-i}{n}$.
  (c) Reset $n = n - 1$.
  (d) If $i < p$, then go to step $i+1$. If $i = p$, terminate with output V.

  If $m < \binom{p-i}{n}$, then execute the following steps.

  (a) Set $V_{i-1} = 0$.
  (b) If $i < p$, then go to step $i+1$. If $i = p$, terminate with output V.

  (Note that if any value $\binom{m}{n}$ with $0 \le m < n$ is encountered, then this value is considered to be 0.)

(4) Compute $c \equiv \sum_{j=0}^{p-1} V_j c_j \pmod{p^n - 1}$, and send c to B.

Once B receives c, then the following is performed.

deciphering stage:

B performs the following steps.

(1) Compute $z \equiv (c - nd) \pmod{p^n - 1}$.

(2) Compute $h(z) \equiv \ell(x)^z \pmod{r(x)}$.

(3) Compute $k(x) = h(x) + r(x)$, where $deg_{\mathbb{Z}/p\mathbb{Z}}(k) = n$ (see Definition A.42 on page 290).

(4) Factor $k(x)$ into linear factors over $\mathbb{Z}/p\mathbb{Z}$:

$$k(x) = \prod_{j=1}^{n} (x + r_j) \qquad (r_j \in \mathbb{Z}/p\mathbb{Z}),$$

which may be accomplished by computing $k(x)$ for all $x \in \mathbb{F}_p$.

(5) Since $V_{\sigma^{-1}(r_j)} = 1$ for $j = 1, 2, \ldots, n$, we recover m as:

$$m = \sum_{j=1}^{n} V_{\sigma^{-1}(r_j)} \left(\frac{p - 1 - \sigma^{-1}(r_j)}{n - \sum_{i=0}^{\sigma^{-1}(r_j)-1} V_i} \right),$$

where we set $\sum_{i=0}^{\sigma^{-1}(r_j)-1} V_i = 0$ if $\sigma^{-1}(r_j) = 0$.

By Exercise 3.31, the deciphering stage must indeed recover m. In the following illustration, we keep the parameters small for simplicity.

Example 3.27 Suppose that entity B chooses $p = 5$, $n = 3$, and randomly selects $r(x) = x^3 + x + 1 = r(x)$, which is irreducible over $\mathbb{F}_5$, and randomly chooses $\ell(x) = 2x^2 + 2$ which generates $\mathbb{F}_{5^3}$ (see Theorem A.47 on page 292). To see the latter, note that

$$\ell(x)^{(5^3-1)/2} = \ell(x)^{124/2} = \ell(x)^{62} \equiv -1 \,(\text{mod } 5)$$

in

$$\mathbb{F}_{5^3} \cong (\mathbb{Z}/5\mathbb{Z})[x]/(r(x)).$$

Then B computes discrete logs as follows using the Pohlig-Hellman Algorithm. $a_0 = \log_{\ell(x)}(x) = 30$, $a_1 = \log_{\ell(x)}(x + 1) = 28$, $a_2 = \log_{\ell(x)}(x + 2) = 106$, $a_3 = \log_{\ell(x)}(x + 3) = 50$, and $a_4 = \log_{\ell(x)}(x + 4) = 23$. Then B randomly selects $\sigma : \{0, 1, 2, 3, 4\} \mapsto \{4, 0, 2, 1, 3\}$, and $d = 29$, after which B computes

$$c_0 \equiv a_{\sigma(0)} + d \equiv a_4 + 29 \equiv 23 + 29 \equiv 52 \,(\text{mod } 124),$$

$$c_1 \equiv a_{\sigma(1)} + d \equiv a_0 + 29 \equiv 30 + 29 \equiv 59 \,(\text{mod } 124),$$

$$c_2 \equiv a_{\sigma(2)} + d \equiv a_2 + 29 \equiv 106 + 29 \equiv 11 \,(\text{mod } 124),$$

$$c_3 \equiv a_{\sigma(3)} + d \equiv a_1 + 29 \equiv 28 + 29 \equiv 57 \,(\text{mod } 124),$$

and

$$c_4 \equiv a_{\sigma(4)} + d \equiv a_3 + 29 \equiv 50 + 29 \equiv 79 \,(\text{mod } 124),$$

so B's public key is $((c_0, c_1, c_2, c_3, c_4), p, n) = ((52, 59, 11, 57, 79), 5, 3)$ and B's private key is $(r(x), \ell(x), \sigma, d) = \left(x^3 + x + 1, 2x^2 + 2, \begin{pmatrix} 0 & 1 & 2 & 3 & 4 \\ 4 & 0 & 2 & 1 & 3 \end{pmatrix}, 29 \right)$.

Now A gets B's public key and represents the message $m = 5$ as a bitstring $m = 101$ of length $\lfloor \log_2 \binom{5}{3} \rfloor = 3$. A then replaces bitstring m with 5-tuple V determined as follows. For $i = 1$, $V_0 = 1$ since $m = 5 > \binom{p-i}{n} = \binom{4}{3} = 4$. For $i = 2$, $V_1 = 0$ since $m = 5 - \binom{4}{3} = 1 < \binom{p-i}{n} = \binom{3}{2} = 3$. For $i = 3$, $V_2 = 1$ since $m = 1 = \binom{p-i}{n} = \binom{2}{2} = 1$. For $i = 4$, $V_3 = 0$ since $m = 0 < \binom{p-i}{n} = \binom{1}{1} = 1$. For $i = 5$, $V_4 = 1$ since $m = 0 = \binom{p-i}{n} = \binom{0}{1}$. Thus, $V = (1, 0, 1, 0, 1)$ and A computes

$$c \equiv \sum_{j=0}^{4} V_j c_j = c_0 + c_2 + c_4 = 52 + 11 + 79 \equiv 18 \,(\text{mod } 124),$$

which A sends to B, who computes $z \equiv (c - nd) = 18 - 3 \cdot 29 \equiv 55 \pmod{124}$, $h(z) = \ell(x)^z = (2x^2 + 2)^{55} \equiv 4x^2 + 3 \pmod{x^3 + x + 1}$, and $k(x) = h(x) + r(x) = x^3 + 4x^2 + x + 4$, which factors over $\mathbb{F}_5$ as $k(x) = (x+2)(x+3)(x+4)$, so $r_1 = 2$, $r_2 = 3$, and $r_3 = 4$. Since $\sigma^{-1}(r_1) = \sigma^{-1}(2) = 2$, $\sigma^{-1}(r_2) = \sigma^{-1}(3) = 4$, and $\sigma^{-1}(r_3) = \sigma^{-1}(4) = 0$, B can recover m as follows.

$$m = \sum_{j=1}^{n} V_{\sigma^{-1}(r_j)} \binom{p - 1 - \sigma^{-1}(r_j)}{n - \sum_{i=0}^{\sigma^{-1}(r_j)-1} V_i} =$$

$$V_2 \binom{2}{2} + V_4 \binom{0}{1} + V_0 \binom{4}{3} = 1 + 0 + 4 = 5 = 101.$$

When the Chor-Rivest Algorithm is properly set up, it was thought to be secure against known attacks. However, as we noted earlier, it has been cryptanalyzed. Nevertheless, to be properly set up means that *none* of $r(x)$, d, $\ell(x)$, or σ can be revealed. Furthermore, the algorithm we have described above over the base field $\mathbb{F}_p$ can be shown to work over any base field $\mathbb{F}_q$ where q is a prime power. The major problem with the Chor-Rivest Cryptosystem is the huge size of the public key. For instance, in practice the recommended values for p and n are $p = 197$ and $n = 24$, and in this case, the public key has approximately $36,000$ bits. In 1991, Hendrik Lenstra introduced a modified version of the algorithm in [123], which does not require step (4) in the key generation. In other words, computation of discrete logs is avoided. However, Lenstra's Algorithm, called the *powerline system*, is not a Knapsack Cryptosystem. The reader may test knowledge of the Chor-Rivest Cipher by solving Exercises 3.29–3.30.

Now that we have looked at both Symmetric-key and Public-key Ciphers, it is time to compare and contrast. From the discussions already given in Chapters Two and Three, we have seen that public-key cryptography ensures efficient key management and digital signatures, whereas symmetric-key cryptography ensures efficient enciphering and data integrity. It may be argued that much of the knowledge concerning symmetric-key cryptography has been developed primarily since the introduction of the DES Cryptosystem. We have already seen that different Symmetric-key Cryptosystems can be combined to produce a new cryptosystem that is stronger than its constituent parts. Weaknesses of Symmetric-key Cryptosystems include the symmetric key having to be secret at both ends of the communication·channel. Thus, the symmetric key should be changed frequently, perhaps as often as for each communication session. Weaknesses of public-key cryptography include the pace of enciphering data being several magnitudes slower than symmetric-key enciphering algorithms (for instance, DES vs. RSA). Also, the private key in a Public-key Cryptosystem must be larger than the secret key in Symmetric-key Cryptosystems. Hence, to ensure the same level of security for Public-key vs. Symmetric-key Cryptosystems, the symmetric keys have bitlength typically ten times smaller than the private keys in Public-key Cryptosystems. For instance, the private key in RSA should be $1,024$ bits whereas in symmetric-key ciphers generally 128 bits will suffice.

Today, there is a practical merging of these cryptosystems. Typically, public-key enciphering algorithms can be used to obtain a key for use in Symmetric-key Cryptosystems. In this fashion, communicating entities can take advantage of the long-term security of private/public key pairs, and the efficient performance of Symmetric-key Cryptosystems.

Exercises

3.27. Let n, b_j, r, s, σ be given as in Example 3.26. Show how the message $m = 1111100$ would be enciphered by A and recovered by B under the Merkle-Hellman Knapsack Cryptosystem.

3.28. With the same instructions as given in Exercise 3.27, use the message $m = 1000001$.

3.29. Use the setup in Example 3.27 to encipher and recover $m = 7$.[3.19]

3.30. Follow the instruction in Exercise 3.29 with $m = 6$.

3.31. Prove that in the deciphering stage of the Chor-Rivest Cipher, the roots of $k(x)$ do indeed provide the values of j for which $V_j = 1$, so that m may be recovered.

3.32. If $S = \{a_1, a_2, \ldots, a_n\}$ is a knapsack set, then the *density of* S is defined to be
$$d = \frac{n}{\max\{\log_2(a_j) : 1 \le j \le n\}}.$$
What is the density of the Chor-Rivest Knapsack Set?[3.20]

3.33. Show that of all superincreasing sequences $\{b_1, b_2, \ldots, b_n\}$, the one with the smallest values of $b_j \in \mathbb{N}$ is for all $j = 1, 2, \ldots, n$ to be $b_j = 2^{j-1}$. Conclude that any superincreasing sequence $\{a_1, a_2, \ldots, a_n\}$ must satisfy $b_j \ge 2^{j-1}$ for all $j = 1, 2, \ldots, n$.

3.34. Suppose that a set $S = \{b_1, b_2, \ldots, b_n\}$ satisfies $b_{j+1} > 2b_j$ for all $j = 1, 2, \ldots, n-1$. Prove that S is a superincreasing sequence.

3.35. Find all subsets of the knapsack set $S = \{1, 2, 3, 5, 9, 10, 11\}$ that have 13 as their sum. Is S a superincreasing sequence?

3.36. Find all subsets of $\{2, 4, 8, 9, 11, 12\}$ that sum to 14.

[3.19]The reader will be aided by a software package such as *Maple* where the *Powmod* function will assist in computing $h(z)$.

[3.20]There is a very strong general attack on Knapsack Cryptosystems, called the *Lattice Basis Reduction Algorithm* (see [134]). It is quite successful if the density of the knapsack set is less than 0.9408. Since the Merkle-Hellman Cryptosystem has density less than 1, then this is significant and explains the breakability of the cipher. On the other hand, Chor-Rivest's density is sufficiently high to thwart such attacks assuming, as detailed earlier, that the parameters are properly chosen. Also, there exist two algorithms (see [38] and [110]) which solve almost all Subset Sum Problems of sufficiently low density. Both of the algorithms rely on basis reduction algorithms.

Chapter 4

Primality Testing

4.1 An Introduction to Primitive Roots

In order to study the primality testing algorithms and related phenomena in this chapter, we need to acquaint ourselves with the notion mentioned in the section header. Toward this end, we first need the following concept related to Euler's Theorem 2.25, which tells us that for $m \in \mathbb{Z}$ and $n \in \mathbb{N}$ with $\gcd(m, n) = 1$, we have $m^{\phi(n)} \equiv 1 \pmod{n}$. One may naturally ask for the *smallest* exponent $e \in \mathbb{N}$ such that $m^e \equiv 1 \pmod{n}$.

Definition 4.1 (Modular Order of an Integer)
   Let $m \in \mathbb{Z}$, $n \in \mathbb{N}$ and $\gcd(m, n) = 1$. Then the order of m modulo n is the smallest $e \in \mathbb{N}$ such that $m^e \equiv 1 \pmod{n}$, denoted by $e = \operatorname{ord}_n(m)$, and we say that m belongs to the exponent e modulo n.

Note that the modular order of an integer given in Definition 4.1 is the same as the element order in the group $(\mathbb{Z}/n\mathbb{Z})^*$.

Example 4.2 Clearly 2 has order 2 modulo 3, so $\operatorname{ord}_3(2) = 2 = \phi(3)$. However, 7 has order 1 modulo 3, so $\operatorname{ord}_3(7) = 1$. A more substantial instance is for the prime $p = 3677$, where $7^{1838} \equiv 1 \pmod{p}$ but $7^e \not\equiv 1 \pmod{p}$ for any $e < 1838$, so $\operatorname{ord}_p(7) = 1838$.

Notice in Example 4.2 that the order of each integer divides $\phi(n)$.

Proposition 4.3 (Divisibility by the Order of an Integer)
   If $m \in \mathbb{Z}$, $d, n \in \mathbb{N}$ such that $\gcd(m, n) = 1$, then $m^d \equiv 1 \pmod{n}$ if and only if $\operatorname{ord}_n(m) \mid d$. In particular, $\operatorname{ord}_n(m) \mid \phi(n)$.

Proof. If $\mathfrak{d} = \operatorname{ord}_n(m)$, and $d = \mathfrak{d}x$ for some $x \in \mathbb{N}$, then

$$m^d = (m^{\mathfrak{d}})^x \equiv 1 \pmod{n}.$$

161

Conversely, if $m^d \equiv 1 \pmod{n}$, then $d \geq \mathfrak{d}$ so there exist integers q and r with $d = q \cdot \mathfrak{d} + r$ where $0 \leq r < \mathfrak{d}$ by the Division Algorithm. Thus, $1 \equiv m^d \equiv (m^{\mathfrak{d}})^q m^r \equiv m^r \pmod{n}$, so by the minimality of $\mathfrak{d}$, $r = 0$. In other words, $\mathfrak{d} \mid d$. In particular, since $m^{\phi(n)} \equiv 1 \pmod{n}$ by Euler's Theorem 2.25, then $\mathfrak{d} \mid \phi(n)$. $\square$

Note that we may rephrase Proposition 4.3 in terms of the group theoretic language surrounding $(\mathbb{Z}/n\mathbb{Z})^*$, namely that if $\mathfrak{d}$ is the order of an element $m \in (\mathbb{Z}/n\mathbb{Z})^*$, then for any $d \in \mathbb{N}$, if $m^d = 1 \in (\mathbb{Z}/n\mathbb{Z})^*$, d must be a multiple of $\mathfrak{d}$. We use this language to prove the next fact.

Corollary 4.4 *If $d, n \in \mathbb{N}$, and $m \in \mathbb{Z}$ with $\gcd(m, n) = 1$, then*

$$\operatorname{ord}_n(m^d) = \frac{\operatorname{ord}_n(m)}{\gcd(d, \operatorname{ord}_n(m))}.$$

Proof. With $\mathfrak{d}$ as above, set $f = \operatorname{ord}_n(m^d)$ (the order of m^d in $(\mathbb{Z}/n\mathbb{Z})^*$) and $g = \gcd(d, \mathfrak{d})$. Thus, by Proposition 4.3, $\mathfrak{d} \mid df$, so $(\mathfrak{d}/g) \mid fd/g$. Therefore, by part (f) of Theorem A.24 on page 280, $(\mathfrak{d}/g) \mid f$. Also, since

$$(m^{\mathfrak{d}})^{d/g} = (m^d)^{\mathfrak{d}/g} = 1 \in (\mathbb{Z}/n\mathbb{Z})^*,$$

then by our above proposition applied to m^d this time, $f \mid (\mathfrak{d}/g)$. Hence, $f = (\mathfrak{d}/g)$, which is the intended result. $\square$

Those integers m for which $\operatorname{ord}_n(m) = \phi(n)$ are of special importance and are the main topic of this section.

Definition 4.5 (Primitive Roots)
If $m \in \mathbb{Z}$, $n \in \mathbb{N}$ and

$$\operatorname{ord}_n(m) = \phi(n),$$

then m is called a primitive root modulo n. *In other words, m is a primitive root if it belongs to the exponent $\phi(n)$ modulo n.*

Example 4.6 If we look back to Example 3.4, we see that

$$\operatorname{ord}_{37}(2) = 36,$$

so 2 is a primitive root modulo the prime 37. In Example 3.15, we see that

$$\operatorname{ord}_{1777}(5) = 1776,$$

so 5 is a primitive root modulo the prime 1777. Also, in Example 3.20, we see that

$$\operatorname{ord}_{3677}(2) = 3676,$$

so 2 is a primitive root modulo the prime 3677. However, 15 has no primitive roots (see Theorem 4.10).

It is handy to have a methodology for computing primitive roots. In [80, Art. 73–74, pp. 47–49], Gauss developed a method for computing primitive roots modulo a prime p as follows.

◆ **Gauss's Algorithm for Computing Primitive Roots** (mod p)

(1) Let $m \in \mathbb{N}$ such that $1 < m < p$ and compute m^t for $t = 1, 2, \ldots$ until $m^t \equiv 1 \pmod{p}$. In other words, compute powers until $\mathrm{ord}_p(m)$ is achieved. If $t = \mathrm{ord}_p(m) = p - 1$, then m is a primitive root and the algorithm terminates. Otherwise, go to step (2).

(2) Choose $b \in \mathbb{N}$ such that $1 < b < p$ and $b \not\equiv m^j \pmod{p}$ for any $j = 1, 2, \ldots, t$. Let $u = \mathrm{ord}_p(b).^{4.1}$ If $u \neq p - 1$, then let $v = \mathrm{lcm}(t, u)$. Therefore, $v = ac$ where $a \mid t$ and $c \mid u$ with $\gcd(a, c) = 1$. Let m_1 and b_1 be the least nonnegative residues of $m^{t/a}$ and $b^{u/c}$ modulo p, respectively. Thus, $g = m_1 b_1$ has order $ac = v$ modulo p. If $v = p - 1$, then g is a primitive root and the algorithm is terminated. Otherwise go to step (3).

(3) Repeat step (2) with v taking the role of t and $m_1 b_1$ taking the role of m. (Since $v > t$ at each step, the algorithm terminates after a finite number of steps with a primitive root modulo p.)

Gauss used the following to illustrate his algorithm.

Example 4.7 Let $p = 73$. Choose $m = 2$ in step (1), and we compute $t = \mathrm{ord}_p(m) = 9$ with

$$m^j \equiv 1, 2, 4, 8, 16, 32, 64, 55, 37, 1 \pmod{p}$$

for $j = 0, 1, 2, 3, 4, 5, 6, 7, 8, 9 = t = \mathrm{ord}_p(m)$, respectively. Now we go to step (2) since $m = 2$ is not a primitive root modulo $p = 73$. Since $3 \not\equiv 2^j \pmod{73}$ for any natural number $j \leq 9$, we choose $b = 3$. Compute b^j for $j = 1, 2, \ldots u$, where $3^u = 3^{12} \equiv 1 \pmod{73}$, where

$$3^j \equiv 3, 9, 27, 8, 24, 72, 70, 64, 46, 65, 49, 1 \pmod{p}$$

for $j = 1, 2, 3, 4, 5, 6, 7, 8, 9, 10, 11, 12 = u = \mathrm{ord}_p(b) = \mathrm{ord}_{73}(3)$, respectively. Since $u \neq p - 1$, then set $v = \mathrm{lcm}(t, u) = 36 = ac = 9 \cdot 4$. Then $m_1 = 2^{t/a} = 2$ and $b_1 = 3^{u/c} = 3^3 = 27$, so $m_1 b_1 = 54$, but $v = \mathrm{ord}_{73}(54) = 36 \neq p - 1$. Thus, we repeat step (2) with $v = 36$ replacing t and choose a value of b not equivalent to any power of the new $m = 54 = m_1 b_1$ modulo 73. Since $b = 5$ qualifies for the role and it is a primitive root modulo 73, the algorithm terminates.

$^{4.1}$Observe that we cannot have $u \mid t$ since if it did, then $b^t \equiv 1 \pmod{p}$. However, it follows from (1) and Exercise 4.12 that m^j for $0 \leq j \leq t - 1$ are all the incongruent solutions of $x^t \equiv 1 \pmod{p}$, so $b \equiv m^j \pmod{p}$ for some such j, a contradiction to the choice of b.

The reader should now solve Exercises 4.1–4.3 on pages 177–178, which are designed to test an understanding of the notion of primitive roots, and set the stage for our existence theorem below.

Example 4.6 suggests a natural question. Is 2 a primitive root of infinitely many primes? This is unknown, but a positive answer is conjectured. In fact, there is a more general famous conjecture as follows.

Conjecture 4.8 (Artin's[4.2] Conjecture)

Every nonsquare integer $m \neq -1$ is a primitive root modulo infinitely many primes.

Although this conjecture remains open, Heath-Brown proved in 1986 that, with the possible exception of at most two primes, it is true that for each prime p there are infinitely primes q such that p is a primitive root modulo q. For example, there are infinitely many primes q such that one of 2, 3, or 5 is a primitive root modulo q (see [94, p. 249]).

The following is a prelude to our first major goal in this section, which is the determination of precisely which integers actually have primitive roots.

Lemma 4.9 (Primitive Roots Modulo a Prime)

Let p be a prime and let $e \in \mathbb{N}$ such that $e \mid (p-1)$. In any reduced residue system modulo p, there exist either 0 or $\phi(e)$ distinct $m \in \mathbb{Z}$, $0 \leq m \leq p-1$, with $\mathrm{ord}_p(m) = e$. In particular, there exist $\phi(p-1)$ primitive roots modulo p.

Proof. Assume that there exists an integer with order e modulo p, and let $r(e)$ be the number of incongruent integers that belong to the exponent e modulo p. Since every natural number $e < p$ must belong to *some* exponent modulo p, then

$$\sum_{e \mid (p-1)} r(e) = p - 1.$$

By Exercise 4.12, $x^e \equiv 1 \pmod{p}$ has exactly e incongruent solutions modulo p, and by Exercise 4.2, these solutions are $m, m^2 \ldots, m^{p-1}$, where m is a primitive root modulo p. Of these, the ones with $\mathrm{ord}_p(m^j) = e$ are those for which

---

[4.2]Emil Artin (1898–1962) was born in Vienna, Austria in 1898. In World War I, he served in the Austrian army. After the war, he obtained his Ph.D. from the University of Leipzig in 1921. In 1937, he emigrated to the U.S.A., and taught at the University of Notre Dame for one year. Then he spent eight years at Indiana, and in 1946 went to Princeton where he remained for the next twelve years. In 1958, he returned to Germany where he remained for the rest of his life. He was reappointed to the University of Hamburg, which he had left two decades before. Artin contributed to finite group theory, the theory of associative algebras, as well as number theory. His name is attached to numerous deep mathematical entities. For instance, there are the *Artin Reciprocity Law*, *Artin L-functions*, and *Artinian Rings* (see [145]). Among Artin's students were Serge Lang, John Tate, and Max Zorn. Artin had interests outside of mathematics including astronomy, biology, chemistry, and music. In the latter, he excelled as an accomplished musician in his own right playing not only the flute, but also the harpsichord and the clavichord. He died in Hamburg on December 20, 1962.

$\gcd(j, e) = 1$ by Exercise 4.1, which means there are $\phi(e)$ of them. By Exercise 4.4,

$$\sum_{e \mid (p-1)} \phi(e) = p - 1,$$

but

$$\sum_{e \mid (p-1)} r(e) = p - 1,$$

and $r(e) \le \phi(e)$, so $r(e) = \phi(c)$ for all $e \mid (p - 1)$.

The last statement of the lemma follows from Theorem A.47 on page 292, which guarantees that p has a primitive root, and Exercise 4.3, which tells us that there exist $\phi(p - 1)$ incongruent integers of order $p - 1$ modulo p. $\square$

The reader who solves Exercise 4.5 will have set the stage for the following existence result.

Theorem 4.10 (Primitive Root Theorem)

An integer $m > 1$ has a primitive root if and only if m is of the form $2^a p^b$ where p is an odd prime, $0 \le a \le 1$, and $b \ge 0$ or $m = 4$. Also, if m has a primitive root, then it has $\phi(\phi(m))$ of them.

Proof. Suppose that $m > 1$ has a primitive root, and let $p^b \| m$ where p is prime,[4.3] and $b \in \mathbb{N}$. By Lemma 4.9, and Exercise 4.5, if g is a primitive root modulo p, then either g or $g + p$ is a primitive root modulo p^b. If g is a primitive root modulo p^b, then either g or $g + p^b$ is odd, so there is an odd primitive root modulo p^b. If h is such a primitive root, then h must also be a primitive root modulo $2p^b$. To see this, let $\text{ord}_{2p^b}(h) = c$, then $c \mid \phi(2p^b) = \phi(p^b)$, by Proposition 4.3. Hence, $c = \phi(p^b) = \phi(2p^b)$.

We have shown that each of $m = 2, 4$ or $m = 2^a p^b$ where p is an odd prime, $0 \le a \le 1$, and $b \ge 0$ has a primitive root. To show that no other moduli have primitive roots, suppose that $m = m_1 m_2$ where $m_1 > 2$ and $m_2 > 2$, and $\gcd(m_1, m_2) = 1$. By Exercise 2.25 on page 75, $\phi(m_j)$ is even for $j = 1, 2$. Therefore, for any $n \in \mathbb{N}$, $n^{\phi(m)/2} \equiv (n^{\phi(m_1)})^{\phi(m_2)/2} \equiv 1 \pmod{m_1}$, and similarly, $n^{\phi(m)/2} \equiv 1 \pmod{m_2}$. Therefore, $n^{\phi(m)/2} \equiv 1 \pmod{m}$, so no $n \in \mathbb{N}$ can be a primitive root modulo m. The last type of modulus to consider is $m = 2^a$. For $a \ge 3$, $n^{2^{a-2}} \equiv 1 \pmod{2^a}$ for all odd $n \in \mathbb{Z}$, by Exercise 4.6. Lastly, by Exercise 4.3, there are exactly $\phi(\phi(m))$ primitive roots modulo m. $\square$

Example 4.11
If $m = 22$, then $7, 13, 17, 19$ are the four incongruent primitive roots modulo m. Note that $\phi(\phi(m)) = 4$.

We now are in a position to introduce the next important concept of this section, which will have cryptographic applications in this chapter. If $n \in \mathbb{N}$ has a primitive root m, then by Exercise 4.2, the values $1, m, m^2, \ldots, m^{\phi(n)-1}$ form

---

[4.3]This symbol $\|$ means that $p^b | m$, but p^{b+1} does not. We say that p^b *exactly* divides m.

a complete set of reduced residues modulo n. It follows from Exercise 4.3 that, given any $b \in \mathbb{N}$, there is exactly one nonnegative integer $e \le \phi(n)$ for which $b \equiv m^e \pmod{n}$. This value has a distinguished name.

Definition 4.12 (Index)

   Let $n \in \mathbb{N}$ with primitive root m, and $b \in \mathbb{N}$ with $\gcd(b, n) = 1$. Then for exactly one of the values $e \in \{0, 1, \ldots, \phi(n) - 1\}$, $b \equiv m^e \pmod{n}$ holds. This unique value e modulo $\phi(n)$ is the index of b to the base m modulo n, denoted by $\operatorname{ind}_m^n(b)$.

Example 4.13 If $n = 29$, then $m = 2$ is a primitive root modulo n. Also, $\operatorname{ind}_2^{29}(5) = 22$, since $5 \equiv 2^{22} \pmod{29}$, so 5 has index 22 to base 2 modulo 29.

Definition 4.12 gives rise to an arithmetic of its own, the *index calculus*. The following are some of the properties.

Theorem 4.14 (Index Calculus)

   If $n \in \mathbb{N}$ and m is a primitive root modulo n, then for any $c, d \in \mathbb{Z}$ each of the following holds.

(1) $\operatorname{ind}_m^n(cd) \equiv \operatorname{ind}_m^n(c) + \operatorname{ind}_m^n(d) \pmod{\phi(n)}$.

(2) For any $t \in \mathbb{N}$, $\operatorname{ind}_m^n(c^t) \equiv t \cdot \operatorname{ind}_m^n(c) \pmod{\phi(n)}$.

(3) $\operatorname{ind}_m^n(1) = 0$.

(4) $\operatorname{ind}_m^n(m) = 1$.

(5) $\operatorname{ind}_m^n(-1) = \phi(n)/2$ for $n > 2$.

(6) $\operatorname{ind}_m^n(n - c) \equiv \operatorname{ind}_m^n(-c) \equiv \phi(n)/2 + \operatorname{ind}_m^n(c) \pmod{\phi(n)}$.

   Proof. Let $\operatorname{ind}_m^n(cd) = x$, $\operatorname{ind}_m^n(c) = y$, and $\operatorname{ind}_m^n(d) = z$.
   (1) Since $cd \equiv m^x \pmod{n}$, $c \equiv m^y \pmod{n}$ and $d \equiv m^z \pmod{n}$, then

$$m^{y+z} \equiv cd \equiv m^x \pmod{n},$$

so the result follows from Euler's Theorem 2.25.
   (2) Since $c \equiv m^y \pmod{n}$, then

$$c^t \equiv m^{yt} \pmod{n},$$

the result follows by applying ind_m^n to both sides.
   (3) If $\operatorname{ind}_m^n(1) = w$, then $1 \equiv m^w \pmod{n}$. Since m is a primitive root modulo n, and $0 \le w < \phi(n)$, by Definition 4.12, then $w = 0$.
   (4) Let $\operatorname{ind}_m^n(m) = v$. Since $m \equiv m^v \pmod{n}$, then $v = 1$ by Definition 4.12.
   (5) Since $m^{\phi(n)/2} \equiv -1 \pmod{n}$ for $n > 2$, then the result follows as in (2).

(6) Since $m^{\phi(n)/2} \equiv -1 \pmod{n}$, then

$$-c \equiv n - c \equiv -m^y \equiv m^{\phi(n)/2}m^y \equiv m^{\phi(n)/2+y} \pmod{n},$$

so the result follows from Euler's Theorem. $\qquad\square$

The reader will recognize that the properties of the index mimic those of logarithms. Hence, if n is a prime p, the index of b to the base p is often called the *discrete logarithm* of b to the base p. For further such properties, see Exercises 4.7–4.8 on page 178. Moreover, part (1) of Theorem 4.14 provides us with a tool for finding indices by reducing it to solving linear congruences

$$cx \equiv b \pmod{n}$$

for $x \in \mathbb{Z}$. To see this note that when this congruence holds, then

$$\mathrm{ind}_m^n(c) + \mathrm{ind}_m^n(x) \equiv \mathrm{ind}_m^n(b) \pmod{\phi(n)},$$

for any primitive root m modulo n. The process is illustrated as follows.

Example 4.15 Suppose that we wish to solve the congruence $7x \equiv 3 \pmod{29}$. Since 2 is a primitive root modulo 29, then

$$\mathrm{ind}_2^{29}(7) + \mathrm{ind}_2^{29}(x) \equiv \mathrm{ind}_2^{29}(3) \pmod{\phi(29)},$$

so

$$\mathrm{ind}_2^{29}(x) \equiv \mathrm{ind}_2^{29}(3) - \mathrm{ind}_2^{29}(7) = 5 - 12 = -7 \equiv 21 \pmod{28}.$$

Therefore, by raising to appropriate exponents we get, $x \equiv 2^{21} \equiv 17 \pmod{29}$.

Part (2) of Theorem 4.14 also provides us with a mechanism for solving power congruences

$$c^x \equiv b \pmod{n}$$

for $x \in \mathbb{Z}$. To see this, note that if this congruence holds, then

$$x \cdot \mathrm{ind}_m^n(c) \equiv \mathrm{ind}_m^n(b) \pmod{\phi(n)}.$$

This methodology is illustrated as follows.

Example 4.16 Suppose we want to solve $3^x \equiv 7 \pmod{29}$. Since

$$x \cdot \mathrm{ind}_2^{29}(3) \equiv \mathrm{ind}_2^{29}(7) \pmod{28},$$

then $5x \equiv 12 \pmod{28}$. Therefore,

$$x \equiv 5^{-1} \cdot 12 \equiv 17 \cdot 12 \equiv 8 \pmod{28}.$$

Thus, $3^8 \equiv 7 \pmod{29}$.

Exercises 4.9–4.10 are designed to test the methods illustrated in Examples 4.15–4.16.

The above is clearly related to the notion of power residues introduced in Definition 3.13. It is valuable to have a criterion for the solvability of such congruences. With the index calculus as a tool, we may now present such a result. It is essential that the reader solve Exercise 4.11 since this tells us *when* the following power congruence actually has a solution and how many there are.

Theorem 4.17 (Euler's Criterion for Power Residue Congruences)
Let $e, c \in \mathbb{N}$ with $e \geq 2$, $b \in \mathbb{Z}$, $p > 2$ is prime with $p \nmid b$, and $g = \gcd(e, \phi(p^c)) \mid b$. Then the congruence

$$x^e \equiv b \,(\mathrm{mod}\ p^c) \tag{4.18}$$

is solvable if and only if
$$b^{\phi(p^c)/g} \equiv 1 \,(\mathrm{mod}\ p^c).$$

Proof. Suppose that
$$x^e \equiv b \,(\mathrm{mod}\ p^c).$$

Then
$$b^{\phi(p^c)/g} \equiv (x^e)^{\phi(p^c)/g} \equiv (x^{e/g})^{\phi(p^c)} \equiv 1 \,(\mathrm{mod}\ p^c),$$

by Euler's Theorem 2.25.

Conversely, assume that
$$b^{\phi(p^c)/g} \equiv 1 \,(\mathrm{mod}\ p^c).$$

By Exercise 4.5, there exists a primitive root m modulo p^c. Therefore,
$$b \equiv m^n \,(\mathrm{mod}\ p^c), \tag{4.19}$$

for some $n \in \mathbb{N}$, by Exercise 4.2. Hence,
$$1 \equiv b^{\phi(p^c)/g} \equiv m^{n\phi(p^c)/g} \,(\mathrm{mod}\ p^c).$$

Since m is a primitive root modulo p^c, this implies that
$$\phi(p^c) | n\phi(p^c)/g,$$

so $g | n$. Thus, there exists some $k \in \mathbb{N}$ such that $n = kg$. By Corollary A.23 on page 280, there exist $x, y \in \mathbb{Z}$ such that
$$ye - x\phi(p^c) = g,$$

so
$$n = kg = kye - kx\phi(p^c). \tag{4.20}$$

By Euler's Theorem again,
$$m^{kx\phi(p^c)} \equiv (m^{kx})^{\phi(p^c)} \equiv 1 \,(\mathrm{mod}\ p^c),$$

so by (4.19)–(4.20),

$$b \equiv bm^{kx\phi(p^c)} \equiv m^{n+kx\phi(p^c)} \equiv m^{kye-kx\phi(p^c)+kx\phi(p^c)} \equiv$$
$$m^{kye} \equiv (m^{ky})^e \,(\mathrm{mod}\ p^c).$$

Hence, b is an eth power residue modulo p^c, namely $b \equiv x^e\,(\mathrm{mod}\ p^c)$ with $x = m^{ky}$. $\square$

Example 4.21 If $x^5 \equiv 5\,(\mathrm{mod}\ 27)$, with $g = \gcd(e, \phi(p^c)) = \gcd(5, 18) = 1$, then $\mathrm{ind}_2^{27}(5) = 5$, and

$$5 \cdot \mathrm{ind}_2^{27}(x) \equiv \mathrm{ind}_2^{27}(5) \equiv 5\,(\mathrm{mod}\ 18),$$

since $m = 2$ is a primitive root mod 3^3. Therefore, $\mathrm{ind}_2^{27}(x) \equiv 1\,(\mathrm{mod}\ 18)$, so $\mathrm{ind}_2^{27}(x) \equiv 1, 19\,(\mathrm{mod}\ 27)$. Thus, $x \equiv 2\,(\mathrm{mod}\ 27)$ is the only distinct congruence class that satisfies the given congruence, since $2^{19} \equiv 2\,(\mathrm{mod}\ 27)$.

Example 4.22 Consider $x^3 \equiv 4\,(\mathrm{mod}\ 27)$. Then $g = \gcd(3, 18) = 3 \nmid 4$, so by Theorem 4.17, there are no solutions to this congruence.

The above examples suggest the following consequence of Theorem 4.17.

Corollary 4.23 (Power Residues at Prime Powers)
Suppose that p is an odd prime, $b \in \mathbb{Z}$, $p \nmid b$ and $e \in \mathbb{N}$. If

$$x^e \equiv b\,(\mathrm{mod}\ p)$$

has a solution, then so does

$$x^e \equiv b\,(\mathrm{mod}\ p^c)$$

for all $c \in \mathbb{N}$.

Proof. We prove this by induction on c. We are given $x^e \equiv b\,(\mathrm{mod}\ p)$, and we assume that $x^e \equiv b\,(\mathrm{mod}\ p^c)$ has a solution. We need only establish that $x^e \equiv b\,(\mathrm{mod}\ p^{c+1})$ has a solution. If $g = \gcd(e, \phi(p^c))$, then

$$b^{\phi(p^c)/g} \equiv 1\,(\mathrm{mod}\ p^c),$$

by Theorem 4.17. Thus, there is an integer f such that $b^{\phi(p^c)/g} = 1 + fp^c$. Therefore, by Claim 2.23 in the proof of Theorem 2.22 on page 65, $(1 + p^c f)^p = b^{p\phi(p^c)/g} = b^{\phi(p^{c+1})/g}$. However, by the Binomial Theorem,

$$(1 + p^c f)^p = \sum_{j=0}^{p} p^{cj} f^j \binom{p}{j} = 1 + p^{c+1} h,$$

for some $h \in \mathbb{Z}$ since $cj \geq c + 1$ for all $j \geq 2$. Hence, $b^{\phi(p^{c+1})/g} \equiv 1\,(\mathrm{mod}\ p^{c+1})$, so by Theorem 4.17, $x^e \equiv b\,(\mathrm{mod}\ p^{c+1})$ has a solution as required. $\square$

Furthermore, we have the following consequence.

Corollary 4.24 (The Number of Power Residues)

 Under the hypothesis of Theorem 4.17, there are $\phi(p^c)/g$ incongruent (nonzero) e^{th}-power residues modulo p^e.

 Proof. By Theorem 4.17, we want the number of incongruent solutions of $x^{\phi(p^c)/g} \equiv 1 \,(\text{mod } p^c)$. By Exercise 4.12, this congruence has exactly $\phi(p^c)/g$ incongruent solutions.   □

Example 4.25 Suppose that we want to determine the number of incongruent fourth power residues modulo 27, namely the number of incongruent $b \in \mathbb{N}$ such that $x^4 \equiv b \,(\text{mod } 27)$. Since

$$g = \gcd(\phi(27), e) = \gcd(18, 4) = 2 \mid \text{ind}_2^{27}(7) = 16,$$

then we know there are such solutions by Exercise 4.11. By Corollary 4.24, there must be $\phi(p^c)/g = 18/2 = 9$ incongruent such solutions. They are

$$b \in \{1, 4, 7, 10, 13, 16, 19, 22, 25\}$$

given by $x \in \{1, 5, 11, 10, 4, 2, 8, 13, 7\}$, respectively.

An important consideration in complexity theory is the search for an efficient algorithm which, given a prime p and a primitive root m modulo p, computes $\text{ind}_m^p(x)$ for any given $x \in \mathbb{F}_p^*$. These algorithms have significant ramifications for the construction of secure pseudorandom number generators (see Section Four of Chapter Two). There is no such algorithm known in general. However, the Pohlig-Hellman Algorithm for computing discrete logs, described on pages 131–133, is an efficient such algorithm in the case where the prime factors of $p - 1$ are known and are "small" with respect to p. Note that the α which generates $\mathbb{F}_p^*$ in the description of the Pohlig-Hellman Algorithm on page 131 is a primitive root modulo p. Also, the computation of a modulo $p_j^{a_j}$ described on pages 131–132 is merely the computation of

$$\text{ind}_\alpha^{p_j^{a_j}}(\beta).$$

The reader may now return to this algorithm with a fresh perspective. We will see other cryptographic applications of the results in this section as we proceed through this chapter.

 We now proceed to establish an important result that we will need in the next section, namely Gauss's *Quadratic Reciprocity Law* mentioned in Footnote 1.44. After we introduced power residues in Definition 3.13, we talked about various types of residues including *quadratic residues modulo $n \in \mathbb{N}$*, which are those integers c for which there exists an integer x with

$$x^2 \equiv c \,(\text{mod } n).$$

If no such integer x exists, then c is called a *quadratic nonresidue modulo n*. Non-residues for the cubic, quartic, and higher cases are similarly defined. Whether an integer is a residue or nonresidue modulo $n \in \mathbb{N}$ is called its *residuacity* modulo n. There are symbols for representing the residuacity of a given integer. In particular, we will be interested in the quadratic residuacity for which we introduce the following symbol.

Definition 4.26 (Legendre's Symbol)
If $c \in \mathbb{Z}$ and $p > 2$ is prime, then

$$\left(\frac{c}{p}\right) = \begin{cases} 0 & \text{if } p \mid c, \\ 1 & \text{if } c \text{ is a quadratic residue modulo } p, \\ -1 & \text{otherwise,} \end{cases}$$

and $\left(\frac{c}{p}\right)$ is called the Legendre Symbol *of c with respect to p.*

Example 4.27 A corollary of Euler's Criterion, Theorem 4.17, may now be stated as follows. Let p be an odd prime. Then

$$\text{if } c^{(p-1)/2} \equiv 1 \, (\text{mod } p), \text{ then } \left(\frac{c}{p}\right) = 1,$$

and

$$\text{if } c^{(p-1)/2} \equiv -1 \, (\text{mod } p), \text{ then } \left(\frac{c}{p}\right) = -1.$$

This is called Euler's Criterion for quadratic residuacity.

Theorem 4.28 (Properties of the Legendre Symbol)
If $p > 2$ is prime and $b, c \in \mathbb{Z}$ with $p \nmid bc$, then

(1) $\left(\dfrac{c}{p}\right) \equiv c^{(p-1)/2} \, (\text{mod } p).$

(2) $\left(\dfrac{b}{p}\right)\left(\dfrac{c}{p}\right) = \left(\dfrac{bc}{p}\right).$

(3) $\left(\dfrac{b}{p}\right) = \left(\dfrac{c}{p}\right),$ *provided $b \equiv c \, (\text{mod } p)$.*

Proof. As seen in Example 4.27, part (1) is a corollary of Euler's Criterion. This may now be used to establish part (2) as follows.

$$\left(\frac{b}{p}\right)\left(\frac{c}{p}\right) \equiv b^{(p-1)/2} c^{(p-1)/2} \equiv (bc)^{(p-1)/2} \equiv \left(\frac{bc}{p}\right) \, (\text{mod } p).$$

Part (3) is an immediate consequence of the definition of a quadratic residue. $\square$

To establish Gauss's Quadratic Reciprocity Law, we first need a technical result proved by him.

☞ **Lemma 4.29 (Gauss's Lemma on Residues)**

Let $p > 2$ be a prime and $c \in \mathbb{Z}$ such that $p \nmid c$. Suppose that $\mathfrak{c}$ denotes the cardinality of the set

$$\{\overline{jc} : 1 \leq j \leq (p-1)/2, \overline{jc} > p/2\},$$

where the $\overline{jc}$ denotes reduction of jc to its least positive residue modulo p. Then

$$\left(\frac{c}{p}\right) = (-1)^{\mathfrak{c}}.$$

Proof. For each natural number $j \leq (p-1)/2$, define

$$c_j = \begin{cases} \overline{jc} & \text{if } \overline{jc} < p/2, \\ p - \overline{jc} & \text{if } \overline{jc} > p/2. \end{cases}$$

If $1 \leq j, k \leq (p-1)/2$, then it is a simple verification that $c_j \equiv c_k \pmod{p}$ if and only if $j = k$. Hence, $c_j \not\equiv c_k \pmod{p}$ for all $j \neq k$ with $1 \leq j, k \leq (p-1)/2$. Thus, we have $(p-1)/2$ incongruent natural numbers, all less than $p/2$. Therefore,

$$\prod_{j=1}^{(p-1)/2} c_j \equiv \left(\frac{p-1}{2}\right)! \pmod{p}. \tag{4.30}$$

Also, since $p - \overline{jc} \equiv (-1)(\overline{jc}) \pmod{p}$, then

$$\prod_{j=1}^{(p-1)/2} c_j \equiv (-1)^{\mathfrak{c}} \cdot c^{(p-1)/2} \cdot \left(\frac{p-1}{2}\right)! \pmod{p}. \tag{4.31}$$

By equating the two versions of $\prod_{j=1}^{(p-1)/2} c_j$ in (4.30)–(4.31), and dividing through by $(-1)^{\mathfrak{c}} \cdot (\frac{p-1}{2})!$, we get

$$c^{(p-1)/2} \equiv (-1)^{\mathfrak{c}} \pmod{p},$$

and by Euler's Criterion in Example 4.27,

$$c^{(p-1)/2} \equiv \left(\frac{c}{p}\right) \pmod{p},$$

so the result follows.   □

An important consequence of Gauss's Lemma that we will need to prove his quadratic reciprocity law is contained in the following.

☞ **Corollary 4.32** Let $c \in \mathbb{Z}$ be odd, and $p > 2$ prime such that $p \nmid c$. Then

$$\left(\frac{c}{p}\right) = (-1)^M,$$

where

$$M = \sum_{j=1}^{(p-1)/2} \lfloor jc/p \rfloor.$$

Proof. For each natural number $j \leq (p-1)/2$, we have $jc = q_j p + r_j$, where $r_j \in \mathbb{N}$ with $r_j < p$, by the Division Algorithm. In the notation of the proof of Gauss's Lemma, this means that $r_j = \overline{jc}$, so

$$c_j = \begin{cases} r_j & \text{if } r_j < p/2, \\ p - r_j & \text{if } r_j > p/2, \end{cases}$$

and $q_j = \lfloor jc/p \rfloor$. Arrange the r_j so that $r_j > p/2$ for $j = 1, 2, \ldots, c$, and $r_j < p/2$ for $j = c+1, c+2, \ldots, (p-1)/2$, which is allowed since we know from the proof of Gauss's Lemma that the c_j are just the values $1, 2, \ldots, (p-1)/2$ in some order. Thus, we have

$$\sum_{j=1}^{(p-1)/2} jc = \sum_{j=1}^{(p-1)/2} p\lfloor jc/p \rfloor + \sum_{j=1}^{(p-1)/2} r_j. \tag{4.33}$$

Also, since the c_j are just a rearrangement of the numbers $1, 2, \ldots, (p-1)/2$, then

$$\sum_{j=1}^{(p-1)/2} j = \sum_{j=1}^{c}(p - r_j) + \sum_{j=c+1}^{(p-1)/2} r_j = pc - \sum_{j=1}^{c} r_j + \sum_{j=c+1}^{(p-1)/2} r_j. \tag{4.34}$$

Subtracting (4.34) from (4.33), we get

$$(c-1) \sum_{j=1}^{(p-1)/2} j = p \left(\sum_{j=1}^{(p-1)/2} \lfloor jc/p \rfloor - c \right) + 2 \sum_{j=1}^{c} r_j. \tag{4.35}$$

Now we reduce (4.35) modulo 2 to get

$$0 \equiv \left(\sum_{j=1}^{(p-1)/2} \lfloor jc/p \rfloor - c \right) \pmod{2},$$

since $c \equiv p \equiv 1 \pmod 2$, which means that

$$c \equiv \sum_{j=1}^{(p-1)/2} \lfloor jc/p \rfloor \pmod 2.$$

By Gauss's Lemma, we are now done. $\qquad\square$

The reader may now use Corollary 4.32 to solve Exercises 4.17–4.18. We are now in a position to establish Gauss's famous result, which he first proved in his masterpiece [80].

Theorem 4.36 (The Quadratic Reciprocity Law)
If $p \neq q$ are odd primes, then

$$\left(\frac{p}{q}\right)\left(\frac{q}{p}\right) = (-1)^{\frac{p-1}{2} \cdot \frac{q-1}{2}}.$$

Equivalently,

$$\left(\frac{q}{p}\right) = -\left(\frac{p}{q}\right) \ \textit{if } p \equiv q \equiv 3 \,(\text{mod } 4), \ \textit{and} \ \left(\frac{q}{p}\right) = \left(\frac{p}{q}\right) \ \textit{otherwise.}$$

☞ *Proof.* First, we establish that

$$\frac{p-1}{2} \cdot \frac{q-1}{2} = \sum_{k=1}^{(p-1)/2} \lfloor kq/p \rfloor + \sum_{j=1}^{(q-1)/2} \lfloor jp/q \rfloor. \tag{4.37}$$

Let

$$\mathcal{S} = \{(jp, kq) : 1 \leq j \leq (q-1)/2; 1 \leq k \leq (p-1)/2\}.$$

The cardinality of $\mathcal{S}$ is $\frac{p-1}{2} \cdot \frac{q-1}{2}$. Also, it is an easy check to verify that $jp \neq kq$ for any $1 \leq j \leq (q-1)/2$, or $1 \leq k \leq (p-1)/2$. Furthermore, set

$$\mathcal{S} = \mathcal{S}_1 \cup \mathcal{S}_2,$$

where

$$\mathcal{S}_1 = \{(jp, kq) \in \mathcal{S} : jp < kq\},$$

and

$$\mathcal{S}_2 = \{(jp, kq) \in \mathcal{S} : jp > kq\}.$$

If $(jp, kq) \in \mathcal{S}_1$, then $j < kq/p$. Also, $kq/p \leq (p-1)q/(2p) < q/2$. Therefore, $\lfloor kq/p \rfloor < q/2$, from which it follows that

$$\lfloor kq/p \rfloor \leq (q-1)/2.$$

Hence, the cardinality of $\mathcal{S}_1$ is $\sum_{k=1}^{(p-1)/2} \lfloor kq/p \rfloor$. Similarly, the cardinality of $\mathcal{S}_2$ is $\sum_{j=1}^{(q-1)/2} \lfloor jp/q \rfloor$. This establishes (4.37).

Now set $M = \sum_{k=1}^{(p-1)/2} \lfloor kq/p \rfloor$, and $N = \sum_{j=1}^{(q-1)/2} \lfloor jp/q \rfloor$. If we let $q = c$ in Corollary 4.32, then

$$\left(\frac{q}{p}\right) = (-1)^M.$$

Similarly,

$$\left(\frac{p}{q}\right) = (-1)^N.$$

Hence,

$$\left(\frac{q}{p}\right)\left(\frac{p}{q}\right) = (-1)^{M+N}.$$

The result now follows from (4.37). □

Example 4.38 Let $p = 7$ and $q = 991$. Then by the Quadratic Reciprocity Law,

$$\left(\frac{p}{q}\right)\left(\frac{q}{p}\right) = \left(\frac{7}{991}\right)\left(\frac{991}{7}\right) = (-1)^{3 \cdot 495} = (-1)^{\frac{p-1}{2} \cdot \frac{q-1}{2}} = -1,$$

so

$$\left(\frac{7}{991}\right) = -\left(\frac{991}{7}\right) = -\left(\frac{4}{7}\right) = -\left(\frac{2}{7}\right)^2 = -1.$$

Hence, $x^2 \equiv 7 \pmod{991}$ has no solutions $x \in \mathbb{Z}$.

Exercises 4.19–4.20 are applications of the Quadratic Reciprocity Law. In the next section, we will see applications to primality testing.

We conclude this section with a generalization of the Legendre symbol, which we will require in Section Three of this chapter.

Definition 4.39 (The Jacobi[4.4] Symbol)

Let $n > 1$ be an odd natural number with $n = \prod_{j=1}^{k} p_j^{e_j}$ where $e_j \in \mathbb{N}$ and the p_j are distinct primes. Then the Jacobi Symbol of a with respect to n is given by

$$\left(\frac{a}{n}\right) = \prod_{j=1}^{k} \left(\frac{a}{p_j}\right)^{e_j},$$

for any $a \in \mathbb{Z}$, where the symbols on the right are Legendre Symbols.

The Jacobi Symbol satisfies the following properties.

Theorem 4.40 (Properties of the Jacobi Symbol)

Let $m, n \in \mathbb{N}$, with n odd, and $a, b \in \mathbb{Z}$. Then

(1) $\left(\dfrac{ab}{n}\right) = \left(\dfrac{a}{n}\right)\left(\dfrac{b}{n}\right).$

(2) $\left(\dfrac{a}{n}\right) = \left(\dfrac{b}{n}\right)$ if $a \equiv b \pmod{n}$.

(3) If m is odd, then $\left(\dfrac{a}{mn}\right) = \left(\dfrac{a}{m}\right)\left(\dfrac{a}{n}\right).$

(4) $\left(\dfrac{-1}{n}\right) = (-1)^{(n-1)/2}.$

[4.4]Carl Gustav Jacob Jacobi (1804–1851) was born in Potsdam in Prussia on December 10, 1804 to a wealthy German banking family. In August of 1825, Jacobi obtained his doctorate from the University of Berlin on an area involving partial fractions. The next year he became a lecturer at the University of Königsberg, and was appointed as a Professor there in 1831. Jacobi's first major work was his application of elliptic functions to number theory. Also, he made contributions to analysis, geometry, and mechanics. He died of smallpox on February 18, 1851.

(5) $\left(\dfrac{2}{n}\right) = (-1)^{(n^2-1)/8}$.

(6) *If* $\gcd(a, n) = 1$ *where* $a \in \mathbb{N}$ *is odd, then*

$$\left(\frac{a}{n}\right)\left(\frac{n}{a}\right) = (-1)^{\frac{a-1}{2}\cdot\frac{n-1}{2}},$$

which is the Quadratic Reciprocity Law for the Jacobi Symbol.

☞   *Proof.* Properties (1)–(2) follow from the results for the Legendre symbol given in Theorem 4.28. Property (3) is an easy consequence of Definition 4.39. For part (4), observe that if $n = \prod_{j=1}^{\ell} p_j$ where the p_j are (not necessarily distinct) primes, then

$$n = \prod_{j=1}^{\ell}(p_j - 1 + 1) \equiv 1 + \sum_{j=1}^{\ell}(p_j - 1)\,(\mathrm{mod}\ 4),$$

since all $p_j - 1$ are even. Thus,

$$\frac{n-1}{2} \equiv \sum_{j=1}^{\ell}(p_j - 1)/2\,(\mathrm{mod}\ 2). \tag{4.41}$$

For convenience's sake, we set $S = \sum_{j=1}^{\ell}(p_j - 1)/2$. Therefore, by part (1) of Theorem 4.28 and (4.41),

$$\left(\frac{-1}{n}\right) = \prod_{j=1}^{\ell}\left(\frac{-1}{p_j}\right) = \prod_{j=1}^{\ell}(-1)^{(p_j-1)/2} = (-1)^S = (-1)^{(n-1)/2},$$

which is part (4). For part (5), first observe that

$$n^2 = \prod_{j=1}^{\ell} p_j^2 = \prod_{j=1}^{\ell}(p_j^2 - 1 + 1) \equiv 1 + \sum_{j=1}^{\ell}(p_j^2 - 1)\,(\mathrm{mod}\ 16),$$

since $p_j^2 \equiv 1\,(\mathrm{mod}\ 8)$ for all such j. Therefore,

$$\frac{n^2 - 1}{8} \equiv \frac{\sum_{j=1}^{\ell}(p_j^2 - 1)}{8}\,(\mathrm{mod}\ 2),$$

and we set $T = \sum_{j=1}^{\ell}(p_j^2 - 1)/8$ for convenience. By Exercise 4.17 on page 179,

$$\left(\frac{2}{n}\right) = \prod_{j=1}^{\ell}\left(\frac{2}{p_j}\right) = \prod_{j=1}^{\ell}(-1)^{(p_j^2-1)/8} = (-1)^T = (-1)^{(n^2-1)/8},$$

which secures part (5). For part (6), let $a = \prod_{j=1}^{t} q_j$, where the q_j are (not necessarily distinct) primes. Since $\gcd(a, n) = 1$, then $p_j \neq q_k$ for any j, k. Thus, by properties (1) and (3), established above,

$$\left(\frac{a}{n}\right)\left(\frac{n}{a}\right) = \prod_{j=1}^{\ell}\left(\frac{a}{p_j}\right)\prod_{k=1}^{t}\left(\frac{n}{q_k}\right) =$$

$$\prod_{j=1}^{\ell}\prod_{k=1}^{t}\left(\frac{q_k}{p_j}\right)\prod_{k=1}^{t}\prod_{j=1}^{\ell}\left(\frac{p_j}{q_k}\right)=\prod_{j=1}^{\ell}\prod_{k=1}^{t}\left(\frac{p_j}{q_k}\right)\left(\frac{q_k}{p_j}\right),$$

and by Theorem 4.36, this equals

$$\prod_{j=1}^{\ell}\prod_{k=1}^{t}(-1)^{\frac{p_j-1}{2}\cdot\frac{q_k-1}{2}}=(-1)^U,$$

where

$$U=\sum_{j=1}^{\ell}\sum_{k=1}^{t}\frac{p_j-1}{2}\cdot\frac{q_k-1}{2}=\sum_{j=1}^{\ell}\frac{p_j-1}{2}\sum_{k=1}^{t}\frac{q_k-1}{2}.$$

However, as shown for the p_j in (4.41),

$$\sum_{k=1}^{t}\frac{q_k-1}{2}\equiv\frac{a-1}{2}\;(\bmod\;2),$$

so the result follows. $\qquad\square$

Example 4.42 We have the following for $n=15$,

$$\left(\frac{2}{15}\right)=\left(\frac{2}{3}\right)\left(\frac{2}{5}\right)=(-1)(-1)=1.$$

However, 2 is not a quadratic residue modulo 15. Thus, more caution must be exercised with the interpretation of the use of the Jacobi Symbol. See Exercises 4.21–4.22.

The use of the Jacobi Symbol for primality testing will become apparent when we reach Section Three of this chapter.

Exercises

4.1. Let $m\in\mathbb{Z}$, e, $n\in\mathbb{N}$ and $\gcd(m,n)=1$. Prove that

$$\mathrm{ord}_n(m^e)=\mathrm{ord}_n(m)$$

if and only if

$$\gcd(e,\mathrm{ord}_n(m))=1.$$

In particular, this result says that if m is a primitive root modulo n, then m^e is a primitive root modulo n if and only if $\gcd(e,\phi(n))=1$.

4.2. Let $m\in\mathbb{Z}$ and $n\in\mathbb{N}$ relatively prime to m. Prove that if m is a primitive root modulo n, then $\{m^j\}_{j=1}^{\phi(n)}$ is a complete set of reduced residues modulo n.

4.3. Assuming that $n \in \mathbb{N}$ has a primitive root, show that there are $\phi(\phi(n))$ incongruent primitive roots modulo n.

☆ 4.4. Prove that for any $n \in \mathbb{N}$,

$$\sum_{d|n} \phi(d) = n,$$

where d runs over all positive divisors of n.

4.5. Let g be a primitive root modulo a prime $p > 2$. Prove that one of g or $g + p$ is a primitive root modulo p^a for all $a \in \mathbb{N}$.

4.6. Prove that for any integer for $a \geq 3$,

$$n^{2^{a-2}} \equiv 1 \,(\mathrm{mod}\ 2^a)$$

for all odd $n \in \mathbb{Z}$

4.7. Let p be a prime with primitive roots a and b. Also, let c be an integer relatively prime to p. Prove that

$$\mathrm{ind}_a^p(c) \equiv \mathrm{ind}_b^p(c) \cdot \mathrm{ind}_a^p(b) \,(\mathrm{mod}\ p-1).$$

(This property mimics the change of base formula for logarithms.)

4.8. With the same assumptions as in Exercise 4.7, prove that

$$\mathrm{ind}_a^p(p - c) \equiv \mathrm{ind}_a^p(c) + (p - 1)/2 \,(\mathrm{mod}\ p-1).$$

4.9. Use the index calculus to solve $5x^5 \equiv 3 \,(\mathrm{mod}\ 7)$.

4.10. Using the index calculus, find solutions to $3x^7 \equiv 5 \,(\mathrm{mod}\ 11)$.

4.11. Show that congruence (4.18) has $g = \gcd(e, \phi(p^c))$ incongruent solutions modulo p^c if $g \mid \mathrm{ind}_a^{p^c}(b)$ for any primitive root a modulo p^c, and has none otherwise.

4.12. Let $t, n \in \mathbb{N}$, where $n > 1$ has a primitive root. Prove that if $t \mid \phi(n)$, then $x^t \equiv 1 \,(\mathrm{mod}\ n)$ has exactly t incongruent roots modulo n.

4.13. Prove that if $\mathrm{ord}_p(m)$ is odd and $p > 2$ is prime, then $m^e \equiv -1 \,(\mathrm{mod}\ p)$ has no solution $e \in \mathbb{N}$.

4.14. Let $m, n \in \mathbb{N}$ be relatively prime. Prove that if a is a primitive root modulo mn, then a is a primitive root modulo both m and n.

4.15. Let m be a primitive root modulo an odd prime p. Prove that, for any prime $q \mid (p - 1)$, we must have that $m^{(p-1)/q} \not\equiv 1 \,(\mathrm{mod}\ p)$.

4.16. Let $m \in \mathbb{N}$ and $p > 2$ a prime. Prove that if $m^{(p-1)/q} \not\equiv 1 \,(\mathrm{mod}\ p)$ for all primes $q \mid (p - 1)$, then m is a primitive root modulo p.

☆ 4.17. Let p be an odd prime. Verify the Legendre Symbol identity,

$$\left(\frac{2}{p}\right) = (-1)^{\frac{p^2-1}{8}}.$$

This result is known as the Supplement to the Quadratic Reciprocity Law. *Such supplements also exist for the higher reciprocity laws (see* [145, pp. 273–332]*.*

4.18. Let p be an odd prime. Prove that

$$\left(\frac{2}{p}\right) = \begin{cases} 1 & \text{if } p \equiv \pm 1 \,(\text{mod } 8), \\ -1 & \text{if } p \equiv \pm 3 \,(\text{mod } 8). \end{cases}$$

4.19. Prove that if $p > 2$ is prime, then

$$\left(\frac{3}{p}\right) = \begin{cases} 1 & \text{if } p \equiv \pm 1 \,(\text{mod } 12), \\ -1 & \text{if } p \equiv \pm 5 \,(\text{mod } 12). \end{cases}$$

4.20. Verify the Legendre Symbol identity,

$$\sum_{j=1}^{p-1} \left(\frac{j}{p}\right) = 0,$$

where p is an odd prime by showing that there are $(p-1)/2$ quadratic residues and $(p-1)/2$ quadratic nonresidues modulo p. Then use this fact to establish the Legendre symbol identity,

$$\sum_{j=0}^{p-1} \left(\frac{(j-a)(j-b)}{p}\right) = \begin{cases} p-1 & \text{if } a \equiv b \,(\text{mod } p), \\ -1 & \text{if } a \not\equiv b \,(\text{mod } p). \end{cases}$$

4.21. Let $a \in \mathbb{Z}$, and $n \in \mathbb{N}$ odd. Prove that if

$$\left(\frac{a}{p^t}\right) = 1$$

for all primes p such that $p^t || n$ for some $t \in \mathbb{N}$, then a is a quadratic residue modulo n (see Footnote 4.3).

4.22. Let $n \in \mathbb{N}$ be odd. Prove that

$$\left(\frac{m}{n}\right) = 1$$

for all $m \in \mathbb{N}$ with $m < n$ such that $\gcd(m,n) = 1$ if and only if n is a perfect square.

4.2   True Primality Tests

On page 19, we informally agreed upon a definition of primality testing. However, that definition is only one of the types of primality testing, which we now formalize. The following is distinct from the probabilistic kinds that we will study in the following section of this chapter.

Definition 4.43 (Primality Proofs)
A Primality Proving Algorithm, *also known as a* True Primality Test, *is a deterministic algorithm (see Footnote 1.34) that, given an input n, verifies the hypothesis of a theorem whose conclusion is that n is prime. A* Primality Proof *is the computational verification of such a theorem. In this case, we call n a* provable prime — *a prime which is verified by a Primality Proving Algorithm.*

The classical example of a True Primality Test is the following. First, recall that a Mersenne Number is one of the form

$$M_n = 2^n - 1$$

(see the discussion in Footnote 1.38).

◆ **Lucas-Lehmer[4.5] True Primality Test For Mersenne Numbers**
The algorithm consists of the following steps performed on an input Mersenne Number $M_n = 2^n - 1$ with $n \geq 3$.

(1) Set $s_1 = 4$ and compute $s_j \equiv s_{j-1}^2 - 2 \,(\text{mod } M_n)$ for $j = 1, 2, \ldots, n-1$.

(2) If $s_{n-1} \equiv 0 \,(\text{mod } M_n)$, then conclude that M_n is prime. Otherwise, conclude that M_n is composite.

Example 4.44 Input $M_7 = 127$. Then we compute $\overline{s_j}$, the least nonnegative residue of s_j modulo M_7 as follows. $\overline{s_2} = 14$, $\overline{s_3} = 67$, $\overline{s_4} = 42$, $\overline{s_5} = 111$, and $\overline{s_6} = 0$. Thus, M_7 is prime by the Lucas-Lehmer Test.

[4.5] Derrick Henry Lehmer (1905–1991) was born in Berkeley, California on February 23, 1905. After graduating with his bachelor's degree from Berkeley in 1927, he went to the University of Chicago to study under L. E. Dickson, but he left after only a few months. Neither the Chicago weather nor the working environment suited him. Brown University offered him a better situation with an instructorship, and he completed both his master's degree and his Ph.D., the latter in 1930. He had brief stints at the California Institute of Technology, Stanford University, Lehigh University, and Cambridge, England, the latter on a Guggenheim Fellowship, during the period 1930–1940. In 1940, he accepted a position at the University of California at Berkeley where he remained until his retirement in 1972. He was a pioneering giant in the world of computational number theory, and was widely respected in the mathematical community. The reader is advised to look into his contributions given in his collected works [115]. He was also known for his valued sense of humour, as attested by John Selfridge in the forward to the aforementioned collected works, as well as by one of Lehmer's students, Ron Graham. In particular, Selfridge concludes with an apt description of Lehmer's contributions, saying that he "has shown us this beauty with the sure hand of a master."

We now verify that the Lucas-Lehmer Test is a True Primality Test, as an optional result (for the more adventurous reader). The reader will need to solve Exercise 4.23 first. This relies on abstract algebra, which is described for the benefit of the reader on pages 295–297 of Appendix A. The following proof is due to H.W. Lenstra.[4.6]

☞ **Theorem 4.45 (Lucas-Lehmer is True)**
Let $M_n = 2^n - 1$ with $n \geq 3$ odd. Then M_n is prime if and only if $M_n \mid s_{n-1}$ defined in the above Lucas-Lehmer Algorithm.

Proof. Let a be the least positive residue of $2^{(n+1)/2}$ modulo n. Thus,

$$a^2 \equiv 2^{n+1} = (2^n - 1) + (2^n + 1) \equiv 2 \,(\mathrm{mod}\ M_n).$$

In this fashion, a may be considered as a square root of 2 modulo M_n. Let $I = (x^2 - ax - 1)$ be the ideal generated by the polynomial $f(x) = x^2 - ax - 1$ and let α be the image of x under the natural map which takes $(\mathbb{Z}/M_n\mathbb{Z})^*[x]$ to

$$S = (((\mathbb{Z}/M_n\mathbb{Z})^*)/I)[x]$$

(see Definition A.59 on page 295). Given that $f(x)$ is quadratic, then

$$S = \{a + b\alpha : a, b \in (\mathbb{Z}/M_n\mathbb{Z})^*\},$$

where

$$\alpha^2 = a\alpha + 1. \tag{4.46}$$

Therefore, $\beta = a - \alpha = -\alpha^{-1}$ is the other root of $f(x)$ in S. Clearly, $\alpha + \beta = a$ and $\alpha\beta = -1$.[4.7] By Exercise 4.34, these facts imply that

$$\alpha^{2^j} + \beta^{2^j} \equiv s_j \,(\mathrm{mod}\ M_n), \tag{4.47}$$

for any $j \in \mathbb{N}$. Now we use (4.47) to establish the theorem.

If M_n is prime, then S is a field by Theorem 2.13 (see Example A.63 on page 296). By Fermat's Little Theorem, $\alpha^{M_n} \equiv \alpha^{2^n - 1} \equiv \alpha \,(\mathrm{mod}\ M_n)$, so $\alpha^{2^n - 2} \equiv 1 \,(\mathrm{mod}\ M_n)$. Thus, by (4.46):

$$f(\alpha^{2^n - 1}) = f(\alpha) = \alpha^2 - a\alpha - 1 = 0,$$

[4.6]Hendrik Willem Lenstra Jr. (1949–), was born in 1949 in Zaandam, Netherlands. His father, and his two brothers, Arjen and Jan, are also well-known mathematicians. He studied at the University of Amsterdam, and obtained his Ph.D. under the direction of Frans Oort in 1977. When he was twenty-eight, he was appointed full professor at the University of Amsterdam. In 1987, he emigrated to the United States, and was appointed a full professor at Berkeley. Among his honours include the Fulkerton Prize in 1985, plenary lecturer at the International Congress of Mathematicians at Berkeley in 1986, an honourary doctorate at the Université de Franche-Comté, Besançon in 1995, and Kloosterman-lecturer at the University of Leiden in 1995. Professor Lenstra resides at Berkeley.

[4.7]The reader should now compare this setup with the Lucas-Lehmer Theory developed in Exercises 1.20–1.30 on page 30. In fact, there is a means of proving Theorem 4.45 using Lucas-Lehmer Theory. For a complete description of how that is done, see [208, Theorem 8.4.9, p. 198] or [11, Theorem 9.2.4, p. 273].

so we must have that $\alpha^{2^n-1} \equiv \beta \,(\mathrm{mod}\ M_n)$. Therefore,

$$\alpha^{2^n} \equiv \alpha\beta \equiv -1\,(\mathrm{mod}\ M_n).$$

From this fact and (4.47) we get,

$$s_{n-1} \equiv \alpha^{2^{n-1}} + \beta^{2^{n-1}} \equiv \alpha^{2^{n-1}} + \alpha^{-2^{n-1}} \equiv (\alpha^{2^n} + 1)\alpha^{-2^{n-1}} \equiv 0\,(\mathrm{mod}\ M_n).$$

Conversely, assume that $s_{n-1} \equiv 0\,(\mathrm{mod}\ M_n)$. Since $\beta^{2^{n-1}} = \alpha^{-2^{n-1}}$, then by (4.47), $\alpha^{2^n} \equiv -1\,(\mathrm{mod}\ M_n)$, and $\alpha^{2^{n+1}} \equiv 1\,(\mathrm{mod}\ M_n)$. Now we may invoke Exercise 4.23 (with $s = 2^{n+1}$, $m = M_n$, and $R = \mathbb{Z}/M_n\mathbb{Z}$ therein). Thus, for any $r \mid s$, there exists a nonnegative j such that $r \equiv M_n^j\,(\mathrm{mod}\ s)$. However,

$$M_n^2 \equiv (2^n - 1)^2 \equiv 1\,(\mathrm{mod}\ 2^{n+1}).$$

Therefore, for any $r \mid M_n$, either $r \equiv 1\,(\mathrm{mod}\ 2^{n+1})$ or $r \equiv M_n\,(\mathrm{mod}\ 2^{n+1})$. Hence, M_n is prime.   □

Exercises 4.24–4.25 pertain to the Lucas-Lehmer Test.

We now look at a primality test for which a partial factorization of $n - 1$ needs to be known in order to determine if it is prime.

Theorem 4.48 (Pocklington's Theorem[4.8])

Let $n = ab + 1 \in \mathbb{N}$ with $a, b \in \mathbb{N}$, $b > 1$ and suppose that for any prime divisor q of $b > 1$, there exists an integer m such that $m^{n-1} \equiv 1\,(\mathrm{mod}\ n)$, and $\gcd(m^{(n-1)/q} - 1, n) = 1$. Then $p \equiv 1\,(\mathrm{mod}\ b)$ for every prime $p \mid n$. Furthermore, if $b > \sqrt{n} - 1$, then n is prime.

Proof. Let $p \mid n$ be prime and set $c = m^{(n-1)/q^e}$ where q is a prime and $e \in \mathbb{N}$ with $q^e \| b$. Therefore, since $\gcd(m^{(n-1)/q} - 1, n) = 1$, then $c^{q^e} \equiv 1\,(\mathrm{mod}\ p)$, but $c^{q^{e-1}} \not\equiv 1\,(\mathrm{mod}\ p)$. Thus, $\mathrm{ord}_p(c) = q^e$, so $q^e \mid (p - 1)$ by Proposition 4.3. Since q was arbitrarily chosen, then $p \equiv 1\,(\mathrm{mod}\ b)$. For the last assertion of the theorem, assume that $b > \sqrt{n} - 1$, but n is composite. Let p be the smallest prime dividing n. Then $p \le \sqrt{n}$, so $\sqrt{n} \ge p > b \ge \sqrt{n}$, a contradiction. Hence, n is prime.   □

Example 4.49 Suppose that we wish to test $n = 57283$ for primality using Pocklington's Theorem knowing that $n - 1 = 6 \cdot 9547$, where 9547 is prime. Since $2^{n-1} \equiv 1\,(\mathrm{mod}\ n)$ but $2^6 \not\equiv 1\,(\mathrm{mod}\ n)$, then n is prime.

[4.8]Henry Cabourn Pocklington (1870–1952) worked mainly in physics, the discoveries in which got him elected as a Fellow of the Royal Society. His professional career was spent as a physics teacher at Leeds Central Higher Grade School in England up to his retirement in 1926. Nevertheless, his six papers in number theory were practical and innovative. See [171] for more detail.

Another figure, whom we discussed at the end of Section Two of Chapter One, involved in the development of primality testing was Proth.[4.9]

Theorem 4.50 (Proth's Theorem)

Let $k, t \in \mathbb{N}$ with t odd and $2^k > t$. Then $n = 2^k t + 1$ is prime if and only if $c^{(n-1)/2} \equiv -1 \pmod{n}$, where c is a quadratic nonresidue modulo n.

Proof. If n is prime, then by Euler's Criterion given in Theorem 4.17, $c^{(n-1)/2} \equiv -1 \pmod{n}$. Conversely, we may invoke Theorem 4.48 with $a = t$, $b = 2^k$ and $m = c$ to conclude that if $c^{(n-1)/2} \equiv -1 \pmod{n}$, then n is prime.$\square$

As the proof of Theorem 4.50 shows, Pocklington's result is more general. In fact, it turns out that it was Pocklington who generalized Proth's result, but he did so without being aware of Proth's work. Thus, one may say that Pocklington was an enlightened amateur who was not aware of the history of number theory that was, after all, merely a hobby for him.

Example 4.51 Suppose that we wish to use Proth's Primality Test on $n = 6529$. Since $n = 6529 = 2^7 \cdot 51 + 1$ and $2^7 = 128 > t = 51$, then we have satisfied the hypothesis of Theorem 4.50. Since 7 is a quadratic nonresidue modulo n, and $7^{3264} = 7^{(n-1)/2} \equiv -1 \pmod{n}$, then n is prime. See Exercises 4.30–4.31.

The next result involves Fermat Numbers which we introduced on page 24.

Theorem 4.52 (Pepin's Primality Test[4.10])

For $n \in \mathbb{N}$, $\mathfrak{F}_n = 2^{2^n} + 1$ is prime if and only if $5^{(\mathfrak{F}_n - 1)/2} \equiv -1 \pmod{\mathfrak{F}_n}$.

Proof. Using Proth's Theorem with $t = 1$, the result will follow if we can show that 5 is a quadratic nonresidue modulo $\mathfrak{F}_n$ for any $n \geq 2$. By a simple induction argument one may verify that $2^{2^n} \equiv 1 \pmod{5}$ for all $n \geq 2$. Thus, $\mathfrak{F}_n \equiv 2 \pmod{5}$ for any $n \geq 2$. Hence, using the Quadratic Reciprocity Law, Theorem 4.36, we have the Legendre Symbol equality,

$$\left(\frac{5}{\mathfrak{F}_n} \right) = \left(\frac{\mathfrak{F}_n}{5} \right) = \left(\frac{2}{5} \right) = -1,$$

where the last equality comes from Exercise 4.17 on page 179. $\square$

Example 4.53 $\mathfrak{F}_3 = 257$ is prime since $5^{(\mathfrak{F}_3 - 1)/2} = 5^{128} \equiv -1 \pmod{\mathfrak{F}_3}$.

[4.9]François Proth (1852–1879) was a self-taught farmer, who lived in the village of Vaux devant Damloup near Verdun, France. The theorem that we prove here is one of four results, which he produced, that can be used for primality testing. Proth probably had proofs of his results, but he did not produce them. For more information see [81] and [164].

[4.10]Pepin's Test was generalized by Hurwitz [97] in 1896 and by Carmichael [44] in 1913 as follows. If $r \in \mathbb{N}$ and $\Phi_r(x)$ is the rth cyclotomic polynomial, then r is prime if and only if there exists an $s \in \mathbb{N}$ such that $\Phi_{r-1}(s) \equiv 0 \pmod{r}$. Pepin's Test is the special case where $r = \mathfrak{F}_n$ and $\Phi_{r-1}(x) \equiv x^{(\mathfrak{F}_n - 1)/2} + 1 \equiv 0 \pmod{\mathfrak{F}_n}$. See [145, Exercise 2.23, p. 87].

We now engage in a discussion of what unifies the algorithms presented in this section, and what contrasts them. The Lucas-Lehmer Primality Test for Mersenne Numbers, Pepin's Primality Test, and Proth's Test are all polynomial time algorithms, which are True Primality Tests. However, they lack generality since they only provide provable primes for special numbers, namely Mersenne Numbers, Fermat Numbers, and numbers of the form $2^k t + 1$ where $2^k > t$, respectively. On the other hand, Pocklington's Theorem relies on a knowledge of at least a partial factorization of $n - 1$. This is also a True Primality Test, since it produces provable primes. However, since the test requires knowledge of factorization (and as we have seen there are no known fast factorization algorithms), then this algorithm does not run in polynomial time . In fact, there is no such test (requiring a knowledge of factorization) that runs in polynomial time.

What underlies the above discussion is essentially a result, encountered in Section One of Chapter Two, namely Fermat's Little Theorem 2.18, which says that if p is prime then $a^{p-1} \equiv 1 \pmod{p}$ for all $a \in \mathbb{Z}$ with $\gcd(a, p) = 1$. The converse of this theorem, if it were true, would provide a simple and fast method for obtaining provable primes. However, the converse fails in general and counterexamples abound. In fact, it has recently been shown that it fails in the worst possible way, infinitely often. What we mean by the "worst possible way" is that there exist composite integers n such that

$$a^{n-1} \equiv 1 \pmod{n} \text{ for } any \text{ integers } a \text{ relatively prime to } n, \qquad (4.54)$$

such as $n = 561$ (see Exercise 4.36–4.37). Composite integers n satisfying (4.54) are called *Carmichael Numbers.*[4.11] In 1994, it was proved that there exist infinitely many Carmichael Numbers, (see [5]). Nevertheless, one can use the converse of Fermat's Little Theorem to prove primality of n if one can find an element of order $n - 1$ in $(\mathbb{Z}/n\mathbb{Z})^*$, namely find a primitive root modulo n (see Exercises 4.15–4.16). Credit for the following result is essentially due to both Lehmer and Kraitchik, although seeds of it may be found in the work of Proth and Lucas.

Theorem 4.55 (Proofs Via the Converse of Fermat's Little Theorem)
If $n \in \mathbb{N}$ with $n \geq 3$, then n is prime if and only if there is an $m \in \mathbb{N}$ such that $m^{n-1} \equiv 1 \pmod{n}$, but $m^{(n-1)/q} \not\equiv 1 \pmod{n}$ for any prime $q \mid (n - 1)$.

Proof. See Exercise 4.26.   □

Of course, the major pitfall with Theorem 4.55 is that it requires a knowledge of the factorization of $n - 1$. However, as we have seen with the algorithms in this section, it works well with special numbers such as the Fermat Numbers

---

[4.11] Robert Daniel Carmichael (1879–1967) was born in Goodwater, Alabama. In 1911, he received his doctorate from Princeton under the direction of G.D. Birkhoff. In 1912, he conjectured that there are infinitely many of the numbers that now bear his name. Carmichael Numbers were generalized to Lucas Sequences by Williams [205] in 1977.

to which Pepin's Test applies. Also, one can get a test with only knowledge of a partial factorization as in Pocklington's Theorem. In the next section, we will see how probabilistic methods are used in an effort to approach primality testing using the converse of Fermat's Little Theorem as enunciated in Theorem 4.55.

The preceding discussion shows us that the True Primality Tests, that are based upon Theorem 4.55, are broken down into two distinct categories — those which sacrifice speed, and those which sacrifice generality. In the next section, which deals with "Probabilistic Primality Tests," we will see that both speed and generality are maintained, but *correctness*, obtaining provable primes, is sacrificed. Thus, in each type of primality test, *exactly* one of the following properties is sacrificed: speed, correctness, or generality.

We conclude this section with a discussion of the complexity of primality testing. In 1971, Pollard [154], demonstrated the existence of an algorithm for determining the primality of n with $O(n^{1/4})$ basic operations, and improved the algorithm somewhat in 1974. In 1981, Adleman and Leighton [4] established the existence of an $O(n^{1/10.89})$ deterministic algorithm for deciding the primality of n. However, these results made use of rather deep methods, beyond the scope of this book, and the results are not effective, which means that these algorithms are impractical as True Primality Tests for the simple reason that there is no easy way of knowing when to terminate the algorithm.

In 1979, Shamir [178] demonstrated that $n!$ can be computed using $O(\ln^2 n)$ arithmetic operations. Using this, he showed that there is an algorithm which takes any composite n as input and finds a nontrivial divisor of it using no more that $O(\ln^3 n)$ arithmetic operations. However, this seemingly simple solution to the problem of factoring and primality testing is deceptive. Shamir's result does not take into account the size of the numbers involved in computing $n!$. For large n this number becomes astronomical. To see this, note that $(2^n + 1)^n$ requires $O(n^2)$ bits of storage on a computer. Hence, Shamir's result is certainly impractical. Nevertheless, it is mentioned here as an interesting fact that there exists a "simple" computer program for factoring and primality testing.

In 1975 Pratt [163] showed that if n is prime, its primality can be proved by executing only $O(\ln^2 n)$ modular multiplications modulo n. However, his paper requires that one know in advance the *complete prime decomposition* of $n-1$. In 1979, this was improved by Plaisted [152], who showed that a Primality Proof (at least under the assumption of a certain likely hypothesis) can be accomplished with only $O((\ln n)(\ln \ln n))$ modular multiplications modulo n. In 1992, Fellows and Koblitz [75] demonstrated that it is possible to provide a proof of the primality of n in deterministic polynomial time given a complete factorization of $n - 1$. This was improved in 1996 by Konyagin and Pomerance [107], who showed that a proof of the primality of n can be obtained in deterministic polynomial time when we know only a *partial* factorization of $n - 1$. The run time for their algorithm is $O((\ln n)^{1+5/(4e)}/\ln \ln n)$, where $0 < \varepsilon < 3/4$.

There is a term underlying the above discussion, which we mentioned in Footnote 1.72 on page 52, namely that of a *certificate*. We now generalize and formalize this notion for later use.

Definition 4.56 (Certificates)

Given a set S, a sequence of computations, which verifies that a given x is an element of S, is called a certificate *for the membership of x in S. When $n \in \mathbb{N}$ and S is some specified set of integers, then a certificate is called* short *if it requires no more than a polynomial in $\ln(n)$ basic operations[4.12] to verify membership of n in S.*

Example 4.57 The Lucas-Lehmer Test described at the outset of this section provides a short certificate for the primality of Mersenne Numbers since only $O(\ln(M_n))$ basic operations need to be executed to verify the primality of M_n. The shortest certificate that can be issued is for Fermat Numbers via Pepin's Test.

Pratt's aforementioned result, in terms of Definition 4.56, may be restated as saying that a certificate for the primality of n can be issued after $O(\ln^2 n)$ basic operations. In 1987, Pomerance [159] invented a method for producing certificates of primality which need only $O(\ln n)$ basic operations to provide a Primality Proof for n. Thus, from Pomerance's work we are guaranteed that any prime has a very short certificate for its primality.

In conclusion, we know that for any prime n there is a polynomial time algorithm for demonstrating that n is indeed prime. Moreover, if a factorization (or even a partial factorization) of $n - 1$ is known, then there is a deterministic polynomial time algorithm for verifying the primality of n.[4.13]

Exercises

☆ 4.23. Let R be a commutative ring with identity containing $\mathbb{Z}/m\mathbb{Z}$ as a subring for a given fixed $m \in \mathbb{N}$ and assume that there is an $\alpha \in R$ such that $\alpha^s = 1$ but for all primes $p \mid s$, $\alpha^{s/p} - 1$ is a unit in R, namely it is invertible in R. Prove that if there exists a $k \in \mathbb{N}$ such that $f(x) = \prod_{j=0}^{k-1}(x - \alpha^{m^j}) \in (\mathbb{Z}/m\mathbb{Z})[x]$, then for any $r \mid s$, there exists a $j \geq 0$ such that $r \equiv m^j \pmod{s}$.

4.24. Use the Lucas-Lehmer Test to verify that M_{13} is prime.

[4.12]If $a, b \in \mathbb{Z}$ and $|a|, |b| < n$, then *basic operations* are defined as follows: $a \pm b$, ab, $\lceil a/b \rceil$, $\lfloor \sqrt{|a|} \rfloor$, and $c \equiv ab \pmod{n}$. These are all polynomial time operations since there are deterministic algorithms to compute: $a \pm b$ in $O(\ln n)$ bit operations, as well as ab, $\lceil a/b \rceil$, $\lfloor \sqrt{|a|} \rfloor$, and $= c \equiv ab \pmod{n}$ in $O(\ln^2 n)$ bit operations. It is sometimes more convenient to measure complexity in terms of these basic operations rather than bit operations (see page 47). The reason is that the *exact* magnitude of the number of bit operations required to multiply two n-bit numbers is not known. Moreover, it depends upon the computing model adopted for the analysis of the complexity (see [208] and [11]).

[4.13]Although it is well beyond the scope of this book, it is worth mentioning that if one assumes what is called the *Extended Riemann Hypothesis* (ERH) then there is a deterministic polynomial time algorithm for establishing the primality of n. However, the ERH is not yet proved, albeit widely believed to be true (see [208]).

4.25. Prove that all Mersenne Numbers M_p are relatively prime for distinct primes p.

4.26. Prove that $n \in \mathbb{N}$ with $n \geq 3$ is prime if and only if there exists an integer m such that $m^{n-1} \equiv 1 \pmod{n}$ but $m^{(n-1)/q} \not\equiv 1 \pmod{n}$ for any prime $q \mid (n-1)$.

4.27. Prove that $n \in \mathbb{N}$ with $n \geq 3$ is prime if and only if, for each prime $q \mid (n-1)$, there exists an integer m_q such that $m_q^{n-1} \equiv 1 \pmod{n}$ but $m_q^{(n-1)/q} \not\equiv 1 \pmod{n}$. (This is an improvement of the result in Exercise 4.26 given by Brillhart and Selfridge in [41].)

4.28. Use Pocklington's Theorem to test $n = 296987$ for primality, armed with the knowledge that the prime $q = 911 \mid (n-1)$.

4.29. Test $n = 3446473$ for primality using Pocklington's Theorem if you know that $b = 8 \cdot 881$ divides $n - 1$.

4.30. Use Proth's Theorem to test $n = 14081$ for primality.

4.31. Test $n = 26113$ for primality using Proth's Theorem.

4.32. By Theorems 2.15 and 2.17, we know that $n \in \mathbb{N}$ is prime if and only if $(n-1)! \equiv -1 \pmod{n}$. Thus, this is a True Primality Test, called *Wilson's Primality Test*. Explain why it is is not a *practical* primality test.

4.33. Given $n \in \mathbb{N}$, set $(n-1)! = q(n)n(n-1)/2 + r(n)$, where $q(n), r(n) \in \mathbb{N}$ with $0 \leq r(n) < n(n-1)/2$. In other words, $r(n)$ is the remainder after dividing $(n-1)!$ by $n(n-1)/2$. Prove that

$$\{r(n) + 1 : r(n) > 0\} = \{p : p > 2 \text{ is prime}\}.$$

(*Hint: Use Wilson's Primality Test described in Exercise 4.32.*) This result was first proved by J. de Barinaga in 1912 (see [60, p. 428]).

4.34. Use the facts given in the outset of the proof of Theorem 4.45 to verify that (4.47) holds.

4.35. Use Pepin's Test to verify that $\mathfrak{F}_4 = 65537$ is prime.

4.36. Prove that if $n \in \mathbb{N}$ is a squarefree,[4.14] odd, composite number such that $(p-1) \mid (n-1)$ for all primes $p \mid n$, then n is a Carmichael Number — see the congruence (4.54).

4.37. Prove that 561, 1729, 10585, and 294409 are all Carmichael Numbers.

4.38. Let $n \in \mathbb{N}$ and assume that $a^{n-1} \equiv 1 \pmod{n}$ for all $a \in (\mathbb{Z}/n\mathbb{Z})^*$ with $\gcd(a, n) = 1$. Prove that n is squarefree.

[4.14]Here n is *squarefree* is defined to mean that there is no prime p such that $p^2 | n$. Thus, in particular 1 is squarefree.

4.3   Probabilistic Primality Tests

The primality tests in this section will be based upon *randomized algorithms* (see Footnote 1.34 on page 20), in contrast to the primality tests in the previous section that were essentially based upon deterministic algorithms. In order to define the first of these algorithms, we need the following notion. The reader should be familiar with the contents of Exercise 4.43 on page 193 in order to see the larger picture.

Definition 4.58 (Euler Liars, Pseudoprimes, and Witnesses)
   Let $n \in \mathbb{N}$ be odd. If $a \in (\mathbb{Z}/n\mathbb{Z})^*$ such that

$$\left(\frac{a}{n}\right) \equiv a^{(n-1)/2} \,(\mathrm{mod}\ n),$$

where the symbol on the left is the Jacobi Symbol given in Definition 4.39, and n is composite, then a is called an Euler Liar *(to the primality of n), or simply an n-Euler Liar, and*

$$E(n) = \left\{ a \in (\mathbb{Z}/n\mathbb{Z})^* : \left(\frac{a}{n}\right) \equiv a^{(n-1)/2} \,(\mathrm{mod}\ n) \right\}$$

is the group of Euler Liars *(see Exercise 4.39). If $a \in (\mathbb{Z}/n\mathbb{Z})^*$, but $a \notin E(n)$, then a is called an* Euler Witness *(to the compositeness of n). If a is an n-Euler Liar, then n is called an* Euler Pseudoprime[4.15] *to base a.   The set of all Euler Pseudoprimes to base a is denoted by* epsp(a).

Example 4.59 $E(15) = \{1, 14\}$ and $E(85) = \{1, 13, 16, 38, 47, 69, 72, 84\}$.

The following result is needed for our first test, and it is of importance in its own right.

Proposition 4.60 (Euler Liars and Primes)
   If $n \in \mathbb{N}$, $n \geq 3$ is odd, then n is prime if and only if

$$E(n) = (\mathbb{Z}/n\mathbb{Z})^* . \tag{4.61}$$

[4.15] According to Erdös [71], the term "Pseudoprime" is due to D.H. Lehmer. The notion of a Pseudoprime has been generalized to linear recurrence sequences and elliptic curves in a sequence of papers beginning with E. Lehmer [116] in 1964 and more recently by Mo and Jones [140] in 1995. The reader is cautioned that some authors allow a Pseudoprime to be prime (for instance see [182, Definition 43, p. 226]). However, this leads to some unfortunate and unnecessary terminology (such as "composite pseudoprime" and "prime pseudoprime"), and potential confusion, so we avoid this interpretation. The term "Euler Pseudoprime" was introduced by Shanks [182] in 1978. Also, in the literature, the term "false witness" is sometimes used for the term "liar."

Proof. If n is prime, then by part (1) of Theorem 4.28,

$$a^{(n-1)/2} \equiv \left(\frac{a}{n}\right) \pmod{n}$$

for all a relatively prime to n. Conversely, if n is composite and (4.61) holds, then

$$a^{n-1} = \left(a^{(n-1)/2}\right)^2 \equiv \left(\frac{a}{n}\right)^2 \equiv 1 \pmod{n},$$

for all $a \in (\mathbb{Z}/n\mathbb{Z})^*$. Therefore, n is a Carmichael Number — see (4.54). Thus, by Exercise 4.38, n is squarefree, so we may set $n = pm$ where p is prime $m > 1$ and $p \nmid m$. Let g be a quadratic nonresidue modulo p. By the Chinese Remainder Theorem, there is a solution $x = a$ to the congruences

$$x \equiv g \pmod{p} \text{ and } x \equiv 1 \pmod{m}.$$

Hence, using the properties of the Jacobi Symbol developed in Theorem 4.40 we get the following.

$$\left(\frac{a}{n}\right) = \left(\frac{a}{pm}\right) = \left(\frac{a}{p}\right)\left(\frac{a}{m}\right) = \left(\frac{g}{p}\right)\left(\frac{a}{m}\right) = \left(\frac{g}{p}\right)\left(\frac{1}{m}\right) = (-1) \cdot 1 = -1.$$

Since $\left(\frac{a}{n}\right) \equiv a^{(n-1)/2} \pmod{n}$, for all $a \in (\mathbb{Z}/n\mathbb{Z})^*$, $a^{(n-1)/2} \equiv -1 \pmod{m}$. However, by the choice of a, we have $a \equiv 1 \pmod{m}$, a contradiction. Hence, n is prime. $\square$

Corollary 4.62 *If $n \in \mathbb{N}$ is composite, then $|E(n)| \leq \phi(n)/2$.*

Proof. By Exercise 4.39, $E(n)$ is a subgroup of $(\mathbb{Z}/n\mathbb{Z})^*$. Since the cardinality of $(\mathbb{Z}/n\mathbb{Z})^*$ is $\phi(n)$ by Example 2.21, and when n is composite, $E(n) \neq (\mathbb{Z}/n\mathbb{Z})^*$ by Proposition 4.60, then $E(n)$ is a proper subgroup of $(\mathbb{Z}/n\mathbb{Z})^*$. So by Lagrange's Theorem for groups (see page 296) $|E(n)| \mid \phi(n)$. Hence, $|E(n)| \leq \phi(n)/2$ since $|E(n)| \neq \phi(n)$ and $\phi(n)$ is even. $\square$

We may now present the first Probabilistic Primality Test.

◆ **Solovay-Strassen Probabilistic Primality Test**
Given $n \in \mathbb{N}$, $n \geq 5$ odd as input and a parameter $r \in \mathbb{N}$, perform the following steps.

(1) Choose a random integer a such that $2 \leq a \leq n - 2$.

(2) If $\gcd(a, n) > 1$, then conclude that n is composite and terminate the algorithm. Otherwise, go to step (3).

(3) Compute $a^{(n-1)/2}$ and $\left(\frac{a}{n}\right)$ modulo n. If

$$a^{(n-1)/2} \not\equiv \left(\frac{a}{n}\right) \pmod{n}, \tag{4.63}$$

conclude that n is composite and terminate the algorithm. Otherwise, set r to $r - 1$ and go to step (1) if $r > 0$. Otherwise, terminate the algorithm.

Notice that if the Solovay-Strassen Test declares n to be composite upon verification of (4.63), then n is indeed composite by part (1) of Theorem 4.28. Similarly, if n is prime, then certainly the test will declare it to be so. However, if n is composite, then there is no certainty that n will *not* be declared to be "prime." Carmichael Numbers provide infinitely many reasons for this to be true. Nevertheless, by Corollary 4.62, the probability that a given odd, composite integer will be declared to be prime, in general, by the Solovay-Strassen Test is less than $(1/2)^r$ if r is randomly chosen. Hence, the larger the parameter r, the greater the probability that the test will *not* declare a composite to be prime. Also, there is a means of avoiding the computation of the greatest common divisor in step (2) since if $g = \gcd(a, n) > 1$, then $\left(\frac{a}{n}\right) = 0$, so g need not be computed. Moreover, since g divides $a^{(n-1)/2}$ modulo n, then testing whether $a^{(n-1)/2} \not\equiv 1 \,(\mathrm{mod}\ n)$ will eliminate the need to test whether $g = 1$ (see Exercise 4.40).

Example 4.64 Let $n = 1121$ and $r = 3$. We choose $a = 2$ and compute $2^{(n-1)/2} = 2^{560} \equiv 137 \,(\mathrm{mod}\ 1121)$, whereas, $\left(\frac{2}{n}\right) = 1 \not\equiv 137 \,(\mathrm{mod}\ n)$ so n is composite.

Example 4.65 Let $n = 1123$ and $r = 5$. We compute $a^{(n-1)/2} - \left(\frac{a}{n}\right) \,(\mathrm{mod}\ n)$ for $a = 2, 3, 5$ and get 0 in each case so the algorithm concludes that n is prime and indeed it is.

Example 4.66 Let $n = 2821$ and $r = 10$. We compute $a^{(n-1)/2} - \left(\frac{a}{n}\right) \,(\mathrm{mod}\ n)$ for $a \in \{3, 4, 9, 12, 16, 17, 25, 27, 36\}$, and get 0 each time. So the Solovay-Strassen Test says that n is prime, but it is not. It is a Carmichael Number and $n = 7 \cdot 13 \cdot 31$. See Exercises 4.41–4.42.

The Solovay-Strassen Test was first devised by Solovay and Strassen [192] in 1978. It was modified in 1982 by Atkin and Larson [7].

Perhaps the most widely used Probabilistic Primality Test is the next one that we will describe. First, we need the following notion.

Definition 4.67 (Strong Pseudoprimes, Liars, and Witnesses)
Let $n = 1 + 2^t m$ where $m \in \mathbb{N}$ is odd, $t \in \mathbb{N}$, and set

$$S(n) = \{a \in (\mathbb{Z}/n\mathbb{Z})^* : a^m \equiv 1 \,(\mathrm{mod}\ n) \text{ or } a^{2^j m} \equiv -1 \,(\mathrm{mod}\ n) \text{ for } 0 \le j < t\},$$

which is called the set of strong liars for n.[4.16] If $a \in (\mathbb{Z}/n\mathbb{Z})^*$, $a \notin S(n)$, and n is composite, then a is called a strong witness (to compositeness) for n. If $a \in S(n)$ for composite n, then n is called a Strong Pseudoprime to base a. The set of all Strong Pseudoprimes is denoted by spsp(a).

[4.16]It is known that if $n \ne 9$, then $|S(n)| \le \phi(n)/4$ (see [134] or [11]). The term "Strong Pseudoprime" was introduced by Selfridge (see Footnote 4.17) in an unpublished manuscript in the mid 1970s. However, it did get published in a paper of Williams [206] in 1978.

The following analogue of Proposition 4.60, also of interest in its own right, is needed for the second primality test of this section. Before proceeding, the reader will need to solve Exercise 4.44 on page 194, which is used in the following proof. Also, Exercise 4.54 will be of interest since it shows that the antithesis of the following can hold.

Proposition 4.68 (Strong Pseudoprimes and Primes)
Let $n = 1 + 2^t m$ be squarefree where $m \in \mathbb{N}$ is odd and $t \in \mathbb{N}$. Then n is prime if and only if

$$S(n) = (\mathbb{Z}/n\mathbb{Z})^* . \tag{4.69}$$

Proof. If n is prime and a is relatively prime to n, then $a^{n-1} \equiv 1 \,(\mathrm{mod}\, n)$, so by Exercise 2.8 on page 72, $a^{(n-1)/2} \equiv \pm 1 \,(\mathrm{mod}\, n)$. If $a^{(n-1)/2} \equiv 1 \,(\mathrm{mod}\, n)$, then $a^{(n-1)/4} \equiv \pm 1 \,(\mathrm{mod}\, n)$ by Exercise 2.8 again. Continuing in this fashion, we see that either $a^m \equiv 1 \,(\mathrm{mod}\, n)$ or $a^{2^j m} \equiv -1 \,(\mathrm{mod}\, n)$ for some natural number $j < t$.

Conversely if (4.69) holds, and p is the smallest prime dividing n, then we may select a primitive root a modulo p with $1 < a < p$. Therefore, $\gcd(a, n) = 1$, so $a \in S(n)$. Thus, one of $a^m \equiv 1 \,(\mathrm{mod}\, n)$ or $a^{2^j m} \equiv -1 \,(\mathrm{mod}\, n)$ must hold so either $a^m \equiv 1 \,(\mathrm{mod}\, p)$ or $a^{2^j m} \equiv -1 \,(\mathrm{mod}\, p)$. In the former case, $(p-1) \mid m$ and $m \mid (n-1)$. In the latter case, $(p-1) \mid 2^j m$ and $2^j m \mid (n-1)$, by Proposition 4.3. Hence, by Exercise 4.36 on page 187, n is a Carmichael Number. Now we may proceed exactly as in the proof of Proposition 4.60, since Exercise 4.44 tells us that n is an Euler Pseudoprime. $\square$

The following test for primality is also known as the *Strong Pseudoprime Test* in the literature.

◆ **The Miller-Rabin-Selfridge Primality Test**[4.17]
Let $n - 1 = 2^t m$ where $m \in \mathbb{N}$ is odd and $t \in \mathbb{N}$. With n and a parameter $r \in \mathbb{N}$ as input, execute the following steps.

(1) Choose a random integer a with $2 \leq a \leq n - 2$.

(2) Compute $x \equiv a^m \,(\mathrm{mod}\, n)$ using the repeated squaring method described on pages 71–72. If $x \equiv 1 \,(\mathrm{mod}\, n)$, then go to step (4). Otherwise, go to step (3).

(3) Compute $x \equiv a^{2^j m} \,(\mathrm{mod}\, n)$, using the aforementioned method for $j \geq 1$ until either $x \equiv \pm 1 \,(\mathrm{mod}\, n)$ or $j = t - 1$. If $x \not\equiv 1 \,(\mathrm{mod}\, n)$ for any such j in the computation, then terminate the algorithm with "n is composite." Otherwise, go to step (4).

[4.17]This test is most often called the *Miller-Rabin Test* in the literature. However, John Selfridge was using the test in 1974 before Miller first published the result, so we credit Selfridge here with this recognition. John Selfridge was born in Ketchican, Alaska, on February 17, 1927. He received his doctorate from U.C.L.A. in August of 1958, and became a Professor at Pennsylvania State University six years later. He is a pioneer in computational number theory and continues to produce significant contributions to the area.

(4) Set r to $r - 1$ and go to step (1) if $r > 0$. Otherwise, terminate the algorithm with "n is prime."

Example 4.70 Let $n = 3539$ and $r = 1$. Since $n - 1 = 2 \cdot 1769 = 2^t m$, then we choose $a = 2$ and compute $x = 2^m \equiv -1 \pmod{n}$ in step (2). Since $t - 1 = 0$ and $x \not\equiv 1 \pmod{n}$ in step (3), we set $r = 0$ and terminate with "n is prime."

The above algorithm tests whether or not a randomly chosen a is in $S(n)$. If $x \equiv 1 \pmod{n}$ in step (3), then $a^{2^j m} \equiv 1 \pmod{n}$ but $a^{2^{j-1} m} \not\equiv \pm 1 \pmod{n}$. Thus, $\gcd(n, a^{2^{j-1} m} - 1) > 1$ is a nontrivial factor of n (see (5.3) and the discussion surrounding it on page 198). If, in addition, $x \not\equiv -1 \pmod{n}$ in step (3), then $a \in S(n)$. Therefore, if the Miller-Rabin-Selfridge Algorithm declares n to be composite, then indeed it is composite. In other words, if n is prime, then the algorithm declares n to be prime (see Exercise 4.45). However, if n is composite, then the above discussion can be used to show that the algorithm declares n to be prime with probability at most $(1/4)^r$.

If a composite $n \in \mathbb{N}$ passes the Miller-Rabin-Selfridge Test, then $n \in \text{spsp}(a)$ for at least r values of a. If $n \in \mathbb{N}$ such that $a^n \equiv a \pmod{n}$ for *all* $a \in \mathbb{Z}$, then n is called an *Absolute Pseudoprime*. By Exercise 4.46, the Absolute Pseudoprimes are exactly the Carmichael Numbers — see (4.54).

Of the Solovay-Strassen and the Miller-Rabin-Selfridge Tests, the latter is computationally less expensive and easier to implement than the former, which requires the computation of Jacobi Symbols. Moreover, given an upper bound on the error probability of $(1/2)^r$ for the former and $(1/4)^r$ for the latter, the probability of *correctness of outcome* is higher with the latter. These facts are augmented by Exercise 4.44, which tells us that the latter can never err *more* than the former since strong liars must be Euler Liars. The complexity of both Solovay-Strassen and Miller-Rabin-Selfridge is $O(\ln^3 n)$ for input n with parameter $r = 1$.

In the literature, the term *Monte Carlo*[4.18] Algorithm is used to mean a *Probabilistic Algorithm* (or an algorithm that uses random numbers — see the discussion of randomized algorithms in Footnote 1.34 on page 20) that achieves a correct answer at least fifty percent of the time. Thus, both Solovay-Strassen and Miller-Rabin-Selfridge are Monte Carlo Algorithms for compositeness. In other words, these tests, with a high probability, provide a proof that a given input is composite, but only provide (good) evidence of primality. The reason is that if input n is declared to be composite, then indeed it is composite for reasons cited above. However, if n is declared to be prime, then we have only a specified probability that it *is* prime since some composite numbers can be declared to be prime by these tests (see Exercise 4.42 for instance).

A *yes-biased* Monte Carlo Algorithm is a Monte Carlo Algorithm in which a "yes" answer is always correct, but a "no" answer may be incorrect. A *no-biased* Monte Carlo Algorithm is defined as one for which a "no" answer is

---

[4.18] Use of the term *Monte Carlo Algorithm* for an algorithm using random numbers goes back to the turn of the century (see [101]).

always correct, but a "yes" answer may be incorrect. A yes-biased Monte Carlo Algorithm is said to have *error probability* equal to $\alpha \in \mathbb{R}^+$ with $0 \leq \alpha < 1$ if, for any occurrence in which the answer is "yes," the algorithm will give the incorrect answer "no" with probability at most α (where the probability is computed over all possible random choices made by the algorithm for a given input). Thus, the Solovay-Strassen Algorithm is a yes-biased Monte Carlo Algorithm for the decision problem "Is n composite?" with error probability $\alpha = (1/2)^r$. The Miller-Rabin-Selfridge Algorithm for the same decision problem is a yes-biased Monte Carlo Algorithm with error probability $\alpha = (1/4)^r$.

Other randomized algorithm types are given as follows. An *Atlantic City Algorithm*[4.19] is a Probabilistic Algorithm that answers correctly at least seventy-five percent of the time. A *Las Vegas*[4.20] Algorithm is a Probabilistic Algorithm that either produces "no answer" as output or produces a correct answer as output. There is an assigned *failure probability* α such that the algorithm will produce "no answer." Thus, if a Las Vegas Algorithm is run $r \in \mathbb{N}$ times, then the probability that the algorithm will declare "no answer" r times in a row is α^r.

Exercises

4.39. Prove that a nonempty set H is a subgroup of a multiplicative group G if and only if $a \cdot b^{-1} \in H$ for all $a, b \in H$. Use this to show that $H - E(n)$ given in Definition 4.58 is a subgroup of $G = (\mathbb{Z}/n\mathbb{Z})^*$. (*Hint: Use Exercise 2.39 on page 100.*)

4.40. Let $n = 1 + p^t m$, p an odd prime, $t, m \in \mathbb{N}$, and suppose that there exists an $a \in \mathbb{Z}$ such that $a^{n-1} \equiv 1 \pmod{n}$. Prove that if $q \mid n$ is prime, then either $a^{(n-1)/p} \equiv 1 \pmod{q}$ or $q \equiv 1 \pmod{p^t}$.

4.41. Test $n = 7487$ for primality using the Solovay-Strassen Test.

4.42. Run the Solovay-Strassen Test on $n = 188461$.

4.43. An odd composite number n for which there exists an integer a such that

$$a^{n-1} \equiv 1 \pmod{n},$$

is called a *Pseudoprime to base a*, and a is called a *Fermat Liar*. The set of all Pseudoprimes to base a is denoted by $\mathrm{psp}(a)$.[4.21] Prove that the set of all Fermat Liars,

$$F(n) = \{a \in (\mathbb{Z}/n\mathbb{Z})^* : a^{n-1} \equiv 1 \pmod{n}\}$$

is a multiplicative group. Also, show that $\mathrm{epsp}(a) \subseteq \mathrm{psp}(a)$ (see Definition 4.58).

[4.19]The term "Atlantic City" was first introduced in 1982 by J. Finn in an unpublished manuscript entitled *Comparison of probabilistic tests for primality*.

[4.20]The term "Las Vegas" Algorithm was introduced by L. Babai in 1982 (see [99]).

[4.21]In the literature, these are often called *Fermat Pseudoprimes* or *Ordinary Pseudoprimes* — see (4.54) on page 184.

☆ 4.44. Prove that spsp(a) ⊆ epsp(a) (see Definitions 4.58 and 4.67).

4.45. Prove that if n is prime, then the Miller-Rabin-Selfridge Algorithm declares it to be so.

4.46. Prove that the set of Absolute Pseudoprimes is equal to the set of Carmichael Numbers (see the discussion on page 192).

4.47. Use the Miller-Rabin-Selfridge Algorithm to test $n = 6601$ for primality with input parameter $r = 1$.

4.48. With input parameter $r = 2$, test $n = 1093$ for primality using the Miller-Rabin-Selfridge Algorithm.

4.49. Show that $n \in$ epsp(a) does not imply $n \in$ spsp(a) in general by exhibiting a counterexample.

4.50. Prove that if $n \equiv 3 \,(\mathrm{mod}\ 4)$, then $n \in$ epsp(a) if and only if $n \in$ spsp(a). (This was first proved by Malm in 1977 — see [125].)

4.51. Prove that there exist infinitely many Pseudoprimes to base 2.

4.52. Prove that if $n \in \mathbb{N}$ such that $2^n \equiv 1 \,(\mathrm{mod}\ n)$, then $n = 1$.

4.53. Let $n \in \mathbb{N}$ with $n > 1$ odd. Show that $a \in S(n)$ if and only if $n - a \in S(n)$.

4.54. Let n be a product of the first $\ell \geq 2$ odd primes. Prove that $S(n) = \{1, n-1\}$.

4.55. Let $a \in \mathbb{Z}$, $n \in \mathbb{N}$ such that $a^{(n-1)/2} \equiv -1 \,(\mathrm{mod}\ n)$, and let $|x|_2$ denote the highest power of 2 dividing $x \in \mathbb{N}$. In other words, if $x = 2^b c$ where c is odd, then $|x|_2 = 2^b$. Prove that $|\phi(r)|_2 \geq |n-1|_2$ for all divisors r of n and that equality holds if and only if $\left(\frac{a}{r}\right) = -1$.

4.56. For $a \in \mathbb{N}$, show that spsp$(a) \subseteq$ psp(a). Then show that if p is a prime and $b \in \mathbb{N}$, then $p^b \in$ spsp(a) if and only if $p^b \in$ psp(a).

4.57. Given $D, n, k \in \mathbb{N}$ such that $\gcd(n, D) = 1$, with n odd having canonical prime factorization $n = \prod_{j=1}^{k} p_j^{a_j}$, define

$$\psi_D(n) = \frac{1}{2^{k-1}} \prod_{j=1}^{k} p_j^{a_j-1} \left(p_j - \left(\frac{D}{p_j} \right) \right),$$

where the symbol on the right is the Legendre Symbol. Prove that n is prime if and only if $\psi(n) = n - (D/n)$.

4.58. With reference to Exercise 4.57, prove that if n is odd and

$$(n - (D/n)) \mid \psi_D(n),$$

then n is prime.

Chapter 5

☞Factoring

This optional chapter deals with factoring algorithms as they pertain to cryptography. The first section deals with three such methods for factoring. We first present Pollard's $p-1$ Algorithm, then the Continued Fraction Method, and conclude with the Quadratic Sieve.

5.1 Three Algorithms

Given the fact that certain cryptosystems such as the RSA and Rabin Public-key Ciphers depend, for their security, upon the intractability of factoring large integers, it is valuable to have a look at some factorization algorithms, by which we mean an algorithm that solves the problem of determining the complete factorization of an integer $n > 1$ guaranteed by the Fundamental Theorem of Arithmetic (Theorem A.28 on page 282). In other words, the algorithm should (*ultimately*, as described below) find distinct primes p_j and $a_j \in \mathbb{N}$ such that $n = \prod_{j=1}^{k} p_j^{a_j}$. We observe that it suffices for such algorithms to merely find $r, s \in \mathbb{N}$ such that $1 < r \leq s < n$ with $n = rs$ (called *splitting n*), since we can then apply the algorithm to n/r and to s, thereby recursively splitting each composite number until a complete factorization is found. Furthermore, given that *deciding* whether a given $n > 1$ is composite or prime is easier, in general, than factoring, one should always check first that n is composite before applying a factorization algorithm. We have already dealt with the notion of *primality testing* in Chapter Four. Furthermore, if n is a perfect power, namely of the form $n = m^d$ where $m, d \in \mathbb{N}$ such that $d > 1$, then one can efficiently determine m and d as follows. For each prime $p \leq \log_2 n$, do a binary search for $r \in \mathbb{N}$ such that $n = r^p$ where attention is restricted to $2 \leq r \leq 2^{\lfloor (\log_2 n)/p \rfloor + 1}$. This can be accomplished in $O((\log_2 n)^3(\log_2(\log_2(\log_2 n))))$ bit operations. Hence, when looking at factorization algorithms below, we may assume that we have already tested n to be composite and not a perfect power, thereby ensuring that we are looking at a product of (at least) two (or, as in the case of RSA, *exactly* two) distinct primes.

This section is devoted to looking at three factorization algorithms in detail. This provides the briefest of overviews, since we could easily devote the entire text to the numerous factorization algorithms that exist in the literature. One of the most primitive algorithms is, of course, *trial division*. For a given $n \in \mathbb{N}$, this means that one divides n by all primes up to $\lfloor \sqrt{n} \rfloor$. For "small" natural numbers, say $n < 10^{12}$, this is not an unreasonable method. However, for "large" integers, we need more elaborate methods. Today, the most effective factorization algorithms are the Quadratic[5.1] and Number Field Sieves as well as the elliptic curve algorithm. In Chapter Six, we will learn about the Elliptic Curve Method. In Section Two of this chapter we will learn about the Number Field Sieve, and in this section, we will learn about the Quadratic Sieve.

A predecessor of these three powerhouses is the first topic of discussion. To understand what it says, we first need the following concept.

Definition 5.1 (Smooth Numbers)
If $B \in \mathbb{N}$, then $n \in \mathbb{N}$ is said to be smooth with respect to B *or simply* B-smooth, *provided that all primes dividing n are no bigger than B. The bound B is called a* smoothness bound.

Smooth numbers satisfy the triad of properties:

(1) They are fairly numerous (albeit sparse).

(2) They enjoy a simple multiplicative structure.

(3) They play an essential role in discrete logarithm algorithms.

In 1974, Pollard [155] introduced the following algorithm.

◆ **Pollard's p − 1 Factoring Algorithm**
Suppose that we wish to factor $n \in \mathbb{N}$, and that a smoothness bound B has been selected.[5.2] Then we execute the following.

(1) Choose a base $a \in \mathbb{N}$ where $2 \leq a < n$ and compute $g = \gcd(a, n)$. If $g > 1$, then we have a factor of n. Otherwise, go to step (2).

(2) For all primes $p \leq B$, compute $m = \left\lfloor \frac{\ln(n)}{\ln(p)} \right\rfloor$ and replace a by $a^{p^m} \pmod{n}$ using the repeated squaring method given on page 71. (Note that this iterative procedure ultimately gives $a^{\prod_{p \leq B} p^m}$ modulo n for the base a chosen in (1).)

(3) Compute $g = \gcd(a - 1, n)$. If $g > 1$, then we have a factor of n, and the algorithm is successful. Otherwise, the algorithm fails.

[5.1] A Quadratic Sieve (in general) is one in which about half of the possible numbers being sieved are removed from consideration. Such a sieve has been used for centuries to eliminate impossible cases from consideration.

[5.2] In practice, B is usually selected such that $10^5 < B < 10^6$. See the discussion following the description of the algorithm.

The reasoning behind Pollard's Algorithm is given as follows.

Let $\ell = \text{lcm}(p_1^{a_1}, \ldots, p_t^{a_t})$, where $p_j^{a_j}$ runs over all prime powers such that $p_j \leq B$. Since $p_j^{a_j} \leq n$, then $a_j \ln(p_j) \leq \ln(n)$, so $a_j \leq \left\lfloor \frac{\ln(n)}{\ln(p_j)} \right\rfloor$. Hence, $\ell \leq \prod_{j=1}^{t} p_j^{\lfloor \ln(n)/\ln(p_j) \rfloor}$. Now, if $p \mid n$ is a prime such that $p - 1$ is B-smooth, then $(p-1) \mid \ell$. Therefore, for any $a \in \mathbb{N}$ with $p \nmid a$, $a^\ell \equiv 1 \pmod{p}$, by Fermat's Little Theorem (2.18). Thus, if $g = \gcd(a^\ell - 1, n)$, then $p \mid g$. If $g = n$, then the algorithm fails. Otherwise, it succeeds.

Example 5.2 Let $n = 13193$, and choose a smoothness bound $B = 13$, then select $a = 2$. We know that a is relatively prime to n so we proceed to step (2). The table shows the outcome of the calculations for step (2).

p	2	3	5	7	11	13
m	13	8	5	4	3	3
a	6245	1365	1884	3133	5472	396

Then we go to step (3) and check $\gcd(a - 1, n) = \gcd(395, 13193) = 79$. Thus, we have factored $n = 79 \cdot 167$. Observe that $p = 79$ is B-smooth since $p - 1 = 2 \cdot 3 \cdot 13$, but $q = 167$ is not since $q - 1 = 2 \cdot 83$. See Exercises 5.1–5.2.

In practice, factors found by a single iteration of the $p-1$ method delineated above will be smaller than those found by the next two algorithms that we will describe in this section.

The running time for Pollard's $p-1$ Algorithm is $O(B \ln(n)/\ln(B))$ modular multiplications, assuming that $n \in \mathbb{N}$ and there exists a prime $p \mid n$ such that $p - 1$ is B-smooth. This is of course the drawback to this algorithm, namely that it requires n to have a prime factor p such that $p-1$ has only "small" prime factors. A generalization of the $p - 1$ method was given by Lenstra using elliptic curves, which we will study in Chapter Six. In the Elliptic Curve Algorithm, we will see that success in factoring depends upon an integer "close" to p having only small prime factors, which is less demanding than the $p - 1$ algorithm and therefore more likely to occur. Another improvement was given by Williams in [207], called the $p + 1$ method of factoring, which is efficient if n has a prime factor p such that $p + 1$ is B-smooth. There have been other refinements and improvements. One may even vary the original algorithm by choosing a "large" smoothness bound, say $B > 10^6$, but executing step (2) in the algorithm for only a few primes p and computing the gcd condition in step (3) after each prime chosen. Pollard also developed another method for factoring in 1975, called the *Monte Carlo Factoring Method*, also known as the *Pollard Rho Method* (see [144, pp. 127–130]).

In Chapter One, we discussed Fermat's Difference of Squares Method for factoring (see page 24).[5.3] On the other hand, Legendre looked at congruences $x^2 \equiv \pm py^2 \pmod{n}$ for primes p. A solution to this congruence means that $\pm p$

[5.3]Fermat's Factoring Method is greatly improved by constructing a sieve on the equation

is a quadratic residue modulo all prime factors of n. Hence, if the residue is $+2$, for instance, then we know (see Exercise 4.17 on page 179) that all prime factors of n are congruent to ± 1 modulo 8, so already we have halved the search for factors of n. Legendre applied this method repeatedly for various primes p. Thus, what Legendre was essentially doing was to construct a quadratic sieve by getting lots of residues modulo n, thereby eliminating potential prime divisors of n that sit in various linear sequences. He found that if you could get enough of these, then one could eliminate primes up to $\sqrt{n}$ as prime divisors and thus show n was prime. What underlies this method of Legendre is the continued fraction expansion of $\sqrt{n}$ since essentially what Legendre was doing was finding *small* residues modulo n (see [199] for more historical background). It should also be noted that, in retrospect, what Euler did could be used to get equation (5.3) below as follows. Euler considered two representations of n:

$$n = x^2 + ay^2 = z^2 + aw^2,$$

so

$$(xw)^2 \equiv (n - ay^2)w^2 \equiv nw^2 - ay^2w^2 \equiv -ay^2w^2 \equiv (z^2 - n)y^2 \equiv (zy)^2 \,(\mathrm{mod}\ n),$$

and we are back to a potential factor for n. The basic idea in the above, for a given $n \in \mathbb{N}$, is simply that if we can find $x, y \in \mathbb{Z}$ such that

$$x^2 \equiv y^2 \,(\mathrm{mod}\ n), \tag{5.3}$$

and $x \not\equiv \pm y \,(\mathrm{mod}\ n)$, then $\gcd(x - y, n)$ is a nontrivial factor of n. This idea is currently exploited by numerous algorithms. For instance, the algorithms that we study in the balance of this section use the construction of these (x, y) pairs.

Legendre was only concerned with building the sieve on the prime factors of n, and so he was unable to *predict*, for a given prime p, a second residue to yield a square. In other words, if he found a solution to $x^2 \equiv py^2 \,(\mathrm{mod}\ n)$, he could not predict a different solution $w^2 \equiv pz^2 \,(\mathrm{mod}\ n)$. If he would have been able to do this, then he would have been able to combine the two as

$$(xw)^2 \equiv (pzy)^2 \,(\mathrm{mod}\ n),$$

so if $xw \not\equiv \pm pzy \,(\mathrm{mod}\ n)$, then $\gcd(xw - pzy, n)$ would be a nontrivial factor of n, thereby putting us back in the situation given in (5.3).

This idea, mentioned above, of trying to match the primes to create a square

---

$n = x^2 - y^2$, by writing it as $y^2 = x^2 - n$, as Fermat did, and then constructing the Quadratic Sieve on $y^2 \equiv x^2 - n \,(\mathrm{mod}\ m_j)$ for various choices of m_j. In the 1960s, John Brillhart programmed this method and got it running at 10^5 of the x values per second. He was able to factor relatively large numbers in this fashion. For a description of the details of how this was done, see [41]. The reader will also benefit from looking at [115, Volume III, pp. 886–893], wherein the (then) twenty-two year old Dick Lehmer talks about his construction of a bicycle-chain sieve, which he used to factor numbers.

is what Kraitchik[5.4] did. Kraitchik, in the early 1920s,[5.5] reasoned that it might suffice to find a *multiple* of $n \in \mathbb{N}$ as a difference of squares. He chose a quadratic polynomial of the form $kn = ax^2 \pm by^2$ for some $k \in \mathbb{N}$. In its simplest form with $k = a = b = 1$, he would sieve over $x^2 - n$ for $x \geq \lfloor \sqrt{n} \rfloor$. This is the basic idea behind the Quadratic Sieve method that we will describe in detail below. Thus, what Kraitchik had done was to opt for "fast" generation of quadratic residues, and in so doing abandoned Legendre's Method (meaning that, generally, he did not have residues less than $2\sqrt{n}$), but gained control over finding of two distinct residues at a given prime to form a square (as described above), which Legendre was unable to do. This will become clear when we describe the Quadratic Sieve method below. Thus, Kraitchik could start at values bigger than $\sqrt{n}$ and sieve until "large" residues were found.

Prior to the inception of the Quadratic Sieve in the early 1980s, there was another factorization algorithm that reigned from its birth in mid 1970 until the Quadratic Sieve took over in the early 1980s. In 1931, Lehmer and Powers wrote about the failures of Kraitchik's method in [117], where they show that if (5.3) fails, then the other method they discuss would automatically fail. Their paper is not one about technique, and they viewed Kraitchik's method negatively as being unsuitable even for hand calculations. However, Kraitchik's *success* with the method, even with hand calculations, demonstrates that this is not true. Kraitchik developed the method, refining it in various ways, over a period of some thirty-five years. Lehmer did exploit Kraitchik's method for factoring by getting residues as Legendre had obtained them. Thus, to this point historically, there were three factoring methods: (1) Legendre's Method, using continued fractions of $\sqrt{n}$ to get relatively small factored residues, which he multiplied, suppressed squares, and recovered "small" residues, which were used to create a sieve on the prime factors of n. (2) Kraitchik used two other ideas. He used quadratic polynomials to get the residues, then multiplied them to get squares (not a square times a small number). (3) Lehmer did the same as Kraitchik, except that he got his residues as Legendre had. However, Lehmer did not think this was a good method for hand computing.

An early version of the continued fraction method was described by Knuth

---

[5.4]Maurice Borisovich Kraitchik (1882–1957) was born on April 21, 1882 in Minsk, capital of the former Belorussian Soviet Socialist Republic. Although he graduated from high school in 1903, he was prevented from attending university in Russia due to the two percent upper bound put on enrollment for Jewish students at that time. Thus, he left Russia in 1905, and by 1910, he graduated as an electrical engineer from a university in Liége, Belgium. In 1923 he obtained his doctorate from the University of Brussels, and maintained a high degree of interest in number theory throughout his life. Between 1915 and 1948, he worked both as an engineer in Brussels, and later as a Director at the Mathematical Sciences section of the Mathematical Institute for Advanced Studies there. During the period 1941–1946, he was Associate Professor at the New School for Social Research in New York. In 1946, he returned to Belgium where he spent the rest of his life. He died on August 19, 1957 in Brussels.

[5.5]According to Williams [208, pp. 127–129], Paul Seelhoff (1829–1896), a high school teacher in Mannheim, Germany, had the basic idea for the Quadratic Sieve before Kraitchik, but the import of Kraitchik's work and the details of development ensure that it deserves to be called *Kraitchik's Method.*

in 1967 (see [104] where the idea is attributed to John Brillhart[5.6]). In 1970, the Continued Fraction Method was used to factor the seventh Fermat Number. The factorization that was done in 1970 was announced in 1971 and published in [146] in 1975. This was the first demonstration of the power of the Continued Fraction Method since it doubled the size of the integers, having from twenty-five to fifty digits, that one could factor at that time.

The reader unfamiliar with the theory of continued fractions should consult the review section on pages 299–302 in Appendix A. In any case, each reader should consult those pages in order to be acquainted with the notation used below.

In what follows, a *factor base* is a set of "small" primes that will remain the primes under consideration for the algorithm at hand.

◆ **The Continued Fraction Factoring Algorithm**

Suppose that we wish to factor $n \in \mathbb{N}$ and a smoothness bound B has been selected. Then execute the following steps:

(1)   Choose a factor base of primes $\mathcal{F} = \{p_1, p_2, \ldots, p_k\}$ for some $k \in \mathbb{N}$ determined by B and a large upper index value J.[5.7]

(2)   Set $Q_0 = 1$, $P_0 = 0$, $A_{-1} = 1$, $A_0 = \lfloor \sqrt{n} \rfloor = q_0 = P_1$. For each natural number $j \leq J$, recursively compute Q_j using the following formulas:

$$Q_j = \frac{n - P_j^2}{Q_{j-1}},$$

$$q_j = \left\lfloor \frac{P_j + \lfloor \sqrt{n} \rfloor}{Q_j} \right\rfloor,$$

$$A_j = q_j A_{j-1} + A_{j-2},$$

$$P_{j+1} = q_j Q_j - P_j,$$

and trial divide Q_j by the primes in $\mathcal{F}$ to determine if Q_j is p_k-smooth. If it is, use its factorization $Q_j = \prod_{i=1}^{k} p_i^{a_{i,j}}$ to form the binary $k + 1$-tuple $\mathfrak{v}_j = (v_{0,j}, v_{1,j}, v_{2,j}, \ldots, v_{k,j})$, where $v_{0,j}$ is respectively 0 or 1 according as j is even or odd, and for $1 \leq i \leq k$, $v_{i,j}$ is respectively 0 or 1 according as $a_{i,j}$ is even or odd. If Q_j is not p_k-smooth, discard it and return to calculate Q_{j+1}.

[5.6]The Brillhart-Morrison Algorithm was the first *subexponential* time algorithm arising from a random square method (see page 49). Also, it was the leader in multiprecision factorization methods that had been developed at that time. In a private conversation with this author, John Brillhart stated that the XORing (see Footnote 2.22 on page 88) of bit strings used to produce quadratic residues modulo n of the form $\pm pa^2$ in the computer version of Legendre's Method suggested that the same XORing could just as easily produce a quadratic residue of the form a^2 modulo n, which leads to Kraitchik's Method and avoids the messy sieve in Legendre's Method. This led to the Continued Fraction Algorithm.

[5.7]From knowledge about the distribution of smooth integers close to $\sqrt{n}$, the optimal k is known to be one which is chosen to be approximately $\sqrt{\exp(\sqrt{\log(n) \log \log(n)})}$.

(3) For each set S of the vectors v_j constructed in (2), for which it is discovered that

$$\sum_{j \in S} v_{i,j} \equiv 0 \,(\text{mod } 2), \quad 0 \le i \le k,$$

we have $x^2 \equiv y^2 \,(\text{mod } n)$, where

$$x = \left[\prod_{j \in S} (-1)^j Q_j \right]^{1/2} \quad \text{and} \quad y \equiv \prod_{j \in S} A_{j-1} \,(\text{mod } n).$$

If $x \not\equiv \pm y \,(\text{mod } n)$, then $\gcd(x \pm y, n)$ gives a nontrivial factor of n.[5.8]

By Corollary A.96 on page 302, $A_{j-1}^2 - nB_{j-1}^2 = (-1)^j Q_j$, which is the heart of the algorithm. Thus, we have that $nB_{j-1}^2 \equiv A_{j-1}^2 \,(\text{mod } p)$, for any prime $p \mid Q_j$, so n is a quadratic residue modulo p. Hence, we only put primes p in the factor base for which n is a quadratic residue modulo p. The following gives a small illustration of the continued fraction algorithm, called CFRAC by some of its users.

Example 5.4 Let $n = 6109$. Our factor base will be $\mathcal{F} = \{3, 5, 11, 13, 31, 37\}$. Since $\lfloor \sqrt{n} \rfloor = 78$, then we compute the following table (where $J = 3$).

j	P_j	q_j	A_{j-1}	$(-1)^j Q_j$	v_j
0	0	78	1	1	$(0,0,0,0,0,0,0)$
1	78	6	78	-25	$(1,0,0,0,0,0,0)$
2	72	4	469	37	$(0,0,0,0,0,0,1)$
3	76	17	1954	-9	$(1,0,0,0,0,0,0)$

We have a set S such that $\sum_{j \in S} v_{i,j} \equiv 0 \,(\text{mod } 2)$ for each $i = 0, 1, \ldots, 6$. This set is $S = \{1, 3\}$ for which we have $Q_1 = -5^2$, $Q_3 = -3^2$, $A_0 - 78$ and $A_2 = 1954$. We compute $\prod_{j \in S} A_{j-1} \equiv 5796 \,(\text{mod } 6109)$ and since

$$y^2 = \prod_{j \in S} A_{j-1}^2 \equiv x^2 = \prod_{j \in S} Q_j = 15^2 \,(\text{mod } n),$$

then we check $\gcd(x \pm y, n)$. We compute that both $\gcd(x - y, n) = \gcd(15 - 5796, 6109) = 41$, and $\gcd(x + y, n) = \gcd(15 + 5796, 6109) = 149$. Thus, we have factored $n = 41 \cdot 149$.

[5.8]The computational details for finding the sets S is given in [146]. The description of the algorithm given here is necessarily quite simplified for this level of exposition. Note in practice that the set of bit vectors is checked periodically to see if it is linearly independent and if this leads to a factorization. If not, then one continues until either a factor is found or until one gets tired and gives up. If n is large, the expectation of having the algorithm end by giving a factor is low. Yet, it can end in a reasonably short time and give spectacular factorizations, but this is rare.

It is possible that no subset S as in Example 5.4 can be found, in which case we try a new attack. One may multiply n by a suitable factor, say m, and perform CFRAC on mn. The value of m is usually chosen to be a product of a small set of primes. For instance, consider the following illustration.

Example 5.5 Let $n = 3827$, and select a factor base consisting of some primes for which n is a quadratic residue, $\mathcal{F} = \{2, 17, 23, 29, 31\}$. Since $\lfloor \sqrt{n} \rfloor = 61$ and we compute the following table (where $J = 5$).

j	P_j	q_j	A_{j-1}	$(-1)^j Q_j$	v_j
0	0	61	1	1	$(0,0,0,0,0,0)$
2	45	6	62	17	$(0,0,1,0,0,0)$
3	57	3	433	-34	$(1,1,1,0,0,0)$
5	61	61	3155	-2	$(1,1,0,0,0,0)$

We see that a subset S of the j for which the $v_{i,j}$ sum to zero modulo 2 for each $i = 0,1,2,3,4,5$ is $S = \{2,3,5\}$. We compute

$$\prod_{j \in S} A_{j-1} = 3155 \cdot 433 \cdot 62 \equiv 3793 = y \, (\mathrm{mod} \ 3827),$$

and $\prod_{j \in S}(-1)^j Q_j = 34^2 = x^2$. But, $\gcd(y+x, n) = \gcd(3793+34, 3827) = 3827$, so we have a trivial factorization. Moreover, there is no point in computing for higher values of j than $j = 5$ since the continued fraction repeats given that $P_6 = 61 = P_5$ and $Q_j = Q_{10-j}$ for $j = 0,1,2,3,4,5$ (see Theorem A.98). Hence, we choose $m = 2$ and apply CFRAC to $mn = 7654$. We will use the following factor base consisting of primes for which mn is a quadratic residue, $\mathcal{F} = \{2,3,5,11,17\}$. Since $\lfloor \sqrt{7654} \rfloor = 87$, we compute the following table.

j	P_j	q_j	A_{j-1}	$(-1)^j Q_j$	v_j
0	0	87	1	1	$(0,0,0,0,0,0)$
1	87	2	87	-85	$(1,0,0,1,0,1)$
2	83	18	175	9	$(0,0,0,0,0,0)$

We see that for $S = \{2\}$, we have a square Q_2. Thus, we compute $A_1 = 175 = y$, $Q_2 = 3^2 = x^2$, and $g = \gcd(x - y, mn) = \gcd(3 - 175, 7654) = 86$. Thus, $g/m = 43$ is a nontrivial factor of n, and we have as well that $g_1 = \gcd(x + y, mn) = \gcd(3 + 175, 7654) = 178$, so $g_1/m = 89$. Hence, we have a complete factorization $n = 43 \cdot 89$. (See Exercises 5.5–5.6.)

The expected running time of CFRAC (see [157]) is known to be

$$O(\exp(\sqrt{2} + o(1))\sqrt{(\ln n)(\ln \ln n)})).$$

In the early 1980s, Carl Pomerance was able to fine-tune the parameters in Kraitchik's Quadratic Sieve to show that it was better than the continued fraction algorithm for producing residues, then multiplying the congruences to produce a solution of (5.3) for the purposes of factoring.

◆ **The Quadratic Sieve Algorithm**

(1) Choose a *factor base* $\mathcal{F} = \{p_1, p_2, \ldots, p_k\}$, where the p_j are primes for $j = 1, 2, \ldots, k \in \mathbb{N}$.

(2) For each nonnegative integer j, let $t = \pm j$. Compute $y_t = (\lfloor \sqrt{n} \rfloor + t)^2 - n$ until $k + 2$ such values are found that are p_k-smooth. For each such t,

$$y_t = \pm \prod_{i=1}^{k} p_i^{a_{i,t}}, \tag{5.6}$$

and we form the binary $k + 1$-tuple, $\mathfrak{v}_t = (v_{0,t}, v_{1,t}, v_{2,t}, \ldots, v_{k,t})$, where $v_{i,t}$ is the least nonnegative residue of $a_{i,t}$ modulo 2 for $1 \le i \le k$, $v_{0,t} = 0$ if $y_t > 0$, and $v_{0,t} = 1$ if $y_t < 0$.[5.9]

(3) Obtain a subset $\mathcal{S}$ of the values of t found in step (2) such that for each $i = 0, 1, 2, \ldots, k$,

$$\sum_{t \in \mathcal{S}} v_{i,t} \equiv 0 \,(\text{mod } 2). \tag{5.7}$$

In this case, $x^2 = \prod_{t \in \mathcal{S}} x_t^2 \equiv \prod_{t \in \mathcal{S}} y_t = y^2 \,(\text{mod } n)$, where $x_t = \lfloor \sqrt{n} \rfloor + t$, so $\gcd(x \pm y, n)$ provides a nontrivial factor of n if $x \not\equiv \pm y \,(\text{mod } n)$.

In step (2), we have that $y_t \equiv x_t^2 \,(\text{mod } n)$. Thus, if a prime $p \mid y_t = x_t^2 - n$, we have $x_t^2 \equiv n \,(\text{mod } p)$. Thus, we must exclude from the factor base any primes p for which there is no solution $x \in \mathbb{Z}$ to the congruence $x^2 \equiv n \,(\text{mod } p)$. In other words, we exclude from the factor base any primes p for which n is *not* a quadratic residue modulo p (see Definition 3.13 and the discussion following it).

Example 5.8 Let $n = 30167$. From Footnote 5.7, $k = 11$, so we choose the first eleven primes for which n is a quadratic residue. They comprise our factor base $\mathcal{F} = \{2, 7, 11, 17, 29, 31, 37, 41, 43, 53, 67\}$. We see, by inspection, that a subset $\mathcal{S}$ of the values of t in the table below (which we computed given $\lfloor \sqrt{n} \rfloor = 173$) such that $\sum_{t \in \mathcal{S}} v_{i,t} \equiv 0 \,(\text{mod } 2)$ for each $i = 0, 1, 2, \ldots, 11$ is $\mathcal{S} = \{0, 18, -23\}$. Thus,

$$\prod_{t \in \mathcal{S}} x_t^2 = 2^2 3^2 5^4 173^2 191^2 \equiv 9062^2 \equiv x^2 \,(\text{mod } 30167),$$

and

$$\prod_{t \in \mathcal{S}} y_t = 2^2 7^2 11^2 17^2 41^2 \equiv 16837^2 \equiv y^2 \,(\text{mod } 30167),$$

so $y^2 - x^2 \equiv 16837^2 - 9062^2 \equiv 7775 \cdot 25899 \,(\text{mod } 30167)$.

By computing both of the values, $\gcd(7775, 30167) = 311 = \gcd(y - x, n)$ and $\gcd(25899, 30167) = 97 = \gcd(x + y, n)$, we get that $n = 30167 = 97 \cdot 311$.

[5.9]We can execute this step using trial division. There is also a sieving process to determine (5.6), that we will not explicitly describe here (see [158]) from which the Quadratic Sieve gets its name.

t	x_t	y_t	$\mathfrak{v}_t$
0	173	$-2 \cdot 7 \cdot 17$	$(1,1,1,0,1,0,0,0,0,0,0,0)$
-1	172	$-11 \cdot 53$	$(1,0,0,1,0,0,0,0,0,0,1,0)$
-5	168	$-29 \cdot 67$	$(1,0,0,0,0,1,0,0,0,0,0,1)$
5	178	$37 \cdot 41$	$(0,0,0,0,0,0,0,1,1,0,0,0)$
-6	167	$-2 \cdot 17 \cdot 67$	$(1,1,0,0,1,0,0,0,0,0,0,1)$
7	180	$7 \cdot 11 \cdot 29$	$(0,0,1,1,0,1,0,0,0,0,0,0)$
11	184	$7 \cdot 17 \cdot 31$	$(0,0,1,0,1,0,1,0,0,0,0,0)$
14	187	$2 \cdot 7^4$	$(0,1,0,0,0,0,0,0,0,0,0,0)$
-15	158	$-11 \cdot 43$	$(1,0,0,1,0,0,0,0,0,1,0,0)$
-17	156	$-7^3 \cdot 17$	$(1,0,1,0,1,0,0,0,0,0,0,0)$
18	191	$2 \cdot 7 \cdot 11 \cdot 41$	$(0,1,1,1,0,0,0,0,1,0,0,0)$
-23	150	$-11 \cdot 17 \cdot 41$	$(1,0,0,1,1,0,0,0,1,0,0,0)$
28	201	$2 \cdot 7 \cdot 17 \cdot 43$	$(0,1,1,0,1,0,0,0,0,1,0,0)$

What underlies the solution to a factorization problem using the Quadratic Sieve as depicted in Example 5.8 is some elementary linear algebra. By ensuring that there are $k + 2$ vectors $\mathfrak{v}_t$ in a $k + 1$-dimensional vector space $\mathbb{F}_2^{k+1}$, we guarantee that there is a linear dependence relation among the $\mathfrak{v}_t$.[5.10] In other words, we ensure the existence of the set $\mathcal{S}$ in step (3) of the algorithm such that congruence (5.7) holds.[5.11] There is no guarantee that $x \not\equiv \pm y \pmod{n}$. Nevertheless, there are usually several dependency relations among the $\mathfrak{v}_t$, so there is a high probability that at least one of them will yield an (x, y) pair such that $x \not\equiv \pm y \pmod{n}$ (see Exercise 5.3 for instance). The problem, of course, is that for "large" smoothness bounds B, we need a lot of congruences before we may be able to get these dependency relations.

For k as given in Footnote 5.7, the asymptotic running time[5.12] of the Quadratic Sieve is $O\left(\exp\left((1 + o(1))\sqrt{\ln(n) \ln(\ln(n))} \right) \right)$, (for a definition of $o(1)$ see page 49).[5.13] The Quadratic Sieve has been quite successful at fac-

---

[5.10]In [146], the algorithm for finding a square residue from factored residues by setting up and reducing a bit matrix was introduced. In practice, one finds that only about .9($k + 1$) rows are needed, because the placement of the bits in the matrix reflects a strong dependency among the residues. This is because the modulus n is *not* prime. Therefore, for example, if $n = pq$ is an RSA modulus, it may occur that 6 is a residue of n, which implies that 6 is a residue of both p and q. This does *not* necessarily mean that 2 is a residue of both p and q. Suppose that 2 is a nonresidue of p. Thus, if we were to eliminate 3 by multiplying, then the 2 could not stand alone as a residue and so would disappear with the 3. This means that there is a dependency among the prime factors of the residues, which is what brings down the number of rows needed to get a dependency. In a private communication with this author, John Brillhart stated that he has never commented on the above in print. Thus, it is worth mentioning here. He commented in that communication that "...it is one of the realities in the statistical business of combining residues to get a square."

[5.11]For the reader needing a reminder of these basic facts in linear algebra see pages 297–298.

[5.12]The running time of an algorithm (defined in Footnote 1.77) is sometimes difficult to determine. When this is the case, one may be forced to settle for approximations for the running time, called the *asymptotic running time*, which is a measure of how the running time of the algorithm increases as the size of the input increases without bound.

[5.13]In the literature, $\exp\left(\sqrt{\ln(n) \ln(\ln(n))}\right)$ is usually denoted by $L(n)$ for convenience. It can

toring RSA moduli (those $n = pq$ where p and q are primes of roughly the same size), even more so than the Elliptic Curve Method. This is because the Quadratic Sieve involves single precision operations, whereas elliptic curve operations are more computationally intensive. The first successful implementation of the Quadratic Sieve in which a serious number was factored occurred in 1983 when J. Gerver [83] factored a 47-digit number. Then, in 1984, the authors of [57] factored a 71-digit number.

The Quadratic Sieve has been employed using an approach called *factoring by electronic mail.* This is a term used by Lenstra and Manasse in [121] to mean the distribution of the Quadratic Sieve operations to hundreds of physically separated computers all over the world, and in 1988 they used this approach to factor a 106-digit number.[5.14] In 1994, the authors of [8] factored the RSA-129 number by using the electronic mail factoring technique with over 1600 computers and more than 600 researchers around the globe.[5.15] The unit of time measurement for factoring is called a *mips year*, which is defined as being tantamount to the computational power of a computer rated at one *million instructions per second* (mips) and used for one year, which is equivalent to approximately $3 \cdot 10^{13}$ instructions. For instance, factoring the RSA-129 challenge number required 5000 mips years, and in 1989 the aforementioned factorization of the 106-digit number needed 140 mips years. The Number Field Sieve, mentioned at the outset of this section, also factors n by ultimately constructing a solution to (5.3), but does so in rings of integers of number fields. The Number Field Sieve has asymptotic running time faster than both the Quadratic Sieve and the Elliptic Curve Method. In Section Two we show how the Number Field Sieve was used to factor the ninth Fermat Number.

The Quadratic Sieve tries to generate sufficiently many factored quadratic residues of n that are close to $\sqrt{n}$. (It is the rate of getting the latter that is the key to these methods when applied to large composite numbers.) This is essentially an idea that is generalized by the Number Field Sieve, which tries to generate d^{th} power residues of n close to $\sqrt[d]{n}$. In the next section we will see that Pollard originated the idea by looking at the case $d = 3$, and tried to factor numbers that are "close" to being perfect cubes.

There are numerous sources for the reader to explore factoring algorithms in more depth if desired. Among them are numerous factorizations of Cunningham Numbers (see Footnote 1.47), which may be found in [40]. A general overview of factoring is given by Pomerance in [160], and by Lenstra and Lenstra in [118]. There is also a discussion of factoring b.c. (before computers) by Williams and

---

be shown that for fixed $\alpha, \beta \in \mathbb{R}^+$, a random integer no bigger than n^α is $L(n)^\beta$-smooth with probability $L(n)^{-\alpha/(2\beta)+o(1)}$ for sufficiently large n.

[5.14]It is *parallel computing* that picks up the time.

[5.15]RSA "challenge" numbers for factoring algorithms are published on the Internet (for a copy of the list, send a request to: *challenge-rsa-list@rsa.com*). These numbers are denoted by RSA-n for some $n \in \mathbb{N}$, which are n-digit numbers that are products of two primes of approximately the same length. In [148], Odlyzko stated: "...if somebody asks how large an integer can be factored, a good first answer is to ask the question in return how many friends that person has. There is a huge and growing amount of idle computing power on the Internet, and harnessing it is more a matter of social skills than technology."

Shallit in [209], and this article has over 160 references for further reading.

This section has presented some algorithms that perform more efficiently on certain numbers of a special form such as the $p-1$ method, as well as more general purpose algorithms such as the Quadratic Sieve technique. An overall strategy should go down many avenues. First, one should try to get small prime factors, which may be accomplished using trial division up to some reasonable bound, then use Fermat's Difference of Squares Method, after which one could employ other algorithms for small prime factors such as Pollard's $p-1$ Method or his rho method. When all else fails to get a complete factorization, the big guns may be brought to bear such as the Quadratic Sieve or the Number Field Sieve, about which we will learn in the next section. The Elliptic Curve Method, which we will study in Chapter Six, can also be used in advance of the latter two sieves for finding small prime factors.

Although there is no known polynomial time algorithm for integer factoring, what we have seen in the above presentations is that it is highly unlikely that there is one. Complexity theory in the modern day does not help us since it gives mostly upper bounds and we need lower bounds to determine the amount of time a cryptanalyst would need to break a cryptosystem. There is a certain degree of consensus in the literature that the use of asymmetric cryptosystems in critical areas of secrecy is open to question, since finding the trapdoor in a public-key cryptosystem would be as much of a threat as a cryptanalyst's attack. The problem is that the nonexistence of trap doors (at least other than the ones about which we are already aware) is difficult, possibly intractable, to prove.

Exercises

5.1. Let $n = 26869$ and $a = 7$. Factor n using Pollard's $p-1$ Algorithm.

5.2. Factor $n = 13861$ using Pollard's $p-1$ Factoring Method with $a = 2$.

5.3. Let $n = 39617$. Factor n as in Example 5.8, by setting up an appropriate table and invoking the Quadratic Sieve algorithm.

5.4. Let $n = 34087$. Use the Quadratic Sieve algorithm, as in Example 5.8, to completely factor n.

5.5. Use the Continued Fraction Algorithm, factor $n = 6557$.

5.6. Using the Continued Fraction Algorithm, factor 703891 by setting up a table as in Example 5.4.

5.7. Show that the number of bit operations required to factor a number $n \in \mathbb{N}$ using trial division is $O(\sqrt{n} \log_2^2 n)$.

5.8. Provide an argument to show that Fermat's Difference of Squares Method for factoring, described on page 24, is not much better than trial division when $n = pq$ where p and q are primes that are sufficiently far apart.

5.2 The Number Field Sieve

In 1988, John Pollard[5.16] circulated a manuscript containing the outline of a new algorithm for factoring integers, of which he had a practical version running on his small computer. In 1990, a more general version of his algorithm was developed and published in 1993 in [119], which the authors called the *Number Field Sieve* (integer factoring algorithm). In order to better understand the Number Field Sieve both historically and mathematically, we look at Pollard's original idea for factoring using *cubic integers*, specifically those from $\mathbb{Z}[\sqrt[3]{2}] = \mathbb{Z}[\sqrt[3]{-2}]$, which is the ring of integers of $F = \mathbb{Q}(\sqrt[3]{2}) = \mathbb{Q}(\sqrt[3]{-2})$.[5.17] We will show how to use these cubic integers to factor certain integers by employing the unique factorization property of $\mathbb{Z}[\sqrt[3]{-2}]$ (see [145, Chapter 3]). We begin by showing how the fifth Fermat Number $\mathfrak{F}_5 = 2^{32} + 1$ can be factored using these techniques. Recall from page 25 that Euler found the factor 641 of $\mathfrak{F}_5$ in 1732.

Let $\alpha = \sqrt[3]{-2}$. Then

$$N_F(x - \alpha) = (x - \alpha)(x - \alpha\zeta_3)(x - \alpha\zeta_3^2) = x^3 + 2,$$

where $N_F(x - \alpha)$ is the *norm* of $x - \alpha$ and $\zeta_3 = (-1 + \sqrt{-3})/2$ is a *primitive cube root of unity*. We observe that

$$2\mathfrak{F}_5 = x^3 + 2.$$

If we set $x = 2^{11}$, then $x - \alpha \in \mathbb{Z}[\sqrt[3]{-2}]$, so we may use the uniqueness of factorization to write $x - \alpha$ as a product of distinct primes in $\mathbb{Z}[\alpha]$. Thus, if $\beta \in \mathbb{Z}[\alpha]$ is a prime dividing $(x - \alpha)$, then

$$N_F(\beta) \mid N_F(x - \alpha) = x^3 + 2,$$

and we may be able to find a nontrivial factor of $\mathfrak{F}_5$ via norms of certain elements of $\mathbb{Z}[\alpha]$.

For convenience, we consider elements of the form $a + b\alpha \in \mathbb{Z}[\alpha]$ and sieve over values of a and b testing for

$$\gcd(N_F(a + b\alpha), \mathfrak{F}_5) = \gcd(a^3 - 2b^3, \mathfrak{F}_5) > 1.$$

[5.16]On January 18, 1999, RSA Data Security Inc. announced that John M. Pollard (and John Gilmore, cofounder of the Electronic Frontier Foundation) were named winners of the 1999 RSA award. This award was instituted in 1998 in recognition of individuals and organizations that have made "significant, ongoing contributions to security issues and cryptography in the areas of mathematics, public policy, and industry." This is most appropriate given that Pollard is almost single-handedly responsible for most of the really novel ideas in factorization and discrete logs, such as the rho, and $p - 1$ methods, as well as the basic idea behind the Number Field Sieve and lattice sieving.

[5.17]In order to follow this section, the reader will be assumed to be familiar with, at a bare minimum, the basic notions of algebraic number theory contained in [145, Chapter 1, pp.1–66]. However, a deeper understanding will be gleaned if familiarity with the notions in [145, Chapter 2–3] is also held. In fact, the examples in this section and much of the discussion is taken from [145, Chapter 1, pp. 67–72; Chapter 2, pp. 117–126; and Chapter 3, p. 190–191].

Also, for convenience, we let a range over the values $1, 2, \ldots, 100$ and b range over the values $1, 2, \ldots, 20$. There are formal reasons for this that we will discuss later. In a run with values $1 \leq a \leq 15$ and $1 \leq b \leq 20$, we get $\gcd(a^3 - 2b^3, \mathfrak{F}_5) = 1$. However, at $a = 16$, $b = 5$, we get

$$\gcd(16^3 - 2 \cdot 5^3, \mathfrak{F}_5) = 641.$$

Let's look at what is going on in the ring of integers that allows this to happen. We may factor $16 + 5\alpha$ as follows.

$$16 + 5\alpha = (1 + \alpha)(-1 + \alpha)(\alpha)(-9 + 2\alpha - \alpha^2),$$

where $1 + \alpha$ is a unit with norm -1; $-1 + \alpha$ has norm -3; α has norm -2; and $\beta = -9 + 2\alpha - \alpha^2$ has norm -641. This accounts for $16^3 - 2 \cdot 5^3 = 2 \cdot 3 \cdot 641$, and shows that β is the predicted prime divisor of $x - \alpha$, which gives us the nontrivial factor of $\mathfrak{F}_5$. The above method works well largely because of the small value of $\mathfrak{F}_5$, but it may not be feasible for larger values to check all of the gcd conditions over a much larger range.

We learned in the preceding section that $\mathfrak{F}_7$ was factored in 1970 by using the Continued Fraction Algorithm. However, the following method of Pollard, which he introduced in 1991, and published in [156], uses the above notions of factorizations in $\mathbb{Z}[\alpha]$ to factor $\mathfrak{F}_7$. His method is modeled after the strategy used in the Quadratic Sieve, which we also studied in the preceding section. Pollard's Method may be used as a paradigm for factoring numbers of the form $x^3 + k$ for any $k \in \mathbb{Z}$ such that $\mathbb{Z}[\sqrt[3]{-k}]$ is both the ring of integers of $F = \mathbb{Q}(\sqrt[3]{-k})$, and has unique factorization, called a *Unique Factorization Domain* or *UFD* (see [145, Chapter 3]).[5.18]

The same starting point for Pollard's Method is used for $\mathfrak{F}_7$ as with $\mathfrak{F}_5$ since

$$2\mathfrak{F}_7 = m^3 + 2 \text{ where } m = 2^{43}.$$

Pollard's Method to factor $\mathfrak{F}_7$ involves the use of B-smooth numbers of the form $a + bm$ for a suitable smoothness bound B defined in the algorithm below (see Definition 5.1 on page 196). We use the term B-smooth in reference to algebraic integers $a + b\alpha \in \mathbb{Z}[\alpha]$ to mean that the *norm* of $a + b\alpha$ is B smooth. If we set $f(x) = x^3 + 2$ and $n = \mathfrak{F}_7$, then since $f(m) \equiv 0 \pmod{n}$, we may define the natural ring homomorphism,

$$\psi : \mathbb{Z}[\alpha] \mapsto \mathbb{Z}/n\mathbb{Z}$$

given by $\psi(\alpha) = m$, so

$$\psi\left(\sum_{j=0}^{2} z_j \alpha^j\right) = \sum_{j=0}^{2} z_j m^j \in \mathbb{Z}/n\mathbb{Z}, \text{ where } z_j \in \mathbb{Z}.$$

[5.18]The Number Field Sieve can be made to work when we do *not* have a UFD, but the exposition is simplified when we do have a UFD since we can more readily see what it means for $a + b\alpha$ to factor into "small primes," namely it gives us an idea of what it means for $a + b\alpha$ to be a smooth element (see [161]).

The role of this map ψ in attempting to factor a number n is given by the following.

Suppose that we have a set $\mathcal{S}$ of polynomials

$$g(x) = \sum_{j=0}^{2} z_j x^j \in \mathbb{Z}[x]$$

such that

$$\prod_{g \in \mathcal{S}} g(\alpha) = \beta^2$$

where $\beta \in \mathbb{Z}[\alpha]$, and

$$\prod_{g \in \mathcal{S}} g(m) = y^2,$$

where $y \in \mathbb{Z}$. Then if $\psi(\beta) = x \in \mathbb{Z}/n\mathbb{Z}$, we have

$$x^2 \equiv \psi(\beta)^2 \equiv \psi(\beta^2) \equiv \psi\left(\prod_{g \in \mathcal{S}} g(\alpha)\right) \equiv \prod_{g \in \mathcal{S}} g(m) \equiv y^2 \,(\mathrm{mod}\ n).$$

In other words, this method finds a pair of integers x, y such that

$$x^2 - y^2 \equiv (x - y)(x + y) \equiv 0 \,(\mathrm{mod}\ n),$$

so we may have a nontrivial factor of n by looking at $\gcd(x - y, n)$.

We now describe the algorithm, but give a simplified version of it, since this is meant to be a simple introduction to the ideas behind the Number Field Sieve, which we will present after this example. Also, we use a very small value of n as an example for the sake of simplicity, namely $n = 23329$. Note that $2n = 36^3 + 2 = m^3 + 2$. We will also make suitable references in the algorithm in terms of how Pollard factored $n = \mathfrak{F}_7$.

◆ Pollard's Algorithm

Step 1. Compute a *factor base*.

As we saw on page 200, Brillhart and Morrison used the term "factor base" to mean the choice of a suitable set of rational primes over which they may factor a set of integers. In the case of the Continued Fraction Algorithm, this meant that they would discard those primes for which the proposed value of n, which they were trying to factor, was not a quadratic residue. The reason for this is that the discarded primes could not divide the residue (modulo n) of $a^2 - nb^2$, where a/b is a convergent in the simple continued fraction expansion of $\sqrt{n}$. In the case of cubic integers in $\mathbb{Z}[\alpha] = \mathbb{Z}[\sqrt[3]{-2}]$, we take for $n = 23329$ only the first eleven primes, those up to and including 41 (or for $n = \mathfrak{F}_7$, Pollard chose the first five hundred rational primes) as $\mathfrak{FB}_1$, the first part of the factor base,[5.19] and

---

[5.19] As we will see later in this section, the reasons behind the choice of the number of primes in $\mathfrak{FB}_1$ are largely empirical.

for the second part, $\mathfrak{FB}_2$, we take those primes of $\mathbb{Z}[\alpha]$ with norms $\pm p$, where $p \in \mathfrak{FB}_1$. Also, we include the units: -1, $1 + \alpha$, and $1/(1 + \alpha) = -1 + \alpha - \alpha^2$ in $\mathfrak{FB}_2$.

Step 2. Run the sieve.

In this instance, the sieve involves finding numbers $a + bm$ that are composed of some primes from $\mathfrak{FB}_1$.[5.20] For $n = 23329$, we sieve over values of a from -5 to 5 and values of b from 1 to 10 (or for $n = \mathfrak{F}_7$, Pollard chose values of a from -4800 to 4800, and values of b from 1 to 2000). Save only coprime pairs (a, b).

Step 3. Look for smooth values of the norm, and obtain factorizations of $a + bx$ and $a + b\alpha$.

Here, smooth values of the norm means that

$$N = N_F(a + b\alpha) = a^3 - 2b^3$$

is not divisible by any primes bigger than those in $\mathfrak{FB}_1$. For those (a, b) pairs, factor $a + bm$ by trial division, and eliminate unsuccessful trials. Factor $a + b\alpha$ by computing the norm $N_F(a + b\alpha)$ and using trial division. When a prime p is found, then divide out a $\mathbb{Z}[\alpha]$-prime of norm $\pm p$ from $a + b\alpha$. This will involve getting primes in the factorization of the form $a + b\alpha + c\alpha^2$ where $c \neq 0$. Units may also come into play in the factorizations, and a table of values of $(1 + \alpha)^j$ is kept for such purposes with $j = -2, \cdots, 2$ for $n = 23329$ (or for $\mathfrak{F}_7$, one should choose to keep a record of units for $j = -8, -7, \ldots, 8$). Some data extracted for the run on $n = 23329$ is given as follows.

Table 5.9

$a + b\alpha + c\alpha^2$	N	factorization of $a + b\alpha + c\alpha^2$
$5 + \alpha$	$3 \cdot 41$	$(-1 + \alpha)(-1 - 2\alpha - 2\alpha^2)$
$4 + 10\alpha$	$-2^4 \cdot 11^2$	$-(3 + 2\alpha)^2 \alpha^4 (-1 + \alpha - \alpha^2)^2$
$-1 + \alpha$	-3	$-1 + \alpha$
$-1 - 2\alpha - 2\alpha^2$	-41	$-1 - 2\alpha - 2\alpha^2$
$3 + 2\alpha$	11	$3 + 2\alpha$
α	-2	α
$-1 + \alpha - \alpha^2$	-1	$unit$

Table 5.10

$a + bm + cm^2$	factorization of $a + bm + cm^2$
$5 + m$	41
$4 + 10m$	$2^2 \cdot 7 \cdot 13$
$-1 + m$	$5 \cdot 7$
$-1 - 2m - 2m^2$	$-5 \cdot 13 \cdot 41$
$3 + 2m$	$3 \cdot 5^2$
m	$2^2 \cdot 3^2$
$-1 + m - m^2$	$-13 \cdot 97$

Step 4. Complete the factorization.

[5.20] As we will see in the description of the Number Field Sieve, there is also the allowance for possibly one larger prime not in $\mathfrak{FB}_1$, but we ignore this possibility here in the interests of simplicity of presentation.

By selecting -1 times the first four rows in the third column of Table 5.9, we get a square in $\mathbb{Z}[\alpha]$:

$$\beta^2 = (-1+\alpha)^2(-1-2\alpha-2\alpha^2)^2(3+2\alpha)^2\alpha^4(-1+\alpha-\alpha^2)^2, \qquad (5.11)$$

and correspondingly, since β^2 is also -1 times the first four rows in the first column of Table 5.9, we get:

$$\beta^2 = (5+\alpha)(-4-10\alpha)(-1+\alpha)(-1-2\alpha-2\alpha^2). \qquad (5.12)$$

Then we get a square in $\mathbb{Z}$ from Table 5.10 by applying ψ to (5.12):

$$\psi(\beta^2) = (5+m)(-4-10m)(-1+m)(-1-2m-2m^2) = 2^2 \cdot 5^2 \cdot 7^2 \cdot 13^2 \cdot 41^2 = y^2.$$

Also, by applying ψ to β via (5.11), we get:

$$\psi(\beta) = (-1+m)(-1-2m-2m^2)(3+2m)m^2(-1+m-m^2) \equiv 9348 \pmod{23329},$$

so by setting $x = \psi(\beta)$, we have $x^2 = \psi^2(\beta) = \psi(\beta^2) \equiv y^2 \pmod{n}$. Since

$$y = 2 \cdot 5 \cdot 7 \cdot 13 \cdot 41 \equiv 13981 \pmod{23329},$$

then $y - x \equiv 4633 \pmod{23329}$. However, $\gcd(4633, 23329) = 41$. In fact, $23329 = 41 \cdot 569$.

Pollard used the algorithm in a similar fashion to find integers X and Y for the more serious factorization

$$\gcd(X - Y, \mathfrak{F}_7) = 59649589127497217.$$

Hence, we have a factorization of $\mathfrak{F}_7$ as follows.

$$\mathfrak{F}_7 = 59649589127497217 \cdot 5704689200685129054721.$$

Essentially, the ideas for factoring using cubic integers above is akin to the notion of the strategy used in the Quadratic Sieve Method. There, we try to generate sufficiently many smooth quadratic residues of n close to $\sqrt{n}$. In the cubic case, we try to factor numbers that are close to perfect cubes. Now we will extend these ideas to show how $\mathfrak{F}_9$ was factored using the Number Field Sieve, and $\mathbb{Z}[\sqrt[5]{2}]$. In his initiation of the aforementioned algorithm, Pollard had been motivated by a discrete logarithm algorithm given in 1986, by the authors of [52], which employed quadratic fields. Pollard looked at the more general scenario by outlining an idea for factoring certain large integers using number fields. The special numbers that he considered are those large composite natural numbers that are "close" to being powers, namely those $n \in \mathbb{N}$ of the form

$$n = r^t - s \text{ for small } r, |s| \in \mathbb{N}, \text{ and a possibly much larger } t \in \mathbb{N}. \qquad (5.13)$$

Examples of such numbers, which the Number Field Sieve had some successes factoring, may be found in tables of numbers of the form

$$n = r^t \pm 1, \text{ called } Cunningham \ Numbers,$$

(for instance, see [144, Appendix D, p. 375]). The first really noteworthy success was factorization of the ninth Fermat Number

$$\mathfrak{F}_9 = 2^{2^9} + 1 = 2^{512} + 1 \text{ (having 155 decimal digits)},$$

by the Lenstra brothers, Manasse and Pollard in 1990, the publication of which appeared in 1993 (see [120]). The Number Field Sieve can be made to work for arbitrary integers. However, for the sake of simplicity in this section, we will deal only with what has been called the *Special Number Field Sieve* by the authors of [42], which means the Number Field Sieve applied to the special class of numbers given in (5.13). The Special Number Field Sieve is asymptotically faster than any known algorithm for the class of integers (5.13) to which it applies. The less specialized number field sieve has come to be known as the *General Number Field Sieve*. The Special Number Field Sieve can factor integers of the form (5.13) in time of the form

$$\exp(c(\ln n)^{1/3}(\ln\ln(n))^{2/3}), \tag{5.14}$$

where $c = (32/9)^{1/3} \approx 1.5263$, whereas the General Number Field Sieve factors arbitrary integers with a running time bound of (5.14) with $c = 1.9229\ldots$. The General Number Field Sieve is responsible for factoring both RSA-140 on February 2, 1999 (see [46]), and RSA-155 on August 26, 1999 (see http://cch.loria.fr/actualites/1999/RSA155/tech.html). Moreover, the record for factorization of a 211-digit Cunningham Integer (by the authors of [46]) was accomplished by the Special Number Field Sieve. The latter was announced by email on April 25, 1999.

For $n = r^t - s$, as above, we wish to choose a number field of degree d over $\mathbb{Q}$. The following choice for d is made for reasons (which we will not discuss here), which make it the optimal selection, at least theoretically. (The interested reader may consult [119, Sections 6.2–6.3, pp. 31–32] for the complexity analysis and reasoning behind these choices.) Set

$$d = \left(\frac{(3 + o(1))\log n}{2\log\log n}\right)^{1/3}. \tag{5.15}$$

Now select $k \in \mathbb{N}$, which is minimal with respect to $kd \geq t$. Therefore,

$$r^{kd} \equiv sr^{kd-t} \pmod{n}.$$

Set

$$m = r^k, \text{ and } c = sr^{kd-t}. \tag{5.16}$$

Then

$$m^d \equiv c \pmod{n}.$$

Set

$$f(x) = x^d - c,$$

and let $\alpha \in \mathbb{C}$ be a root of f. Then this leads to a choice of a number field, namely $F = \mathbb{Q}(\alpha)$. Although the Number Field Sieve can be made to work when $\mathbb{Z}[\alpha]$ is *not* a UFD, the assumption that it *is* a UFD simplifies matters greatly in the exposition of the algorithm, so we will make this assumption. See [119] for a description of the modifications necessary when it is not a UFD.

Now the question of the irreducibility of f arises. If f is reducible over $\mathbb{Z}$, we are indeed lucky, since then

$$f(x) = g(x)h(x), \text{ with } g(x), h(x) \in \mathbb{Z}[x],$$

where $0 < \deg(g) < \deg(f)$. Therefore,

$$f(m) - n = g(m)h(m)$$

is a nontrivial factorization of n, and we are done. Use of the Number Field Sieve is unnecessary. However, the probability is high that f is irreducible since *most* primitive polynomials over $\mathbb{Z}$ *are irreducible*. Hence, for the description of the Number Field Sieve, we may assume that f is irreducible over $\mathbb{Z}$.

Since $f(m) \equiv 0 \pmod{n}$, we may define the natural homomorphism,

$$\psi : \mathbb{Z}[\alpha] \mapsto \mathbb{Z}/n\mathbb{Z},$$

given by

$$\alpha \mapsto \overline{m} \in \mathbb{Z}/n\mathbb{Z}.$$

Then

$$\psi\left(\sum_j a_j \alpha^j\right) = \sum_j a_j \overline{m}^j.$$

Now define a set $\mathcal{S}$ consisting of pairs of relatively prime integers (a, b), satisfying the following two conditions:

$$\prod_{(a,b)\in\mathcal{S}} (a + bm) = c^2, \quad (c \in \mathbb{Z}), \tag{5.17}$$

and

$$\prod_{(a,b)\in\mathcal{S}} (a + b\alpha) = \beta^2, \quad (\beta \in \mathbb{Z}[\alpha]). \tag{5.18}$$

Thus, $\psi(\beta^2) = \overline{c}^2$, so $\psi(\beta^2) \equiv c^2 \pmod{n}$. In other words, since $\psi(\beta^2) = \psi(\beta)^2$, then if we set $\psi(\beta) = h \in \mathbb{Z}$,

$$h^2 \equiv c^2 \pmod{n}.$$

This takes us back to the discussion surrounding (5.3) on page 198. Thus, we may have a nontrivial factor of n. In other words, if $h \not\equiv \pm c \pmod{n}$, then $\gcd(h \pm c, n)$ is a nontrivial factor of n.

The above overview of the Number Field Sieve methodology is actually a special case of an algebraic idea, which is described as follows. Let R be a ring with homomorphism

$$\phi : R \mapsto \mathbb{Z}/n\mathbb{Z} \times \mathbb{Z}/n\mathbb{Z},$$

together with an algorithm for computing nonzero diagonal elements (x, x) for $x \in \mathbb{Z}/n\mathbb{Z}$. Then the goal is to multiplicatively combine these elements to obtain squares in R whose square roots have an image under ϕ not lying in $(x, \pm x)$ for nonzero $x \in \mathbb{Z}/n\mathbb{Z}$. The Number Field Sieve is the special case

$$R = \mathbb{Z} \times \mathbb{Z}[\alpha], \text{ with } \phi(z, \beta) = (\overline{z}, \psi(\beta)).$$

Before setting down the details of the formal Number Field Sieve Algorithm, we discuss the crucial role played by *smoothness*. Recall that a smooth number is one with only "small" prime factors. In particular, $n \in \mathbb{N}$ is *B-smooth* for $B \in \mathbb{R}^+$, if n has no prime factor bigger than B.

If $F = \mathbb{Q}(\alpha)$ is a number field, then by definition an algebraic number $a + b\alpha \in \mathbb{Z}[\alpha]$ is *B-smooth* if $|N_F(a+b\alpha)|$ is *B-smooth*. Hence, $a + b\alpha$ is *B-smooth* if and only if all primes dividing $|N_F(a + b\alpha)|$ are less than B. Thus, the idea behind the Number Field Sieve is to look for small relatively prime numbers a and b such that both $a + \alpha b$ and $a + \overline{m}b$ are smooth. Since $\psi(a + \alpha b) = a + \overline{m}b$, then each pair provides a congruence modulo n between two products. Sufficiently many of these congruences can then be used to find solutions to $h^2 \equiv c^2 \pmod{n}$, which may lead to a factorization of n.

The above overview leaves open the demanding questions as to how we choose the degree d, the integer m, and how the set of relatively prime integers a, b such that Equations (5.17)–(5.18) can be found. These questions may now be answered in the following formal description of the algorithm.

✦ The (Special) Number Field Sieve Algorithm

Step 1. (Selection of a Factor Base and Smoothness Bound)

There is a consensus that smoothness bounds are best chosen empirically. However, there are theoretical reasons for choosing such bounds as

$$B = \exp((2/3)^{2/3}(\log n)^{1/3}(\log\log n)^{2/3}),$$

which is considered to be optimal since it is based upon the choice for d as above. See [119, Section 6.3, p. 32] for details. Furthermore, the reasons for this being called a smoothness bound will unfold in the sequel.

Define a set

$$\mathcal{S} = \mathcal{S}_1 \cup \mathcal{S}_2 \cup \mathcal{S}_3,$$

where the component sets $\mathcal{S}_j$ are given as follows.

$$\mathcal{S}_1 = \{p \in \mathbb{Z} : p \text{ is prime and } p \leq B\},$$

$$\mathcal{S}_2 = \{u_j : j = 1, 2, \ldots, r_1 + r_2 - 1, \text{ where } u_j \text{ is a generator of } \mathcal{U}_F\},^{5.21}$$

and

$$\mathcal{S}_3 = \{\beta = a + b\alpha \in \mathbb{Z}[\alpha] : |N_F(\beta)| = p < B_2 \text{ where } p \text{ is prime }\},$$

[5.21] Here $\{r_1, r_2\}$ is the signature of F (see [145, Definition 1.32, p. 20]), and the generators u_j, called *fundamental units*, are the generators of the infinite cyclic groups given by Dirichlet's Unit Theorem (see [145, Theorem 2.78, p. 114]), where $\mathcal{U}_F$ is the unit group of the ring of integers of F.

where B_2 is chosen empirically. Now we set the factor base as

$$\mathcal{F} = \{a_j = \psi(j) \in \mathbb{Z}/n\mathbb{Z} : j \in \mathcal{S}\}.$$

Also, we may assume $\gcd(a_j, n) = 1$ for all $j \in \mathcal{S}$, since otherwise we have a factorization of n and the algorithm terminates.

Step 2. (Collecting Relations and Finding Dependencies)

We wish to collect relations (5.17)–(5.18) in such a way that they occur simultaneously, thereby yielding a potential factor of n. To do so, one searches for relatively prime pairs (a, b) with $b > 0$ satisfying the following two conditions.

(i) $|a + bm|$ is B-smooth except for at most one additional prime factor p_1, with $B < p_1 < B_1$, where B_1 is empirically determined.

(ii) $a + b\alpha$ is B_2-smooth except for at most one additional prime $\beta \in \mathbb{Z}[\alpha]$ such that $|N_F(\beta)| = p_2$ with $B_2 < p_2 < B_3$, where B_3 is empirically chosen.

The prime p_1 in (i) is called the *large prime*, and the prime p_2 in (ii) is called the *large prime norm*. Pairs (a, b) for which p_1 and p_2 do not exist (namely when we set $p_1 = p_2 = 1$), are called *full relations*, and are called *partial relations* otherwise. In the sequel, we will only describe the full relations since, although the partial relations are more complicated, they lead to relations among the factor base elements in a fashion completely similar to the ones for full relations. For details on partial relations, see [120, Section 5].

First, we show how to achieve relations in Equation (5.17), the "easy" part (relatively speaking). (This is called the *rational part*, whereas relations in Equation (5.18) are called the *algebraic part*.) Then we show how to put the two together. To do this, we need the following notion from linear algebra.

Every $n \in \mathbb{N}$ has an exponent vector $v(n)$ defined by

$$n = \prod_{j=1}^{\infty} p_j^{v_j},$$

where p_j is the jth prime, only finitely many of the v_j are nonzero, and

$$v(n) = (v_1, v_2, \ldots) = (v_j)_{j=1}^{\infty}$$

with an infinite string of zeros after the last significant place. We observe that n is a square if and only if each v_j is even. Hence, for our purposes, the v_j give *too much* information. Thus, to simplify our task, we reduce each v_j modulo 2. Henceforth, then $\overline{v_j}$ means v_j reduced modulo 2. We modify the notion of the exponent vector further for our purposes by letting $B_1 = \pi(B)$, where $\pi(B)$ is the number of primes no bigger than B. Then, with $p_0 = -1$,

$$a + bm = \prod_{j=0}^{B_1} p_j^{v_j}$$

is the factorization of $a + bm$. Set

$$v(a + bm) = (\overline{v_0}, \ldots, \overline{v_{B_1}}),$$

for each pair (a, b) with $a + b\alpha \in \mathcal{S}_3$. The choice of B allows us to make the assumption that $|\mathcal{S}_3| > B_1 + 1$. Therefore, the vectors in $v(a+bm)$ for pairs (a, b) with $a + b\alpha \in \mathcal{S}_3$ exceed the dimension of the $\mathbb{F}_2$-vector space $\mathbb{F}_2^{B_1+1}$. In other words, we have *more than* $B_1 + 1$ vectors in a $B_1 + 1$-dimensional vector space. Therefore, there exist nontrivial linear dependence relations between vectors. This implies the existence of a subset $\mathcal{J}$ of $\mathcal{S}_3$ such that

$$\sum_{a+b\alpha \in \mathcal{J}} v(a + bm) = 0 \in \mathbb{F}_2^{B_1+1},$$

so

$$\prod_{a+b\alpha \in \mathcal{J}} (a + bm) = z^2 \quad (z \in \mathbb{Z}).$$

This solves Equation (5.17).

Now we turn to the algebraic relations in Equation (5.18). We may calculate the norm of $a + b\alpha$ by setting $x = a$ and $y = b$ in the homogeneous polynomial

$$(-y)^d f(-x/y) = x^d - c(-y)^d,$$

with $f(x) = x^d - c$. Therefore,

$$N_F(a + b\alpha) = (-b)^d f(-ab^{-1}) = a^d - c(-b)^d.$$

Let

$$R_p = \{r \in \mathbb{Z} : 0 \leq r \leq p - 1, \text{ and } f(r) \equiv 0 \,(\mathrm{mod}\, p)\}.$$

Then for relatively prime pairs (a, b), we have

$$N_F(a + b\alpha) \equiv 0 \,(\mathrm{mod}\, p) \text{ if and only if } a \equiv -br \,(\mathrm{mod}\, p), \text{ for some } r \in R_p$$

and this r is unique. Observe that by the relative primality of a and b, the multiplicative inverse b^{-1} of b modulo p is defined since, for $b \equiv 0 \,(\mathrm{mod}\, p)$, there are no nonzero pairs (a, b) with $N_F(a + b\alpha) \equiv 0 \,(\mathrm{mod}\, p)$.

The above shows that there is a one-to-one correspondence between those $\beta \in \mathbb{Z}[\alpha]$ with $|N_F(\beta)| = p$, a prime and pairs (p, r) with $r \in R_p$. Note that the kernel of the natural map

$$\psi : \mathbb{Z}[\alpha] \longmapsto \mathbb{Z}/p\mathbb{Z}$$

is

$$\ker(\psi) = \langle a + b\alpha \rangle,$$

the cyclic subgroup of $\mathbb{Z}[\alpha]$ generated by $a + b\alpha$. Thus,

$$|\mathbb{Z}[\alpha] : \langle a + b\alpha \rangle| = |N_F(a + b\alpha)| = p,$$

so

$$\mathbb{Z}[\alpha]/\langle a + b\alpha \rangle$$

is a field (see Example A.63). In ideal-theoretic terms (see [145, Chapter 3]) this says that the $\mathbb{Z}[\alpha]$-ideal

$$\mathcal{P} = (a + b\alpha)$$

is a principal, first-degree, prime $\mathbb{Z}[\alpha]$-ideal. In other words, $N_F(\mathcal{P}) = p^1 = p$, where N_F denotes the ideal norm in this case. Hence,

$$\mathbb{Z}[\alpha]/\mathcal{P} \cong \mathbb{F}_p,$$

the finite field of p elements. The above tells us that in Step 1 of the Number Field Sieve Algorithm, the set $\mathcal{S}_3$ essentially consists of the first-degree prime $\mathbb{Z}[\alpha]$-ideals of norm $N_F(\mathcal{P}) \leq B_2$. These are the *smooth, degree one, prime* $\mathbb{Z}[\alpha]$-*ideals*, namely those ideals whose prime norms are B_2-smooth.

In part (ii) of Step 2 of the algorithm on page 215, the additional prime element $\beta \in \mathbb{Z}[\alpha]$ such that

$$|N_F(\beta)| = p_2 \text{ with } B_2 < p_2 < B_3$$

corresponds to the prime $\mathbb{Z}[\alpha]$-ideal $\mathcal{P}_2$ called the *large prime ideal*. Moreover,

$$\mathcal{P}_2 \text{ corresponds to the pair } (p_2, c \,(\mathrm{mod}\ p_2)),$$

where $c \in \mathbb{Z}$ is such that $a \equiv -bc \,(\mathrm{mod}\ p_2)$, thereby enabling us to distinguish between prime ideals of the same norm. If the large prime in Step 2 does not occur, we write $\mathcal{P}_2 = (1)$.

Since

$$|a + bm| = \prod_{p \in \mathcal{S}_1} p^{v_p}, \tag{5.19}$$

and

$$|a + b\alpha| = \prod_{u \in \mathcal{S}_2} u^{t_u} \prod_{s \in \mathcal{S}_3} s^{v_s}, \tag{5.20}$$

for nonnegative $t_u, v_s \in \mathbb{Z}$, and since $\psi(a + bm) = \psi(a + b\alpha)$, then

$$\prod_{p \in \mathcal{S}_1} \psi(p)^{v_p} = \prod_{u \in \mathcal{S}_2} \psi(u)^{t_u} \prod_{s \in \mathcal{S}_3} \psi(s)^{v_s},$$

in $\mathbb{Z}/n\mathbb{Z}$. Therefore, we achieve a relationship among the elements of the factor base $\mathcal{F}$, as follows

$$\prod_{u \in \mathcal{S}_2} \psi(u)^{t_u} \prod_{s \in \mathcal{S}_3} \psi(s)^{v_s} \equiv \prod_{p \in \mathcal{S}_1} \psi(p)^{v_p} \,(\mathrm{mod}\ n).$$

If $|\mathcal{S}_3| > \pi(B)$, then by applying Gaussian elimination for instance, we can find $x(a,b) \in \{0,1\}$ such that simultaneously both

$$\prod_{a+b\alpha \in \mathcal{S}_3} (a + b\alpha)^{x(a,b)} = \left(\left(\prod_{u \in \mathcal{S}_2} u^{\overline{t_u}} \right) \left(\prod_{s \in \mathcal{S}_3} s^{\overline{v_s}} \right) \right)^2,$$

and

$$\prod_{a+b\alpha \in \mathcal{S}_3} (a+bm)^{x(a,b)} = \left(\left(\prod_{p \in \mathcal{S}_1} p^{\overline{v_p}} \right) \right)^2,$$

from which a factorization of n may be gleaned, by the method surrounding the discussion of (5.3) on page 198.

Note that an ideal-theoretic interpretation of (5.20) is:

$$|a+b\alpha| = \prod_{u \in \mathcal{S}_2} u^{t_u} \prod_{\mathcal{P} \in \mathcal{S}_3} \pi_{\mathcal{P}}^{v_{\mathcal{P}}}, \qquad (5.21)$$

where $\mathcal{P}$ ranges over all of the first-degree prime $\mathbb{Z}[\alpha]$-ideals of norm less than B_2, and $\pi_{\mathcal{P}}$ is a generator of $\mathcal{P}$. Thus, given (5.19), we get the following identity from (5.21).

$$\prod_{p \in \mathcal{S}_1} \psi(p)^{w_p} = \prod_{u \in \mathcal{S}_2} \psi(u)^{t_u} \prod_{\mathcal{P} \in \mathcal{S}_3} \psi(\pi_{\mathcal{P}})^{v_{\mathcal{P}}}.$$

From this, we proceed as above to get a potential factorization.

Practically speaking, the Number Field Sieve tasks consist of sieving all pairs (a, b) for $b = b_1, b_2 \ldots, b_n$ for short (overlapping) intervals $[b_1, b_2]$, with $|a|$ less than some given bound. All relations, full and partial, are gathered in this way until sufficiently many have been collected.

The first big prize garnered by the Number Field Sieve was the factorization of $\mathfrak{F}_9$, the ninth Fermat Number, as described in [120]. In 1903, A. E. Western found the prime factor

$$2424833 = 37 \cdot 2^{16} + 1$$

of $\mathfrak{F}_9$. Then in 1967, Brillhart determined that

$$\mathfrak{F}_9/2424833 \text{ (having 148 decimal digits)}$$

is composite by showing that it fails to satisfy Fermat's Little Theorem. Thus, the authors of [120] chose

$$n = \mathfrak{F}_9/2424833 = \left(2^{512} + 1 \right) /2424833.$$

Then they exploited the above algorithm as follows. If we choose d as in (5.15), we get that $d = 5$. The authors of [120] then observed the following. Since $2^{512} \equiv -1 \pmod{n}$, we get $h^5 \equiv 2^{1025} \equiv 2 \cdot \left(2^{512} \right)^2 \equiv 2 \pmod{n}$ for $h = 2^{205}$. This allowed them to choose the map

$$\psi : \mathbb{Z}[\sqrt[5]{2}] \mapsto \mathbb{Z}/n\mathbb{Z}, \text{ given by } \psi : \sqrt[5]{2} \mapsto 2^{205}.$$

Here $\mathbb{Z}[\sqrt[5]{2}]$ is a UFD. Then they chose m and c as in Equation (5.16), namely since $r = 2$, $s = -1$, and $t = 512$, then the minimal k with $5k = dk \geq t = 512$ is $k = 103$, and $m = 2^{103}$, so $c = -8 \equiv 2^{103 \cdot 5} \pmod{n}$. This gives rise to $f(x) = x^5 + 8$ with root $\alpha = -\sqrt[5]{2^3}$, and $\mathbb{Z}[\alpha] \subseteq \mathbb{Z}[\sqrt[5]{2}]$. Observe that

$$8\mathfrak{F}_9 = 2^{515} + 8 = \left(2^{103} \right)^5 + 8.$$

Thus,

$$\psi(\alpha) = m = 2^{103} \equiv -2^{615} \equiv - \left(2^{205} \right)^3 \pmod{n}.$$

Notice that 2^{103} is small in relation to n, and is in fact closer to $\sqrt[5]{n}$. Since

$$\psi(a + b\alpha) = a + 2^{103}b \in \mathbb{Z}/n\mathbb{Z},$$

we are in a position to form relations as described in the above algorithm. In fact, the authors of [120] actually worked only in the subring $\mathbb{Z}[\alpha]$ to find their relations. The sets they chose from Step 1 are

$$\mathcal{S}_1 = \{p \in \mathbb{Z} : p \le 1295377\},$$

$$\mathcal{S}_2 = \{-1, -1 + \sqrt[5]{2}, -1 + \sqrt[5]{2^2} - \sqrt[5]{2^3} + \sqrt[5]{2^4}\},$$

for units $u_1 = -1$, $u_2 = -1 + \sqrt[5]{2}$, and $u_3 = -1 + \sqrt[5]{2^2} - \sqrt[5]{2^3} + \sqrt[5]{2^4}$, and

$$\mathcal{S}_3 = \{\beta \in \mathbb{Z}[\alpha] : |N_F(\beta)| = p \le 1294973, p \text{ a prime}\}.$$

The authors began sieving in mid-February of 1990 on approximately thirty-five workstations at Bellcore. On the morning of June 15, 1990 the first of the dependency relations that they achieved turned out to give rise to a trivial factorization! However, an hour later their second dependency relation gave way to a 49-digit factor. This and the 99-digit cofactor were determined by A. Odlyzko to be primes, on that same day. They achieved:

$$\mathfrak{F}_9 = q_7 \cdot q_{49} \cdot q_{99},$$

where q_j is a prime with j decimal digits as follows: $q_7 = 2424833$,

$$q_{49} = 7455602825647884208337395736200454918783366342657,$$

and

$$q_{99} = 74164006262753080152478714190193747405994078109751902390582131614441575950470500809281871169394 0737.$$

As we mentioned at the outset of this section, the big prize of factoring RSA-155 goes to the General Number Field Sieve, and the honour of factoring a 211-digit Cunningham Number goes to the Special Number Field Sieve. Factoring RSA-140 required 8000 mips years (see page 205), whereas RSA-140 required 2000 mips years. The authors of the factorization of RSA-155 observed that the asymptotic complexity formula for the Number Field Sieve extrapolated from RSA-140 to RSA-155 would predict 14000 mips years for RSA-155. However, the gain was caused by an "improved application of the polynomial search used for RSA-140." As noted by Odlyzko in [148], by comparison the distributed.net project[5.22] has access to approximately 10^6 computers yielding about 10^7 mips

---

[5.22]See: http://www.distributed.net/.

years. Thus, the upshot is that these *distributed computers* can provide 10^7 mips in a year! (See Footnote 5.15 on page 205.) Hence, 2000 mips years to factor RSA-140 is not so formidable a time span. There is even skepticism that Moore's "Law" (see Footnote 1.38 on page 22) will hold indefinitely. The consensus is that by 2004, the standard processor will have a rating of 1000 mips. However, factoring algorithms are typically limited more by memory requirements than by processors. Thus, it is unlikely that Moore's Law will fail in the immediate future.

The Number Field Sieve is generally considered to be the most important development with respect to the discrete log problem. Regarding the Special Number Field Sieve, there has been some concern over the years that an entity A could design a *trapdoor prime*. In other words, there might exist primes that could be manufactured with representation as a value of a polynomial with small coefficients at some integer, for instance, such that the Special Number Field Sieve could be applied to them, but only entity A, knowing the secret design, could do so. If entity B does not know the details of such a trapdoor prime design, then B would ostensibly be forced to use the General Number Field Sieve. However, A's advantage over B is negligible, so there should be no serious concern over such a trapdoor.

Exercises

5.9. Use Pollard's Method to factor $\mathfrak{F}_6$.

In Exercises 5.10–5.14, use the gcd method described before Pollard's Method to find an odd factor of the given integer.

5.10. $2^{373} + 1$. (*Hint: Use* $\mathbb{Z}[\sqrt[3]{-4}]$.)

5.11. $7^{149} + 1$. (*Hint: Use* $\mathbb{Z}[\sqrt[3]{-7}]$.)

5.12. $3^{239} - 1$. (*Hint: Use* $\mathbb{Z}[\sqrt[3]{3}]$.)

5.13. $2^{457} + 1$. (*Hint: Use the hint in Exercise 5.10.*)

5.14. $5^{77} - 1$.

5.15. $2^{332} + 1$.

5.16. Using the Number Field Sieve, find two factors of $2^{153} + 3$.

5.17. Using the Number Field Sieve, find a prime factor of $2^{488} + 1$.

5.18. Let p be a prime and let ζ_p be a primitive p^{th} root of unity, namely $\zeta_p^p = 1$ but $\zeta_p^j \neq 1$ for any natural number $j < p$. Prove that

$$\sum_{j=0}^{p-1} \zeta_p^j = 1 + \zeta_p + \zeta_p^2 + \cdots + \zeta_p^{p-1} = 0.$$

Chapter 6

☞Advanced Topics

This is an optional chapter devoted to three topics that take us substantially beyond the basic notions covered in the first four chapters. They are, in order of appearance: elliptic curves, zero knowledge proofs, and quantum computers. We will discuss their cryptographic implications.

6.1 Elliptic Curves & Cryptography

The discussion on page 128 revealed that, when used as the basis for a cryptosystem, computation of discrete logs is assumed to be intractable. What has focused attention upon Elliptic Curve Cryptosystems is that there is no known subexponential time algorithm (see page 49) to solve the discrete log problem on a general elliptic curve. What makes the cryptosystems based upon elliptic curves more acceptable, than say RSA, is that one can achieve the same level of security with smaller key sizes (see the discussion on page 159). Therefore, Elliptic Curve Cryptosystems can be used in such environments as smart cards since such ciphers need less memory and smaller processor requirements.

Not only do we have cryptographic reasons for looking at elliptic curves, but also some substantial pure mathematical reasons. The famous problem known as Fermat's Last Theorem was solved relatively recently by Andrew Wiles (see [203]). This essentially began with a 1982 conjecture posed by Gerhard Frey that a solution to the Fermat Equation (given in Footnote 1.40 on page 22) would imply the existence of an elliptic curve which is *semi-stable* but not *modular* (see [162]). This conjecture was proved in 1986 by Ken Ribet. In 1993, Wiles claimed to have a proof that all semi-stable elliptic curves with rational coefficients are modular. However, his proof had holes in it. With the aid of Richard Taylor, these gaps were plugged by early 1995. Hence, FLT fell to the contradiction after centuries of attempts to prove it. For a succinct overview of Wiles's proof see [72].

Given the above discussion, we have the greatest of motivations for studying elliptic curves both from a pure mathematical standpoint and from our main perspective in this text — applications to cryptography. We devote this section to an introduction of the basic features of elliptic curves and their applications to cryptography. First, we formally define the notion of *elliptic curves*.

Definition 6.1 (Elliptic Curves)

Let F be a field with characteristic not equal to 2 or 3 (see Footnote A.10 on page 290). If $a, b \in F$ are given such that $4a^3 + 27b^2 \neq 0$ in F, then an elliptic curve E defined over F is given by an equation $y^2 = x^3 + ax + b \in F[x]$. The set of all solutions $(x, y) \in F$ to the equation:

$$y^2 = x^3 + ax + b \qquad (6.2)$$

together with a point o, called the point at infinity, *is denoted by $E(F)$, called the set of F-rational points on E. The value $\Delta(E) = -16(4a^3 + 27b^2)$ is called the* discriminant *of the elliptic curve E.*

Elliptic curves can also be defined for characteristic $2, 3$ by a slightly different equation, but there are good pedagogical reasons for avoiding this case (see the denominators in Cardano's Formula on page 223). *We assume throughout that the fields under consideration have characteristic* not 2 or 3. However, do note that the characteristic 2 case is a cryptographic favorite quite simply because it is *fast*.

We have not explicitly defined what we mean by the "point at infinity," although we will do so shortly, once we have described the geometry of the situation. First of all, it is worth mentioning that the term "elliptic curve" is somewhat of a misnomer in the sense that elliptic curves are not ellipses. The term arose from the fact that elliptic curves made their debut during attempts to calculate the arc length of an ellipse.[6.1]

The term "point at infinity" comes from the representation of elliptic curves in the real plane $\mathbb{R} \times \mathbb{R}$ where the point at infinity is related to the vertical tangents (or third point of intersection of any vertical line with the curve). In what is called the *projective plane* the situation becomes clearer. Projective geometry studies the properties of geometric objects invariant under projection. For instance, projective 2-space over a field F, denoted by $\mathbb{P}^2(F)$, is the set $\{(x, y, z) : x, y, z \in F\} - \{(0, 0, 0)\}$ of all equivalence classes (See Footnote 2.8 on page 72) of *projective points* $(tx, ty, tz) \sim (x, y, z)$ for nonzero $t \in F$. So, if $z \neq 0$, then there exists a unique projective point in the class of (x, y, z) of the form $(x, y, 1)$, namely $(x/z, y/z, 1)$. Thus, $\mathbb{P}^2(F)$ may be identified with all points (x, y) of the ordinary, or *affine*, plane together with points for which $z = 0$. The latter are the points on the *line at infinity*, which one may regard as

---

[6.1] In the area of mathematical inquiry called *algebraic geometry*, elliptic curves are classified as *abelian varieties of dimension one*, and abelian varieties may be viewed as generalizations of elliptic curves. On the other hand, ellipses have genus 0, so the geometries of ellipses and elliptic curves have an almost no common intersection. The theory of *elliptic functions* is related to elliptic curves. These functions are those meromorphic functions having two $\mathbb{R}$-linearly independent periods. These functions are inversions of *elliptic integrals* which are line integrals on elliptic curves over $\mathbb{C}$. One may trace the name *elliptic curves* back to the fact that they are Riemann surfaces associated to the integrals for the arc-length of ellipses.

the *horizon* on the plane. With this definition, one sees that the point at infinity in Definition 6.1 is $(0, 1, 0)$ in $\mathbb{P}^2(F)$. This is the intersection of the y-axis with the line at infinity. Thus, we may think of the point at infinity as lying infinitely far off in the direction of the y-axis.

In Definition 6.1, the discriminant is required to be nonzero, since this gives us the guarantee that $x^3 + ax + b = 0$ has no multiple roots (see Definition A.43).[6.2] The reader may also wonder why we did not take the more general form of the equation $Y^2 = X^3 + AX^2 + BX + C$ in Definition 6.1. The reason is that we may translate from the latter to the one in the definition by letting $X \mapsto x - A/3$. This yields $y^2 = x^3 + ax + b$ with $a = B - A^2/3$ and $b = A^3/9 - AB/3 - A^3/27 + C$. Furthermore, by Theorem A.52, $-A$ is the sum of all the roots of $X^3 + AX^2 + BX + C = 0$. Also, once the translation is made, we may find the real root of $x^3 + ax + b = 0$ via the formula:

$$x = \sqrt[3]{-\frac{b}{2} + c} + \sqrt[3]{-\frac{b}{2} - c}, \quad \text{where} \quad c = \sqrt{\left(\frac{b}{2}\right)^2 + \left(\frac{a}{3}\right)^3},$$

called *Cardano's Formula*[6.3] (see Footnote 1.22 on page 9). The form of equation (6.2) is called the *Weierstrass Form*.

We now begin to look at elliptic curves, especially those with rational integer solutions, namely those $P = (x, y)$ on E such that $x, y \in \mathbb{Q}$ since these will provide the tools for applications to factoring and primality testing. We begin with an example illustrated by the following figure.

Figure 6.3: $y^2 = x^3 + 4x + 16$ and $y = -2(x + 2)$

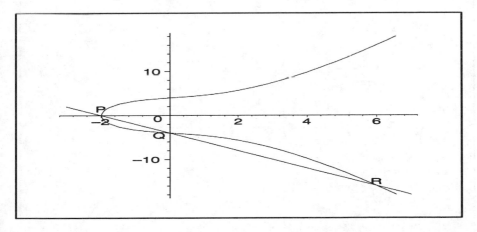

[6.2]We want to avoid multiple roots since if $x = \alpha$ is a multiple root of $x^3 + ax + b = 0$, then $(\alpha, 0)$ is a *singular point* on the curve $y^2 = x^3 + ax + b$ (see the solution of Exercise 6.5 on page 326).

[6.3]The two complex roots are similarly found as part of Cardano's Formula (see [202, p. 364]).

Consider the points $P = (-2,0)$, $Q = (0,-4)$ and $R = (6,-16)$ which are the points of intersection of the straight line and the elliptic curve in Figure 6.3. Notice that we may easily find these points by substituting the equation $y = -2(x+2)$ into $y^2 = x^3 + 4x + 16$ to get, after rewriting and simplifying, $x(x-6)(x+2) = 0$. Thus, the x-values $x = 0, -2, 6$ which determine the points are obtained. If two of the three points (intersecting a straight line, possibly repeated) on an elliptic curve are rational, then the third must be as well. However, if only one of the three points is known to be rational, then the other two may not be. For instance, if $y = 4x+8$, then clearly $(-2,0)$ is a point on both the line $y = 4x + 8$ and the elliptic curve given by $y^2 = x^3 + 4x + 16$. However, neither of the other two points are rational. To see this, notice that when we substitute $y = 4x+8$ into $y^2 = x^3+4x+16$, we get $(x+2)(x^2-18x-24) = 0$ after simplification. We compute the roots $x = 9 \pm \sqrt{105}$ of $x^2 - 18x - 24 = 0$ which are irrational. In fact, there exist elliptic curves having *no* rational points (x,y) with both x and y nonzero. By Exercises 6.1–6.3 on page 248, the elliptic curve given by $y^2 = x^3 - 4x$ is one of these (see Figure 6.4 below). Hence, to generate a rational point on an elliptic curve from two given others on the curve, one can find a straight line intersecting the curve at the two known rational points. Then the third point of intersection is guaranteed to be rational since two already are and the degree 3 polynomial has rational coefficients. Note as well that Figure 6.3 is an example of an elliptic curve with only one real root, whereas the one in Figure 6.4 has exactly three real roots, namely $x = -2, 0, 2$.

Figure 6.4: $y^2 = x^3 - 4x$

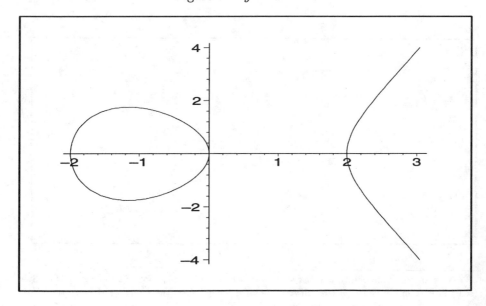

Now we proceed with the important task of determining how points on an

elliptic curve form an additive group. Given the above, it is tempting to define
the meaning of $P + Q$ to be R where P, Q, R are collinear (lie on the same
straight line intersecting the elliptic curve at P, Q, and R). Suppose that this
definition of addition leads to an additive group structure on an elliptic curve
E. If 0 is the additive identity, then $P + 0 = P$ for all points P on E. However,
this means that the straight line through P and 0 must intersect the curve as a
tangent. Thus, by definition $P + P = P$. Given the existence of additive inverses
$-P$, we get $P = P + P - P = P - P = 0$. This means that $P = 0$ for all points on
E. Hence, the assumption of two distinct points on E leads to a contradiction
under this definition of "group" addition — not a very satisfactory outcome.
The valid method for obtaining a nontrivial group structure is developed as
follows.

Suppose that P and Q are points on an elliptic curve E. We need a definition
of $P + Q$. If $P \neq o$ and $P \neq \pm Q$ where $-Q$ means the reflection of the point Q
about the x-axis, then there must be a third point R on E, uniquely determined
as the intersection point of the line through P and Q. However, as we have seen
above, R cannot be $P + Q$. Instead, we take $P + Q = -R$, the reflection of R
about the x-axis.

As a specific example consider the elliptic curve given by $y^2 = x^3 - x + 1$
and look at the two points $P = (0, -1)$ and $Q = (1, 1)$, illustrated in Figure 6.5
below.

Figure 6.5: Addition of distinct points on $y^2 = x^3 - x + 1$

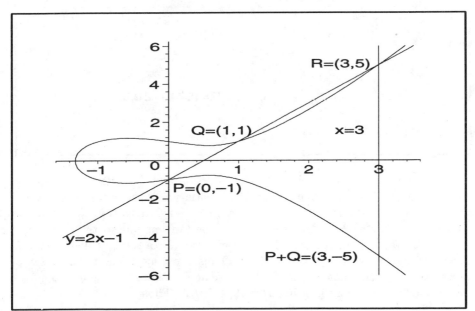

The line joining P and Q is given by $y = 2x - 1$. Thus, we calculate that
the third point of intersection of the line with the curve is $R = (3, 5)$, which is

obtained by substituting $y = 2x - 1$ into $y^2 = x^3 - x - 1$ and solving for x given that we already have two of the values, $x = 0$ and $x = 1$. The reflection of R about the x-axis is $-R = P + Q = (3, -5)$.

Figure 6.5 illustrates the process of adding two distinct points. Now let us assume that $P = Q$ and $P \neq -Q$. Then to form $P + Q = 2P$, we take the tangent line at P, which gives rise to a third point $R = (x_3, y_3)$, uniquely determined as the intersection point of E with that tangent line. Then the reflection $-R$ of R about the x-axis is what we define as $P + P = 2P = -R$. Thus, $-R$ is the other point of intersection of E with the line $x = x_3$, which is also the intersection of the line containing R and o with E. This is illustrated by Figure 6.6 below.

Figure 6.6: Addition of a point to itself on $y^2 = x^3 - x + 4$

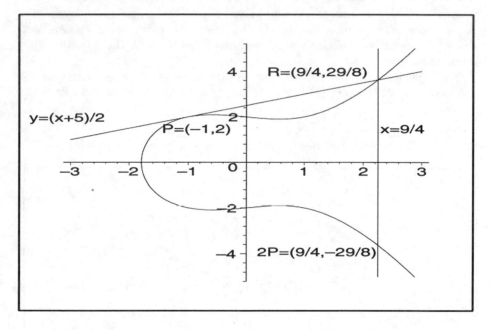

In Figure 6.6, the point $P = (-1, 2)$ is to be added to itself. To get the tangent line at P, we note that implicit differentiation of the curve yields $2yy' = 3x^2 - 1$. Substituting the value of $P = (x, y) = (-1, 2)$ into this derivative, we get that $y' = 1/2$ at P. Thus, the slope of the tangent line to the curve at P is $m = 1/2$. Putting the point P into the equation for a straight line, with this slope, through P we get $y = (x + 5)/2$. Thus, we may calculate the point of intersection R of the tangent line to the curve at P by substituting $y = (x+5)/2$ into $y^2 = x^3 - x + 4$ and get that $R = (9/4, 29/8)$. Thus,

$$2P = -R = (9/4, -29/8).$$

The above discussion motivates the following formal definition for addition of points on elliptic curves and it provides a summary of what we have discussed by giving the addition in parametric form.

Definition 6.7 (Additive Group Structure for Elliptic Curves)

Let E be an elliptic curve over a field (of characteristic not 2 or 3). For any two points $P = (x_1, y_1)$ and $Q = (x_2, y_2)$ on E, define

$$P + Q = \begin{cases} \mathfrak{o} & \text{if } x_1 = x_2 \text{ and } y_1 = -y_2, \\ Q = Q + P & \text{if } P = \mathfrak{o}, \\ (x_3, y_3) & \text{otherwise,} \end{cases}$$

where

$$x_3 = m^2 - x_1 - x_2, \tag{6.8}$$

$$y_3 = m(x_1 - x_3) - y_1, \tag{6.9}$$

and

$$m = \begin{cases} (y_2 - y_1)/(x_2 - x_1) & \text{if } P \neq Q, \\ (3x_1^2 + a)/(2y_1) & \text{if } P = Q. \end{cases} \tag{6.10}$$

To see that Definition 6.7 actually provides us with an additive abelian group structure, we note that additive closure is immediate from the definition as is the commutativity of addition $P + Q = Q + P$. Also, $\mathfrak{o}$ is clearly the additive identity, and $-P = (x, -y)$ is the additive inverse of $P = (x, y)$ for any point P on E. Hence, it remains only to show additive associativity, which is a tedious exercise, at least without some mathematical software package such as *Maple* at one's disposal.[6.4]

In the motivating discussion leading up to Definition 6.7, we provided a look at geometric considerations. Now we provide an algebraic explanation to add to the geometry. Let E be given by

$$y^2 = x^3 + ax + b. \tag{6.11}$$

[6.4]However, if the reader has knowledge of algebraic geometry, then establishing associativity is straightforward as follows. In algebraic geometry, *Bezout's Theorem* says that two projective curves of orders r and s having no component in common intersect in rs points. Here a "component" refers to a curve determined by an irreducible factor of the polynomial equation that defines the original curve (see the solution of Exercise 6.5 on page 326). A consequence of Bezout's Theorem is that if we have two cubic curves (which satisfy certain properties that are in fact satisfied by our elliptic curves), then a third curve passing through eight of their nine points of intersection must pass through the ninth as well. Given an elliptic curve E, with points P and Q, let PQ denote the third point of intersection of E with the straight line through P and Q. It can be shown that an elliptic curve E, with points P_1, P_2, and P_3, determines two other elliptic curves with eight common points: P_1, P_2, P_3, $P_1 + P_2$, $P_2 + P_3$, $P_1 P_2$, $P_2 P_3$, and $\mathfrak{o}$. Hence, $P_1(P_2 + P_3)$ and $(P_1 + P_2)P_3$ must coincide with their ninth point of intersection. This yields the result immediately. There is, however, one caveat here. Bezout's Theorem only holds over algebraically closed fields. Thus, the above proof is carried out over the algebraic closure L of the base field F. Then, since additively closed subsets of groups inherit associativity from any extension, $E(F)$ is a group since it is closed in the group $E(L)$.

If $P = (x_1, y_1) \neq Q = (x_2, y_2)$ on E with $x_1 \neq x_2$, namely $P \neq -Q$, then $-(P + Q)$ is the third point of intersection, $R = (x_3, -y_3)$, of E with the line joining P and Q. The equation of this line has slope

$$m = (y_1 - y_2)/(x_1 - x_2),$$

which is (6.10). This may be rewritten as $y = m(x - x_1) + y_1$, and substituted into (6.11) to get:

$$m^2(x - x_1)^2 + 2m(x - x_1)y_1 + y_1^2 = x^3 + ax + b,$$

which simplifies to

$$x^3 - m^2 x^2 + Ax + B = 0, \qquad\qquad (6.12)$$

where $A = a + 2m^2 x_1 - 2my_1$ and $B = b - y_1^2 + 2m_1 x_1 - m^2 x_1^2$. However, by Theorem A.52, $m^2 = x_1 + x_2 + x_3$, or by rewriting,

$$x_3 = m^2 - x_1 - x_2,$$

which is (6.8). Thus $P + Q = (x_3, y_3)$, where

$$y_3 = m(x_1 - x_3) - y_1,$$

which is (6.9). If $P = Q = (x_1, y_1)$ and $P \neq -Q$, namely $y_1 \neq 0$, then the slope of the tangent at P is given by $2yy' = 3x^2 + a$, namely by

$$m = \frac{3x_1^2 + a}{2y_1},$$

which is the case (6.10) for $P = Q$. Lastly, if $P = -Q$, then the line through P and $-Q$ is vertical, so the third point of intersection is $\mathfrak{o}$, as noted above, and

$$P + Q = -Q + Q = \mathfrak{o}.$$

Remark 6.13 *All of the above can be summarized in a single equation that covers all cases including the possibility that $P = \mathfrak{o}$, and the possibility that the points are indistinct. It is that if P, Q, R are three collinear points (all in the same straight line) on E, then*

$$P + Q + R = 0.$$

The rational points on elliptic curves are important, as we shall see, from the perspective of the elliptic curve cryptography that we will study in this chapter. In particular, rational points on elliptic curves E over $\mathbb{Q}$ are broken down into two classes.

Definition 6.14 (Torsion Points on Elliptic Curves)

If E is an elliptic curve over a field F, and $P \in E(F)$ such that $nP = o$ for some $n \in \mathbb{N}$, then P is called a torsion point *or a* point of finite order. *The smallest such value of n is called the* order *of P. We call o the* trivial torsion point. *If P is not a torsion point, then P is said to be a* point of infinite order. *If P is a torsion point and $P \neq o$, then P is called a* nontrivial torsion point.

Note that the notion of order given in Definition 6.14 coincides with the element order in the group of points on E.

Example 6.15 Given an elliptic curve E defined by $y^2 = x^3 + 1$ and a point $P = (2,3)$, we can use Definition 6.7 to calculate that $2P = (0,1)$, $3P = (-1,0)$, $4P = (0,-1)$, $5P = (2,-3)$, and $6P = o$. Thus, P is a torsion point of order six on E.

The set of torsion points $E(F)_t$ form a subgroup of $E(F)$ by Exercise 6.7 on page 249, called the *torsion subgroup of $E(F)$*. The fundamental result on torsion points was proved in the 1930s by Nagell[6.5] with the proof later refined by Lutz.

Theorem 6.16 (Nagell-Lutz)

If E is an elliptic curve over $\mathbb{Q}$ given by $y^2 = x^3 + ax + b$ with $a, b \in \mathbb{Z}$, and $P = (x_1, y_1) \in E(\mathbb{Q})_t$ then $x_1, y_1 \in \mathbb{Z}$ and either $y_1 = 0$ (in which case P has order 2), or $y_1 \neq 0$ and $y_1^2 \mid (4a^3 + 27b^2)$.

Proof. See [102, Section 4, pp. 144-145].[6.6] $\square$

What Theorem 6.16 says is that *if $P = (x_1, y_1)$ is a rational torsion point*, then it is an integral point of order two (see Exercise 6.6), or it has *ordinate* y_1 whose square divides $|\Delta(E)|$. However, Theorem 6.16 cannot be used to verify that an integral point has finite order, since there exist integral points on elliptic curves that have infinite order. We actually need to find an $n \in \mathbb{N}$ such that $nP = o$ in order to prove that it is indeed torsion. Yet, we can use the *contrapositive*[6.7] of Theorem 6.16 to show that a given point has *infinite* order. In other words, if nP is *not* an integral point for *some* $n \in \mathbb{N}$, then P has infinite order. For instance, if $y^2 = x^3 - x + 4$ and if we take the point $Q = (-1,2)$, then $2Q = (9/4, -29/8)$, and $nQ \neq o$ for any $n \in \mathbb{N}$ by Theorem 6.16. Hence, not all rational (or even integral) points are torsion.

[6.5]Trygve Nagel (changed to Nagell) (1895–1988) was born in Oslo on July 13, 1895. He studied at the University of Oslo and was appointed professor at Uppsala in 1931, where he stayed until his retirement in 1962. He died on January 24, 1988.

[6.6]We state this with result without proof, as we do with numerous other deep results in this section. The intent is to give the reader an overview of the flavour and richness of the subject. References for the proofs are given for those readers interested in looking at these topics in more detail.

[6.7]Logically, the contrapositive of a statement A implies B is *not B* implies *not A*. Thus, for example in Theorem 6.16, if we can show that an integral point has a multiple that is *not* integral, then that point is *not* torsion.

We can also use Theorem 6.16 to check that the torsion points that we found, for instance in Example 6.15, are all that exist for that elliptic curve. Since $4a^3 + 27b^2 = 27$ in that example, then $y = \pm 1$ or $y = \pm 3$ for any torsion point $P = (x, y)$ (with $y \neq 0$) by Theorem 6.16. We have already seen that $(2, 3)$ has order 6, and similarly $(2, -3)$ has order 6. The point $(0, 1)$ has order 3 since $2(0, 1) = (0, -1)$ and $3(0, 1) = \mathfrak{o}$. Finally $(-1, 0)$ has order 2. Thus $E(\mathbb{Q})_t$ is a cyclic group of order 6 generated (see Footnote A.12 on page 292) by $(2, 3)$.

Now we illustrate how to geometrically obtain multiples of torsion points on elliptic curves. The following figure illustrates the next example.

Figure 6.17: Multiples of a Torsion Point on $y^2 = x^3 - 43x + 166$

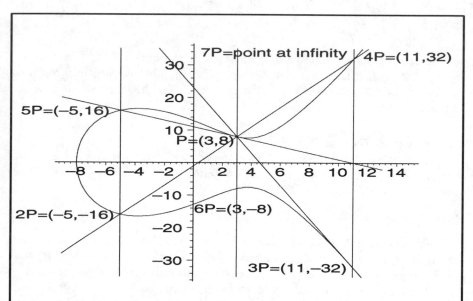

Example 6.18 When we speak about "reflection" in this example, we will mean "reflection about the x-axis." In Figure 6.17, we begin with a tangent line at $P = (3, 8)$ which intersects the curve again at $(-5, 16)$, so the reflection is $2P = (-5, -16)$. Then the line through P and $2P$ intersects the curve at $(11, 32)$, whose reflection is is $3P = (11, -32)$. The line through $P = (3, 8)$ and $3P = (11, -32)$ intersects the curve at $(11, -32)$ again, so the reflection is $4P = (11, 32)$. The line through $4P$ and P intersects the curve at $(-5, -16)$, whose reflection is $5P = (-5, 16)$. Then the line through $5P$ and P intersects the curve at $(3, 8)$, so $6P = (3, -8)$ is the reflection. Finally, the line through $6P$ and P is vertical, so $7P = \mathfrak{o}$.

Example 6.18 shows that every torsion point on E can be obtained from generators of $E(\mathbb{Q})_t$ by repeatedly taking lines through them, intersecting with

E, and reflecting about the x-axis to create new points on E.

The problem of determining the cardinality of $E(\mathbb{Q})_t$ as E varies over all elliptic curves over $\mathbb{Q}$ was solved by B. Mazur in 1976.

Theorem 6.19 (Mazur's Theorem) *If E is an elliptic curve over $\mathbb{Q}$, then either*

$$E(\mathbb{Q})_t \cong \mathbb{Z}/n\mathbb{Z},$$

for some $n \in \{1, 2, 3, 4, 5, 6, 7, 8, 9, 10, 12\}$ or

$$E(\mathbb{Q})_t \cong \mathbb{Z}/2\mathbb{Z} \oplus \mathbb{Z}/2n\mathbb{Z},$$

where $n \in \{1, 2, 3, 4\}$.

Proof. See [131]. □

Theorem 6.19 shows that a torsion group cannot have order bigger than 16 for elliptic curves over $\mathbb{Q}$. For instance, in the above, we showed that the curve in Example 6.15 has $|E(\mathbb{Q})_t| = 6$. An example where $|E(\mathbb{Q})_t| = 4$ is given in Figure 6.4 since the only torsion point are $(\pm 2, 0)$, $(0, 0)$ and $\mathfrak{o}$. In this case,

$$E(\mathbb{Q})_t \cong \mathbb{Z}/2\mathbb{Z} \times \mathbb{Z}/2\mathbb{Z}$$

(see Exercise 6.6 on page 249). An example where $|E(\mathbb{Q})_t| = 1$ is

$$y^2 = x^3 - 2,$$

by Exercise 6.9. Also, by Exercise 6.8, the elliptic curve given by

$$y^2 = x^3 - 432$$

has $|E(\mathbb{Q})_t| = 3$ since the only torsion points are $(12, \pm 36)$ and $\mathfrak{o}$.

The problem of determining $|E(F)_t|$ as E varies over all elliptic curves for a given number field F remains an open problem. However, in 1996, L. Merel [136] proved what is called the *strong uniform boundedness conjecture* or (UBC) of Mazur and Kamienny, which says:

$$|E(F)_t| \leq B_F$$

for an elliptic curve E, where B_F is a constant depending only on $|F : \mathbb{Q}|$. For instance, Mazur's Theorem tells us that $B_\mathbb{Q} = 16$.

Once we have a rational point (that is not integral), then we have a point of infinite order by the Nagell-Lutz Theorem. By Exercise 6.10 on page 249, the rational points on $E(\mathbb{Q})$ form a subgroup thereof. We may naturally ask for a description of the structure of this subgroup. This was answered by Mordell[6.8] in 1922.

[6.8]Louis Joel Mordell (1888–1972) was born in Philadelphia, Pennsylvania on January 28, 1888. He was educated at Cambridge, and lectured at Manchester College of Technology from 1920 to 1922. In 1922, he went to Manchester University where he remained until he succeeded G.H. Hardy (1877–1947) at Cambridge in 1945. Among his honours were being elected as a member of the Royal Society in 1924, winning the De Morgan Medal in 1941, being president of the London Mathematical Society from 1943 to 1945, and winning the Sylvester Medal in 1949. He died in Cambridge on March 12, 1972.

Theorem 6.20 (Mordell's Theorem)

If E is an elliptic curve over $\mathbb{Q}$, then the subgroup of rational points $E(\mathbb{Q})$ is a finitely generated abelian group.

Proof. See [102, Theorem 4.11, p. 95]. $\qquad\qquad\qquad\qquad\qquad\qquad\qquad$ $\square$

What Mordell's Theorem says is that the rational points on any elliptic curve defined over $\mathbb{Q}$ may be obtained from only finitely many of them by adding them in all possible ways. More precisely, we may describe them in terms of the following notion.

Definition 6.21 (Basis, Rank, and Independence)

If G is an additive abelian group, then $g_1, \ldots, g_r \in G$ are called independent *provided that*

$$z_1 g_1 + z_2 g_2 \cdots + z_r g_r = 0,$$

where $z_j \in \mathbb{Z}$ if and only if $z_j = 0$ for all $j = 1, 2, \ldots, r$. The maximum number of independent elements in G is called the rank *of G. If this number is finite, then G is said to be of finite rank (where possibly $r = 0$, in which case G is a finite group, so r is a nonnegative integer in general).*

The fundamental structure theorem for finitely generated additive abelian groups says that such groups are isomorphic to

$$\underbrace{\mathbb{Z} \oplus \cdots \oplus \mathbb{Z}}_{r \text{ copies}} \oplus \mathbb{Z}/p_1^{a_1}\mathbb{Z} \oplus \cdots \oplus \mathbb{Z}/p_k^{a_k}\mathbb{Z},$$

where the p_j are (not necessarily distinct) primes and $a_j \in \mathbb{N}$ for $j = 1, \ldots, k$ (see [96] for instance). The rank r refers to the number of summands of $\mathbb{Z}$ (possibly $r = 0$ if G is finite). Roughly speaking, this says that there are generators $P_1, \ldots, P_r, Q_1, \ldots, Q_k \in G$ such that every point $P \in G$ can be written in the form

$$P = z_1 P_1 + \cdots + z_r P_r + n_1 Q_1 + \cdots + n_k Q_k,$$

where the $z_j \in \mathbb{Z}$ for $j = 1, 2, \ldots, r$ are uniquely determined by P and the $n_j \in \mathbb{Z}$ are uniquely determined modulo $p_j^{a_j}$ for $j = 1, \ldots, k$. In particular, what Mordell's Theorem says in view of Definition 6.21 is that if $G = E(\mathbb{Q})$ is the set of rational points on an elliptic curve E over $\mathbb{Q}$, then

$$E(\mathbb{Q}) \cong \underbrace{\mathbb{Z} \oplus \cdots \oplus \mathbb{Z}}_{r_{\mathbb{Q}}(E) \text{ copies}} \oplus E(\mathbb{Q})_t,$$

where $r_{\mathbb{Q}}(E) \in \mathbb{N}$ is called the *Mordell-Weil rank* of E. Thus, $|E(\mathbb{Q})| < \infty$ if and only if $r_{\mathbb{Q}}(E) = 0$. The study of the rank of elliptic curves is one of the most active research areas in modern mathematics. See Exercise 6.14 on page 250 for an application of rank to the notion of *congruent numbers* defined in Exercise 6.11.

Another way of stating the Mordell Theorem is that the group of rational points on an elliptic curve over $\mathbb{Q}$ is a *free abelian group of finite rank*, since a "free abelian group" is just a sum of copies of the additive group of $\mathbb{Z}$. There is no known general method for computing $r_{\mathbb{Q}}(E)$, for an arbitrary elliptic curve E, that leads to an answer in all cases. Nevertheless, the general consensus is that the method we have does lead to an answer in all cases.

Mordell's Theorem was generalized by Weil[6.9] in 1928.

We have seen instances where an integer point is not a torsion point. The question arises naturally: Are there infinitely many integer points on a given elliptic curve? In 1926, this was answered by C. L. Siegel.[6.10]

Theorem 6.22 (Siegel's Theorem)
The equation
$$y^2 = x^3 + ax + b$$
with $a, b \in \mathbb{Z}$ and $4a^3 + 27b^2 \neq 0$ has only finitely many solutions $x, y \in \mathbb{Z}$.

Proof. See [185]. □

Now that we have some rudimentary knowledge pertaining to elliptic curves, it is time to turn our attention to elliptic curves over finite fields, since this is the final gateway to the applications of elliptic curves to factoring and primality

---

[6.9]André Weil (1906–1998), pronounced *vay*, was born on May 6, 1906 in Paris, France. As he said in his autobiography, *The Apprenticeship of a Mathematician*, he was passionately addicted to mathematics by the age of ten. After graduating from the École Normal in Paris, he eventually made his way to Göttingen, where he studied under Jacques Salomon Hadamard (1865–1963). His doctoral thesis contained a proof of what we call the *Mordell-Weil Theorem*, where he extended Mordell's result to any *abelian variety* defined over a number field F (see [200]). His first position was at Aligarh Muslim University, India (1930–1932), then the University of Strasbourg, France (1933–1940), where he became involved with the controversial *Bourbaki project*, which attempted to give a unified description of mathematics. The name *Nicholas Bourbaki* was that of a citizen of the imaginary state of Poldavia, which arose from a spoof lecture given in 1923. Weil tried to avoid the draft, which earned him six months in prison. In order to be released from prison, he agreed to join the French army. Then he came to the United States to teach at Haverford College in Pennsylvania. He also held positions at São Paulo University, Brazil (1945–1947), the University of Chicago (1947–1958), and thereafter at the Institute for Advanced Study at Princeton. In 1947 at Chicago, he began a study, which eventually led to a proof of the Riemann hypothesis for algebraic curves. He went on to formulate a series of conjectures that won him the Kyoto prize in 1994 from the Inamori Foundation of Kyoto, Japan. His conjectures provided the principles for modern algebraic geometry. His honours include an honorary membership in the London Mathematical Society in 1959, and election as a Fellow of the Royal Society of London in 1966. However, in his own official biography he lists his only honour as *Member, Poldevian Academy of Science and Letters*. He died on August 6, 1998 in Princeton, and is survived by two daughters, and three grandchildren. His wife Eveline died in 1986.

[6.10]Carl Ludwig Siegel (1896–1981) was born in Berlin, Germany in 1896. He achieved his doctorate under Edmund Georg Hermann Landau (1877–1938). From 1922 to 1937, he was a professor at Frankfurt. Then he was at Göttingen from 1938 to 1940, after which he was at Princeton until 1951. He returned to Göttingen in that year. He built upon the work of Eisenstein, Gauss, Lagrange, and Hermite in the theory of quadratic forms. He also extended the work of Poincaré, by essentially introducing the general theory of automorphic functions of several complex variables. He died in Göttingen on April 4, 1981.

testing and their use in cryptography. The standard approach is to begin with an elliptic curve and "reduce it" modulo a given prime. However, to do this we must first formalize the notion of "reduction" of rational points.

Definition 6.23 (Reduction of Rational Points on Elliptic Curves)
  Let $n \in \mathbb{N}$ and $x_1, x_2 \in \mathbb{Q}$ with denominators prime to n. Then

$$x_1 \equiv x_2 \,(\mathrm{mod}\ n)$$

means

$$x_1 - x_2 = a/b \text{ where } \gcd(a, b) = 1, a, b \in \mathbb{Z}, \text{ and } n | a.$$

Remark 6.24 *In Definition 6.23, for any $x = c/d \in \mathbb{Q}$ with $\gcd(d, n) = 1 = \gcd(c, d)$, there exists an $r \in \mathbb{Z}$, with $0 \le r \le n - 1$, uniquely determined modulo n, and denoted by*

$$r \equiv \overline{x} \,(\mathrm{mod}\ n).$$

(When the context is clear, we will suppress the $(\mathrm{mod}\ n)$, simply writing $r = \overline{x}$, and the same convention will apply below.) Note that we may take

$$r = \overline{cd^{-1}},$$

where d^{-1} is the unique multiplicative inverse of d modulo n. The identification of c/d with cd^{-1} modulo n preserves addition and multiplication. Hence, if $P = (x, y)$ is a point on an elliptic curve E over $\mathbb{Q}$, with denominators of x and y prime to n, then

$$\overline{P} \text{ means } (\overline{x}, \overline{y}).$$

Also, $\overline{E}$ denotes the curve reduced modulo n, whenever defined, namely

$$\overline{E} = \{(\overline{x}, \overline{y}) : y^2 \equiv x^3 + ax + b \,(\mathrm{mod}\ n)\},$$

with $x = \overline{x}$, and $y = \overline{y}$. The cardinality of the set $\overline{E}$ is denoted by $|\overline{E}|$. It turns out that $\overline{E}$ may not be a group, since certain elements may not be invertible. However, we may still use it for practical computational purposes, as illustrated below.

Example 6.25 If $x = 3/4$ and $n = 7$, then $\overline{x} = 6 = r$ is the unique integer (least positive residue) modulo 7 such that $x \equiv r \,(\mathrm{mod}\ 7)$, since

$$x - r = c/d - r = 3/4 - 6 \equiv 3 \cdot 4^{-1} - 6 \equiv 3 \cdot 2 - 6 \equiv 0 \,(\mathrm{mod}\ 7).$$

The following result instructs us on how to add and reduce points on rational elliptic curves, and will be the chief tool in the description of the Elliptic Curve Factoring Method.

Theorem 6.26 (Addition and Reduction of Points on Elliptic Curves)
Let $n \in \mathbb{N}$, $\gcd(6, n) = 1$, and let E be an elliptic curve over $\mathbb{Q}$ with equation

$$y^2 = x^3 + ax + b, \quad a, b \in \mathbb{Z},$$

and

$$\gcd(4a^3 + 27b^2, n) = 1.$$

Let P_1, P_2 be points on E where $P_1 + P_2 \neq \mathfrak{o}$, and the denominators of P_1, P_2 are prime to n. Then $P_1 + P_2$ has coordinates having denominators prime to n if and only if there does not exist a prime $p \mid n$ such that (modulo p):

$$\overline{P_1} + \overline{P_2} = \overline{\mathfrak{o}}$$

on the elliptic curve $\overline{E}$ over $\mathbb{F}_p$, with equation

$$y^2 \equiv x^3 + ax + b \,(\mathrm{mod}\, p).$$

Proof. See [189, Chapter IV, pp. 121–123]. □

We are now in a position to describe Lenstra's Factorization Method using elliptic curves (see [122]). In the following algorithm, if $\overline{E} \,(\mathrm{mod}\, n)$ is not a group, then this is not a problem in the algorithm. The reason is that, even if P_1 and P_2 were points on such a curve and if $P_1 + P_2$ were not defined, then n must be composite! The noninvertibility that would result in step 5 of the algorithm would then give us a factor of n. This is indeed the underlying key element in the Elliptic Curve Algorithm. In other words, we *want the group addition law to fail* since this gives us a factor.

◆ **Lenstra's Elliptic Curve Factoring Method**

The following is the algorithm for factoring an odd composite $n \in \mathbb{N}$, which we assume to have checked in advance is neither a perfect power nor a prime, and is relatively prime to 6.

(1) In some random fashion, we generate a pair (E, P), where E is an elliptic curve over $\mathbb{Q}$ with equation $y^2 = x^3 + ax + b$, $(a, b \in \mathbb{Z})$, and P is a point on E.

(2) Check that $\gcd(n, 4a^3 + 27b^2) = 1$. If not, then we have a factor of n, unless $\gcd(n, 4a^3 + 27b^2) = n$, in which case we choose a different pair (E, P).

(3) Choose $M \in \mathbb{N}$ and bounds $A, B \in \mathbb{N}$ such that the canonical prime factorization of M is

$$M = \prod_{j=1}^{\ell} p_j^{a_{p_j}},$$

for small primes $p_1 < p_2 < \ldots < p_\ell \leq B$, where $a_{p_j} = \lfloor \ln(A)/\ln(p_j) \rfloor$ is the largest exponent such that $p_j^{a_{p_j}} \leq A$.

(4) For a sequence of divisors s of M, compute $\overline{sP}$ (modulo n) as follows. First compute

$$\overline{sP} = \overline{p_1^k P},$$

for $1 \le k \le a_{p_1}$, then

$$\overline{sP} = \overline{p_2^k p_1^{a_{p_1}} P},$$

for $1 \le k \le a_{p_2}$, and so on, until all primes p_j dividing M have been exhausted or the following occurs.

(5) If the calculation of either $(x_2 - x_1)^{-1}$ or $(2y_1)^{-1}$ in (6.10), for some $s|M$ in step (4), shows that one of them is *not* prime to n, then there is a prime $p|n$ such that $\overline{sP} = \mathfrak{o}$, (modulo p) by Theorem 6.26. This will give us a nontrivial factor of n unless $\overline{sP} = \mathfrak{o}$ for all primes $p|n$. In that case $\gcd(s,n) = n$, and we go back and try the algorithm with a different (E, P) pair.

The value of B in step 3 of the above algorithm is the upper bound on the prime divisors of s, from which we form $\overline{sP}$. If B is large enough, then we increase the probability that $\overline{sP} = \mathfrak{o} \pmod{p}$ for some prime $p \mid n$. On the other hand, the larger the value of B, the longer the computational time. Hence, we must also choose B to minimize running time. Moreover, A is an upper bound on the prime powers that divide s, so similar considerations apply. Lenstra has some convincing conjectural evidence that $n \in \mathbb{N}$ can be factored by his algorithm in expected running time

$$O\left(e^{\sqrt{(2+\epsilon)\ln(p)(\ln \ln p)}}(\ln n)^2\right),$$

where p is the smallest prime factor of n and ϵ goes to zero as p gets large.[6.11]
In the next example, which illustrates the above algorithm, we will make use of the following result proved by Hasse.[6.12]

[6.11]A corollary of this fact is that the Elliptic Curve Method can be used to factor n in expected time $O(e^{\sqrt{(1+\epsilon)(\ln n)(\ln \ln n)}})$, with ϵ as above.

[6.12]Helmut Hasse (1898–1979) was born in Kassel, Germany on August 25, 1898. His paternal grandmother shared a common ancestor with the composer Felix Mendelssohn-Bartholdy. It is worthy of note that Dirichlet, Kummer, and Hensel were also related to the Mendelssohn family. From 1908 to 1913, he went to the Willhelms-Gymnasium in Kassel, followed by the Fichte-Gymnasium at Berlin-Wilmersdorf. He became a volunteer in the wartime Navy in June 1915. Nevertheless, he was able to matriculate at Christian-Albrecht University in Kiel by the autumn of 1917. After the war, he matriculated at Georg August University in Göttingen. There he attended lectures by E. Hecke (1887–1947), who had the greatest influence on Hasse, D. Hilbert (see Footnote 1.78), E. Landau (See Footnote 1.61), E. Noether (1882–1935), and C. Runge (1856–1927), among others. In 1920, Hasse went to Marburg to study under Hensel. In October of 1920, Hasse discovered the *local-global* principle that now bears his name. This principle shows that a quadratic form that represents zero nontrivially over the p-adic numbers for each prime p represents zero nontrivially over the rational number field $\mathbb{Q}$. This principle formed an important part of his doctoral thesis as well as his Habilitation Thesis. In 1922, Hasse was appointed lecturer at the University of Kiel. Three years later, he was appointed Professor at Halle. In 1930, Hensel retired from Marburg, and Hasse was appointed to fill his

Theorem 6.27 (Hasse's Bound for Elliptic Curves Over $\mathbb{F}_p$)

If E is an elliptic curve over $\mathbb{F}_p$ for a prime $p > 3$, and N is the number of points on E, then

$$|N - p - 1| < 2\sqrt{p}.$$

Proof. See [102, Theorem 10.5, p.296]. □

The reader may find interest in Exercise 6.4, which is an application for the number of elements on an elliptic curve over a finite field using the Legendre Symbol.

Based upon Hasse's Theorem and the above expected running time, Lenstra concludes that if we take

$$A = p + 1 + 2\sqrt{p}, \text{ and } B = e^{\sqrt{(\ln p)(\ln \ln p)/2}}, \tag{6.28}$$

where p is the smallest prime factor of n, then about one out of every B iterations will be successful in factoring n. Of course, we do not know a prime divisor p of n in advance, so we replace p by $\lfloor \sqrt{n} \rfloor$ and look at incremental values up to that bound.

Once the values of A and B have been chosen, then for a given prime p, the set $\overline{E} \pmod{p}$ is a finite abelian group, since these are the rational points on an elliptic curve over a finite field. Also, if the order g of $\overline{E}$ is not divisible by any primes larger than B, and if p is a prime such that $p + 1 + 2\sqrt{p} < A$, then Hasse's Theorem above tells us that $g \mid m$ in the algorithm, so $\overline{mP} = \mathfrak{o} \pmod{p}$.

We now present an example with a small value for pedagogical purposes. This is of course unrealistic since in practice we would be dealing with values the size of RSA, for instance, where the values would have to be about 208 digits for long-term security, as we saw in Chapter Three.

Example 6.29 Let $n = 88639$ and we seek to factor n using Lenstra's Elliptic Curve Method. We choose a family of elliptic curves and points (E, P):

$$(y^2 = x^3 + ax + 4, (0, 2)),$$

compute

$$\Delta = \Delta(E(\mathbb{Q})) = -16(4 \cdot 1^3 + 27 \cdot 4^2) = -6976,$$

and check that $\gcd(\Delta, n) = 1$. Thus, we may begin with $a = 1$. Since $\lfloor \sqrt{n} \rfloor = 297 \geq p$, then by (6.28), we may choose

$$A = 297 + 1 + 2 \cdot \sqrt{297} = 332 \text{ and } B = 9.$$

Thus, $M = 2^8 \cdot 3^5 \cdot 5^3 \cdot 7^2$, where $8 = \lfloor \ln(332)/\ln(2) \rfloor$, $5 = \lfloor \ln(332)/\ln(3) \rfloor$, $3 = \lfloor \ln(332)/\ln(5) \rfloor$, and $2 = \lfloor \ln(332)/\ln(7) \rfloor$. Using (6.8)–(6.10), we tabulate the following.

chair. In 1934, he was appointed to Göttingen. From 1939 to 1945, Hasse was on war leave from Göttingen. In September of 1945, he was dismissed from his position by the British occupation forces. He then moved to Berlin in 1946, where he took a research post at the Berlin Academy. In May of 1949, Hasse was appointed Professor at Humboldt University in East Berlin. In 1950, he was appointed to Hamburg where he remained until his retirement in 1966. He died on December 26, 1979 in Ahrensburg, near Hamburg, Germany.

s	m	$\overline{m}$	$\overline{sP}$
1	$--$	$--$	$(0,2)$
2	$1/4$	22160	$(5540,87252)$
2^2	$92074801/174504$	54712	$(52834,1747)$
2^3	$8374294669/3494$	20128	$(39125,88437)$
2^4	$2296148438/88437$	59461	$(77117,11444)$
2^5	$4460273767/5722$	29627	$(78795,929)$
2^6	$9312978038/929$	73122	$(53453,58500)$
2^7	$2142917407/29250$	63677	$(39646,8237)$
2^8	$4715415949/16474$	87396	$(47533,45014)$
$2^7 \cdot 3$	$12259/2629$	59176	$(28102,67373)$
$2^8 \cdot 3$	$2369167213/134746$	8617	$(5642,60149)$
$2^7 \cdot 3^2$	$1806/5615$	67186	$(72416,36233)$
$2^8 \cdot 3^2$	$15732231169/72466$	545	$(63554,7051)$
$2^7 \cdot 3^3$	$14591/4431$	49954	$(78296,72532)$
$2^8 \cdot 3^3$	$18390790849/145064$	48013	$(34382,3096)$
$2^7 \cdot 3^4$	$34718/21957$	6933	$(112,38294)$
$2^8 \cdot 3^4$	$37633/76588$	87562	$(7398,8496)$
$2^7 \cdot 3^5$	$-14899/3643$	2940	$(38107,30185)$
$2^8 \cdot 3^5$	$2178215174/30185$	75316	$(59476,49173)$
$2^7 \cdot 3^6$	$18988/21369$	23628	$(25018,63875)$
$2^7 \cdot 3^5 \cdot 5$	$-7351/17229$	47604	$(287,16090)$
$2^8 \cdot 3^5 \cdot 5$	$61777/8045$	27905	$(83475,78180)$
$2^7 \cdot 3^6 \cdot 5$	$4435/5942$	46170	$(83105,74671)$
$2^7 \cdot 3^5 \cdot 5^2$	$3509/370$	33309	$(5816,2153)$
$2^8 \cdot 3^5 \cdot 5^2$	$101477569/4306$	77561	$(34076,77818)$
$2^7 \cdot 3^6 \cdot 5^2$	$15133/5652$	85709	$(35664,54433)$
$2^7 \cdot 3^5 \cdot 5^3$	$-23385/1588$	376	$(71636,70501)$
$2^8 \cdot 3^5 \cdot 5^3$	$15395149489/141002$	77759	$(75341,86432)$
$2^7 \cdot 3^6 \cdot 5^3$	$15931/3705$	66657	$(67436,55576)$
$2^7 \cdot 3^5 \cdot 5^4$	$30856/7905$	24011	$(54566,59079)$
$2^7 \cdot 3^5 \cdot 5^3 \cdot 7$	$27353/20775$	69432	$(40702,15268)$
$2^8 \cdot 3^5 \cdot 5^3 \cdot 7$	$4969958413/30536$	63799	$(16756,22421)$
$2^7 \cdot 3^6 \cdot 5^3 \cdot 7$	$-7153/23946$	8221	$(73104,56724)$
$2^7 \cdot 3^5 \cdot 5^4 \cdot 7$	$34303/56348$	61531	$(25133,56816)$
$2^7 \cdot 3^5 \cdot 5^3 \cdot 7^2$	$34395/8377$	75734	$(33094,35927)$
$2^8 \cdot 3^5 \cdot 5^3 \cdot 7^2$	$3285638509/71854$	45301	$(32924,42289)$

Thus, we have exhausted all the divisors of M without getting a factor of n. Notice that in the line for $s = 2^7 \cdot 3 = 2^8 + 2^7$, this is the first line in which we are adding two *distinct* points. We did this since we had exhausted all of the powers of 2 by doubling, namely we reached $a_2 = 8$ for $p = 2$. Thus, this line is a transition step to powers of 3. Also, the line with $2^7 \cdot 3^2 = 2^7 \cdot 3 + 2^8 \cdot 3$ is a transition step to get to $2^8 \cdot 3^2$ by doubling. Similar comments apply for transitions to the other primes 5 and 7. For instance, $2^7 \cdot 3^6 = 2^8 \cdot 3^5 + 2^7 \cdot 3^5$ is a transition to the prime 5 since $2^8 \cdot 3^5 + 2^7 \cdot 3^6 = 2^7 \cdot 3^5 \cdot 5$, which we then double to get $2^8 \cdot 3^5 \cdot 5$. For the prime 7, there are more transition steps for each addition to the exponent. For instance, we first calculate $2^7 \cdot 3^6 \cdot 5^3$, then

$2^7 \cdot 3^5 \cdot 5^4$, which we then add to $2^8 \cdot 3^5 \cdot 5^3$ to get $2^7 \cdot 3^5 \cdot 5^3 \cdot 7$. This is then doubled to get $2^8 \cdot 3^5 \cdot 5^3 \cdot 7$. Sometimes these transitional sums are called *partial sums*.

Since we did not achieve a factorization for $a = 1$, we abandon the elliptic curve given by $y^2 = x^3 + x + 4$ and seek a new one in the family. However, by Exercises 6.15–6.16, the curves with $a = 2, 3$ also lead to an exhaustion of the factors of M without achieving a factorization. Thus, we select $a = 4$ and perform the following calculations for the family

$$(E, P) : (y^2 = x^3 + 4x + 4, (0, 2)).$$

s	m	$\overline{m}$	$\overline{sP}$
1	$--$	$--$	$(0, 2)$
2	1	1	$(1, 88636)$
2^2	$7/177272$	14772	$(71403, 53759)$
2^3	$15295165231/107518$	14807	$(77474, 21529)$
2^4	$9003331016/21529$	8745	$(1898, 86846)$
2^5	$2701804/43423$	66098	$(14137, 36524)$
2^6	$599564311/73048$	77120	$(55143, 42998)$
2^7	$9122251351/85996$	47836	$(45464, 149)$
2^8	$3100462946/149$	78848	$(42633, 25537)$
$2^7 \cdot 3$	$-25388/2831$	51778	$(75271, 35273)$
$2^8 \cdot 3$	$16997170327/70546$	14892	$(23622, 1032)$
$2^7 \cdot 3^2$	$34241/51649$	$--$	$--$

We get a solution given that the process terminates since

$$\gcd(88639, 51649) = 137.$$

The group addition law broke down at the last step since the denominator of m was not relatively prime to n, thereby giving us the factor 137. Since $n/137$ is prime, then we have a complete factorization:

$$n = 88639 = 137 \cdot 647.$$

The above illustrated Elliptic Curve Algorithm is often denoted by *ECM* in the literature for *Elliptic Curve Method*. As noted on page 205, the Quadratic Sieve is more efficient than ECM, in practice, at factoring natural numbers that are RSA moduli. This is true despite the fact that the expected running times (see Footnote 2.30 on page 105) of the ECM and the Quadratic Sieve are essentially the same with such integers as input. The reason is that the ECM operations are substantially more complicated than those used in the Quadratic Sieve. Since the cryptographically most interesting (that is, the hardest to factor) integers are the RSA moduli, then it is not surprising that ECM has proven to be the most successful at factoring one hundred digit *non*-RSA moduli. ECM factors random integers relatively fast, but RSA moduli slow, since a *random* such integer is rare. Thus, one reason why the ECM should be used in advance of the Quadratic Sieve or the Number Field Sieve is that ECM finds "small" prime factors faster than the two aforementioned sieves. In other words (since the running time of ECM depends upon the size of the prime factors of *general* $n \in \mathbb{N}$), ECM tends to find such small factors first. Hence, ECM may be considered to be a special-purpose factoring algorithm, which is the current favorite for finding prime factors that are (no more than) forty digits in large composite numbers. (The present record for the largest prime factor found by ECM is forty-eight decimal digits.) Furthermore, ECM has shown marked success when used as a factorization subroutine in other algorithms (such as the Quadratic Sieve, or Number Field Sieve), since it requires relatively little storage space. Moreover, both of the aforementioned sieves require the factorization of numbers as a subroutine, at which point the ECM may be employed with some success.

The Pollard $p-1$ Factoring Method studied in Section One of Chapter Five depended upon the smoothness of $p - 1$ for some prime divisor of the input n. Thus, the algorithm fails if no such p divides n. The ECM is a generalization of the $p - 1$ Method in the sense that the latter operates over $(\mathbb{Z}/p\mathbb{Z})^*$ (of order $p - 1$), and the ECM operates on a random elliptic curve over $\mathbb{Z}/p\mathbb{Z}$. It can be shown that the order of such an elliptic curve is (roughly) uniformly distributed over the interval $[p + 1 - 2\sqrt{p}, p + 1 + 2\sqrt{p}]$ (see Hasse's Theorem 6.27 above and Theorem 6.30 below). Thus, if the order of the elliptic curve is chosen such that it is smooth relative to B, given in (6.28), then ECM will, with a high probability, find a nontrivial factor of n. By restricting the interval somewhat, we can be more explicit about this probability. It can be shown that for a given prime p, there exists a constant $c \in \mathbb{R}^+$ such that approximately cp^2 of all pairs $(a, b) \in \mathbb{Z}/p\mathbb{Z} \times \mathbb{Z}/p\mathbb{Z}$ with $4a^3 + 27b^2 \neq 0$ yield an elliptic curve given by $y^2 = x^3 + ax + b$ whose order is in the interval $(p - \sqrt{p}, p + \sqrt{p})$. Moreover, the group orders in that interval are roughly uniformly distributed. (See [54].) What this tells us is that for each elliptic curve chosen, the probability that it has order in the interval $(p - \sqrt{p}, p + \sqrt{p})$ is approximately c.

Given the above, the ECM can be modified to serve as a True Primality Test (see Section Two of Chapter Four) based upon the following result.

Theorem 6.30 (Elliptic Curve Primality Test)
   Let $n \in \mathbb{N}$ with $\gcd(n, 6) = 1$, and let E be an elliptic curve over $\mathbb{Q}$. Suppose that

(a) $n + 1 - 2\sqrt{n} \leq |\overline{E}| \leq n + 1 + 2\sqrt{n}$.

(b) $|\overline{E}| = 2p$, *where $p > 2$ is prime,*

where $\overline{E}$ is reduced modulo n. If $P \neq \mathfrak{o}$ is a point on E and $\overline{pP} = \mathfrak{o}$ on $\overline{E}$, then n is prime.

   Proof. See [54, Lemma 14.23, p. 324].   □

Theorem 6.30 is employed by picking random points P_j for $j = 1, 2, \ldots, m \in$ $\mathbb{N}$ on an elliptic curve E and, for a given prime p, calculating $\overline{pP_j}$ for each such j. If the outcome is that $\overline{pP_j} = \mathfrak{o}$ for some $j = 1, 2, \ldots, m$, then n is prime. For instance, a suitable choice for P_1 is $2Q_1$, where Q_1 is randomly chosen. If $P_1 \neq \mathfrak{o}$, but $pP_1 = \mathfrak{o}$, then n is prime. If $P_1 \neq \mathfrak{o} \neq pP_1$, then n is composite.

Example 6.31 We are given $n = 9343$ and we want to test it for primality using the Elliptic Curve Primality Test. We choose the elliptic curve E given by $y^2 = x^3 + 4x + 4$ with which we had success in Example 6.29 and the point $P = (0, 2)$. We first check that

$$\gcd(n, 6) = 1 = \gcd(\Delta(E), n) = \gcd(-1108, 9343).$$

Assuming that n is prime, we use Exercise 6.4 to verify that

$$|\overline{E} \,(\mathrm{mod}\ n)| = 2 \cdot 4721 = 2p$$

and we calculate that

$$\lfloor n + 1 - 2\sqrt{n} \rfloor = 9150 < 9343 < 9537 = \lfloor n + 1 + 2\sqrt{n} \rfloor.$$

Thus, both (a) and (b) of Theorem 6.30 are satisfied. Given that

$$p = 4721 = 2^{12} + 2^9 + 2^6 + 2^5 + 2^4 + 1,$$

we calculate up to 2^{12}, by doubling, then add the required distinct points until we get $\overline{pP} = 4721(0, 2)$. The following table contains these calculations.

s	m	$\overline{m}$	sP
1	$--$	$--$	$(0, 2)$
2	1	1	$(1, 9340)$
2^2	$7/18680$	1556	$(1297, 1515)$
2^3	$5046631/3030$	2199	$(2676, 2539)$
2^4	$10741466/2539$	3318	$(7061, 4425)$
2^5	$149573167/8850$	8488	$(6835, 7891)$
2^6	$140151679/15782$	1468	$(1807, 1586)$
2^7	$9795751/3172$	6231	$(1582, 8282)$
2^8	$1877044/4141$	6791	$(6812, 6217)$
2^9	$69605018/6217$	2164	$(7115, 1444)$
2^{10}	$7993141/152$	7654	$(7562, 6099)$
2^{11}	$85775768/6099$	6887	$(9263, 4579)$
2^{12}	$13547869/482$	5603	$(1289, 4860)$
$2^{12} + 2^9 = 4608$	$-1708/2913$	8114	$(7157, 3459)$
$2^6 + 4608 = 4672$	$1873/5350$	767	$(59, 3081)$
$2^5 + 4672 = 4704$	$2405/3388$	5585	$(7740, 1590)$
$2^4 + 4704 = 4720$	$-405/97$	5775	$(0, 9341)$
$1 + 4720 = 4721$	$--$	$--$	$\mathfrak{o}$

We get that
$$\overline{pP} = \overline{4721(0,2)} = \mathfrak{o},$$
since
$$(x_1, y_1) + (x_2, y_2) = (0, 9341) + (0, 2) = \mathfrak{o},$$
by the addition formula in Definition 6.7 given that

$$x_1 = x_2 = 0 \text{ and } y_1 = 9341 \equiv -2 = -y_2 \pmod{9343}.$$

In other words, they lie on the same vertical line, so they sum to the point at infinity. Hence, by Theorem 6.30, $n = 9343$ is prime.

If part (a) of Theorem 6.30 *fails* to hold, then n is composite by Theorem 6.27. Hence, failure for (a) to hold is a test for compositeness. On the other hand, (b) in Theorem 6.30 is very special and does not hold for many primes. The reader may test this in practice by using Exercise 6.4 on several prime values for n. The value of n in Example 6.31 is quite small, accounting for the relative ease of calculation. However, as n gets large, calculating the corresponding large value of $|\overline{E} \pmod{n}|$ may turn out to be as difficult as proving that n is prime. (Note that n and $|\overline{E} \pmod{n}|$ have the same order of magnitude by part (a) of Theorem 6.30.) These problems were overcome in a primality test developed by Goldwasser and Kilian in 1986.

Theorem 6.32 (Goldwasser-Kilian)

Suppose that $n \in \mathbb{N}$ with $\gcd(n, 6) = 1$ and $n > 1$ and let E be an elliptic curve over $\mathbb{Z}/n\mathbb{Z}$. If there exists an integer m with a prime divisor

$$q > (n^{1/4} + 1)^2$$

and if there exists a point P on $E(\mathbb{Z}/n\mathbb{Z})$ such that both

$$\overline{mP} = \mathfrak{o} \text{ and } \overline{(m/q)P} \neq \mathfrak{o}, \tag{6.33}$$

then n is prime.

Proof. See [89]. □

In Theorem 6.32, if the multiplications in (6.33) both fail to hold, then a nontrivial factor of n has been found, so n is not prime. This is what occurs in the ECM as well.

Now suppose that we are given an elliptic curve E of group order m over $\mathbb{Z}/p\mathbb{Z}$ where p is prime, and assume further that we can indeed find a prime divisor q of m with $q > (p^{1/4} + 1)^2$. Then it can be shown that there exists a point P on E such that $(m/q)P \neq \mathfrak{o}$. Hence, the Goldwasser-Kilian Test says that all we have to do is to keep generating random elliptic curves and calculate their group orders until one is found that will prove n to be prime. Finding a suitable point on the elliptic curve yields the result.

In 1995, Schoof [176] produced a polynomial time algorithm for counting the points on an elliptic curve, namely for the exact determination of the number of rational points on elliptic curves over finite fields. Thus, Schoof's Algorithm can be used in the Goldwasser-Kilian Test to compute the order of $E(\mathbb{Z}/n\mathbb{Z})$. What this means is that we have a probabilistic polynomial time primality test. Moreover, the algorithm is a certificate (see Definition 4.56) of primality for n, and this can even be checked in shorter time than it took to generate it! Nevertheless, the algorithm is largely impractical (see Footnote 6.14).

With all of the above under our belts, it is time to turn our attention to cryptosystems using elliptic curves. In 1985, Miller (see [139]) and Koblitz (see [105]) independently proposed using elliptic curves for public-key cryptosystems. However, they did not invent a cryptographic algorithm for use with elliptic curves, but rather implemented then-existing public-key algorithms in elliptic curves over finite fields. These types of cryptosystems are more appealing than cryptosystems over finite fields since, rather than just the group of a finite field $\mathbb{F}_p^*$, one has many *elliptic curves over* $\mathbb{F}_p$ from which to choose. Also, whenever the elliptic curve is properly chosen, there is no known subexponential time algorithm (see page 49) for cryptanalyzing such cryptosystems. The security of Elliptic Curve Cryptosystems depends upon the intractability of the following problem.

Definition 6.34 (Elliptic Curve Discrete Log Problem)

If E is an elliptic curve over a field F, then the Elliptic Curve Discrete Log Problem *to base $Q \in E(F)$ is the problem of finding an $x \in \mathbb{Z}$ (if one exists) such that $P = xQ$ for a given $P \in E(F)$.*[6.13]

Currently, the Discrete Log Problem in elliptic curve groups is several orders of magnitude more difficult than the Discrete Log Problem in the multiplicative group of a finite field (of similar size). What this means explicitly is that for a suitably chosen elliptic curve E over $\mathbb{F}_q$, the discrete log problem for

---

[6.13]In 1991, Menezes, Okamoto, and Vanstone found a new means of attacking the Discrete Log Problem for elliptic curves (appearing two years later in [133]). Their method, currently called the *MOV attack* in the literature, involves the use of what is called a *Weil Pairing* (see [187, Section 3.8, pp. 95–99]), which embeds an elliptic curve over a finite field into the multiplicative group of some finite extension field of the given finite field. Hence, their method reduces the problem to the discrete log problem in that extension field, called an *MOV reduction*. To be of any use, the degree of the extension field must be small, and essentially the only elliptic curves for which this degree is small are of a special type called *supersingular* (see [187, p. 137]). What they showed was that if we have a supersingular curve, then the discrete log problem in an elliptic curve group can be reduced in expected polynomial time to the discrete log problem in the extension field of degree no more than 6 over the finite field. However, the vast majority of the elliptic curves are *not* supersingular, called *nonsupersingular* or *ordinary*. For the nonsupersingular curves, the MOV reduction virtually never leads to a subexponential time algorithm. What this suggests is that one of the basic open questions in Elliptic Curve Cryptography is whether or not we can find a subexponential time algorithm for the discrete log problem on some set of nonsupersingular elliptic curves — a difficult question at the present time. The MOV attack was generalized by Frey and Rück [79] in 1994 (see also [31]).

the group of $E(\mathbb{F}_q)$ appears to be (given our current state of knowledge) of complexity exponential in the size $\lceil \log_2 q \rceil$ of the field elements, whereas there exist subexponential algorithms in $\lceil \log_2 q \rceil$ for the discrete log problem in $\mathbb{F}_q^*$.

Remark 6.35 *There is also the following valuable result on the group structure of $E(\mathbb{F}_p)$. If $p > 3$ is prime, then there are $m, n \in \mathbb{N}$ such that $E(\mathbb{F}_p)$ is isomorphic to the product of a cyclic group of order m with one of order n, where $m|\gcd(n, p-1)$. See [189].*

Several of the public-key algorithms that we studied in Chapter Three such as Diffie-Hellman and ElGamal can be implemented in elliptic curves over finite fields. We now show how to implement ElGamal in the group E of an elliptic curve. By Remark 6.35, there exist $m, n \in \mathbb{N}$ satisfying the stated property. If we can find them, then we can find a subgroup of $E(\mathbb{F}_p)$ isomorphic to the cyclic group $\mathbb{Z}/n\mathbb{Z}$. In particular if $m = 1$, then $E(\mathbb{F}_p)$ is cyclic. The following is a standard implementation of ElGamal on elliptic curves, where we assume that the discrete log problem on the cyclic subgroup of E is intractable. The following was presented in [69].

◆ **ElGamal Public-Key Elliptic Curve Cryptosystem**

We assume that E is an elliptic curve over $\mathbb{F}_p$ where p is prime and H is a cyclic subgroup of $E(\mathbb{F}_p)$ generated by a point $P \in E(\mathbb{F}_p)$. Entity A wishes to send a message to entity B whose public key is (E, P, aP) and whose private key is the natural number $a < p - 1$. Entity A performs the following.
enciphering stage:

(1) Choose a random natural number $b < p - 1$.

(2) Consider the plaintext message units embedded as points m on E.

(3) Compute $\beta = bP$ and $\gamma = m + b(aP)$.

(4) Send the ciphertext $\mathfrak{C}_e(m) = c = (\beta, \gamma)$ to B.

deciphering stage:

Once B receives the ciphertext, the plaintext m is recovered via the private key as:
$$\mathfrak{D}_d(c) = m = \gamma - a\beta.$$

Example 6.36 Consider the elliptic curve group E given by $y^2 = x^3 + 4x + 4$ over $\mathbb{F}_{13}$. Then by Exercise 6.4 on page 249, $|E(\mathbb{F}_p)| = 15$, which is necessarily cyclic (see Footnote A.12 on page 292). Also, $P = (1, 3)$ is a generator of E. Assuming that B's public key is $(E, P, 4P)$ where $a = 4$ is the private key and $m = (10, 2)$ is the message that A wants to send to B, then A performs the following. A chooses $b = 7$ at random. Then A calculates:

$$\mathfrak{C}_e(m) = \mathfrak{C}_e((10, 2)) = (bP, m + b(aP)) = (7P, (10, 2) + 7(4P)) =$$

$$((0, 2), (10, 2) + 7(6, 6)) = ((0, 2), (10, 2) + (12, 5)) = ((0, 2), (3, 2)) = (\beta, \gamma) = c.$$

Then A sends c to B and B uses the private key to recover m via:

$$\mathfrak{D}_d(c) = (3, 2) - 4(0, 2) = (3, 2) - (12, 5) = (3, 2) + (12, 8) = (10, 2) = m.$$

(See Exercises 6.21–6.22.)

Although Example 6.36 provides an easy illustration (albeit, as usual, a very unrealistic one in terms of the size of the parameters), there are certain technical difficulties in the implementation of ElGamal Cryptosystems on elliptic curves. As we saw in Section Two of Chapter Three, the standard ElGamal Public-key Cipher over $\mathbb{F}_p$ has a message expansion factor of two. However, an elliptic curve implementation of ElGamal has a message expansion factor of approximately four, because there are p plaintexts, but each ciphertext is comprised of four field elements. Moreover, and perhaps more seriously, plaintext message units m lie on E and there does not exist an appropriate (both theoretically and practically) method of deterministically generating such points. One could use the Schoof Algorithm mentioned above, but it turns out to be substantially unwieldy in practice, despite its being of deterministic polynomial time.[6.14] A more efficient version was discovered by Menezes and Vanstone[6.15] [135], which we describe below. This version allows the plaintext and ciphertext pairs to be in $\mathbb{F}_p^* \times \mathbb{F}_p^*$ rather than on the elliptic curve so that we again have a message expansion factor of two as with the original ElGamal system.

◆ Menezes-Vanstone Elliptic Curve Cryptosystem

In what follows, the discrete log problem in the cyclic subgroup of the elliptic curve group is assumed to be intractable.

Let E be an elliptic curve over $\mathbb{F}_p$ where $p > 3$ is prime and let H be a subgroup of $E(\mathbb{F}_p)$ generated by a point $P \in E(\mathbb{F}_p)$. Assume that randomly chosen $k \in \mathbb{Z}_{|H|}$ and $a \in \mathbb{N}$ are secret. If entity A wants to send message

$$m = (m_1, m_2) \in (\mathbb{Z}/p\mathbb{Z})^* \times (\mathbb{Z}/p\mathbb{Z})^*,$$

then A does the following.

enciphering stage:

(1) $\beta = aP$, where P and β are public.

[6.14]In characteristic two, Schoof's Algorithm is not too bad, assuming you are using the Elkies-Atkin improvements (see [177] and [70]), but characteristic p is unworkable. For example, for decent field size, it's about a day's computation on a Sparc 5 to do one curve, and so if one is waiting for prime order curve of reasonable size, it will take approximately a year.

[6.15]Scott Vanstone is one of the founders of *Certicom*, established in 1985 with office locations in California and in Mississauga, Canada. Certicom is the first company to develop Elliptic Curve Cryptography commercially, and has patented implementations that are perhaps the fastest in the industry. Both Menezes and Vanstone are also Professors of Mathematics and Computer Science at the University of Waterloo, Canada.

(2) $(y_1, y_2) = k\beta$.

(3) $c_0 = kP$.

(4) $c_j \equiv y_j m_j \pmod{p}$ for $j = 1, 2$.

Then A sends the following enciphered message to B,

$$\mathfrak{E}_k(m) = (c_0, c_1, c_2) = c,$$

and upon receipt, B calculates the following to recover m.

deciphering stage:

(1) $ac_0 = (y_1, y_2)$.

(2) $\mathfrak{D}_k((c_1, c_2)) = (c_1 y_1^{-1} \pmod{p}, c_2 y_2^{-1} \pmod{p}) = m$.

Example 6.37 Let E be the elliptic curve given by $y^2 = x^3 + 4x + 4$ over $\mathbb{F}_{13}$, and let $P = (1, 3)$. As we saw in Example 6.36, we may choose $E(\mathbb{F}_{13}) = H$ which is cyclic of order 15, generated by P. If the private keys are $k = 5$ and $a = 2$, then given a message

$$m = (12, 7) = (m_1, m_2),$$

entity A computes

$$\beta = aP = 2(1, 3) = (12, 8),$$

$$(y_1, y_2) = k\beta = 5(12, 8) = (10, 11),$$

$$c_0 = kP = 5(1, 3) = (10, 2),$$

$c_1 \equiv y_1 m_1 = 10 \cdot 12 \equiv 3 \pmod{13}$, and $c_2 \equiv y_2 m_2 = 11 \cdot 7 \equiv 12 \pmod{13}$.

Then A sends

$$\mathfrak{E}_k(m) = \mathfrak{E}_5(12, 7) = (c_0, c_1, c_2) = ((10, 2), 3, 12) = c$$

to B. Upon receipt, B computes

$$ac_0 = 2(10, 2) = (10, 11) = (y_1, y_2).$$

and

$$\mathfrak{D}_k((c_1, c_2)) = \mathfrak{D}_5(3, 12) = (3 \cdot 10^{-1} \pmod{13}, 12 \cdot 11^{-1} \pmod{13}) = (12, 7) = m.$$

(See Exercises 6.23–6.24.)

We abbreviate the term *Elliptic Curve Cryptosystem* in what follows to *ECC* for the sake of convenience. Some other cryptographic uses for elliptic curves, which are beyond the scope of this book, are nevertheless worthy of note. A special class of algebraic curves called *hyperelliptic curves* may be viewed as generalizations of elliptic curves. They were first proposed for cryptographic use by Koblitz [106] in 1989. Essentially, the cryptography is the same as for elliptic curves. The difference is that the points on elliptic curves are replaced by points on the *Jacobian* of a hyperelliptic curve (elliptic curves are Jacobians of dimension one — the simplest case of a Jacobian, see [31]).[6.16] There has been cryptographic interest in these curves, as with ordinary elliptic curves, since there exist lots of Jacobian groups of hyperelliptic curves. Moreover, arithmetic in the Jacobian of a hyperelliptic curve is straightforward since it is closely related to composition of binary quadratic forms (see [31, pp. 171–180]).

In 1992, the four authors of [108] proposed an analogue of RSA by using a special class of elliptic curves over the ring $\mathbb{Z}/n\mathbb{Z}$ where $n \in \mathbb{N}$ is composite. In 1993, Demytko [58] presented another analogue of RSA, in this case, where there is minimal restriction on the types of elliptic curves to be used. In 1997, Vanstone and Zuccherato [198] developed a new cryptosystem based on elliptic curves over $\mathbb{Z}/n\mathbb{Z}$ in which the message is contained in the exponent instead of the group element. The security of all of the above RSA analogues is based upon the presumed difficulty of factoring. In 1997, Anshel and Goldfeld [6] presented an explicit construction of a pseudorandom number generator arising from an elliptic curve, which can be effectively computed at low computational cost. They introduced a new intractable problem, distinct from integer factorization or the discrete log problem, that leads to a new class of one-way functions based on the arithmetic theory of zeta functions, and against which there is currently no known attack.

Let us summarize and conclude our comments on elliptic curves to bring this section to a close. The lack of subexponential attacks on ECCs gives us the advantage of allowing smaller message units, as well as smaller amounts of processing time and electrical power. Moreover, the consensus is that a suitably chosen elliptic curve in a given ECC over a finite field of size approximately 160 bits ensures security equivalent to RSA with modulus of 1024 bits (or for any cryptosystem relying on intractability of discrete log over the base field for their security). However, there is a certain scepticism in the community about ECCs for two reasons. First, the amount of deep mathematics, as we have seen

---

[6.16]The group of rational divisors of degree zero modulo principal divisors forms the Jacobian of a hyperelliptic curve over a finite field. This is a finite abelian group, which forms the starting point for cryptosystems that are based upon hyperelliptic curves (see [31]). Also, hyperelliptic curves were a fundamental component in the random polynomial time algorithm for primality proving given in [3] as well as in integer factorization algorithms such as that presented in [124]. Moreover, hyperelliptic curves correspond to degree two function field extensions, a fact that was exploited by Flassenberg and Paulus in [78] who proposed a working method which could easily be extended to solving the discrete log problem for Jacobians of hyperelliptic curves of small genus. Furthermore, the authors of [2] have shown that on higher genus hyperelliptic curves (see [31]), there exist subexponential discrete log algorithms. See also *www.cacr.math.uwaterloo.ca/techreports/1999/corr99-04.ps*

in this section, necessary to implement and work with ECCs ensures that there is a small set of people qualified to deal with them. Secondly, ECCs have had a short life thus far, so many believe that there has been insufficient time for ECCs to undergo the careful scrutiny thought to be required. Also, the MOV attack cited above, showing that the discrete log problem can be solved on supersingular curves, is worrying to some, since other classes of elliptic curves have similarly fallen victim. For example, there is also a class of elliptic curves called *anomalous* (see [31]), which have recently been shown to have highly efficient discrete log algorithms. If elliptic curves are to be used, then it is considered prudent to use key sizes of (an absolute minimum of) 300 bits for even the most modest security requirements, and 500 bits for more sensitive communication. Generous safety margins such as these (and say 2048 bits for decade-long security of RSA , for instance) is recommended since the real threat to the presumed intractability of both discrete logs and factoring is unexpected new mathematical discoveries.

We are already aware, from Section One of Chapter Five, that there is a history of seriously overestimating the difficulty of factoring.[6.17] We cannot be certain what algorithm will be discovered in the future (such as the powerful number field sieve studied in Chapter Five). This is highlighted by the fact that relatively few people have worked on the problem of discrete logs and integer factorization, which implies that the area has just not been examined as exhaustively as frequently alleged. Lastly, there is the ever-present threat of quantum computers (see Section Three of this chapter), even though they are currently only a "theoretical possibility."

Exercises

6.1. Prove that $Y^2 = X^3 - 4X$ has nonzero solutions $X, Y \in \mathbb{Q}$ if and only if $x^4 + y^4 = z^2$ has nonzero solutions $x, y, z \in \mathbb{Z}$.

☆ 6.2. A *Pythagorean triple* is a solution $x, y, z \in \mathbb{N}$ to $x^2 + y^2 = z^2$. A Pythagorean triple is called *primitive* if $\gcd(x, y, z) = 1$. Prove that x, y, z is a primitive Pythagorean triple with y even if and only if there exist relatively prime integers u and v of different parity such that $u > v$ and $x = u^2 - v^2$, $y = 2uv$, and $z = u^2 + v^2$. (This is called the *Pythagorean Triples Theorem*.)

[6.17]Some further evidence is given by claims made over the years. For instance, in 1874 Jevons [98], an English economist and logician, stated that nobody would ever be able to factor 8616460799. However, Bancroft Brown did just that in 1925, with the factorization $8616460799 = 96079 \cdot 89681$. In 1967, Brillhart and Selfridge [41] stated that one could only expect frustration whenever attempting to factor "...a number of 25 or more digits, even with the speeds available in modern computers." However, in 1970, Brillhart and Morrison factored the thirty-nine digit seventh Fermat Number (see page 200). In 1976, Richard Guy [95] stated that he would "...be surprised if anyone regularly factors numbers of size 10^{80} without special form during the present century." Yet, in 1994, RSA-129 was factored (see page 205). Also, as mentioned on page 212, the RSA-140 challenge number was factored into a product of two seventy digit integers on February 2, 1999, then RSA-155 was factored on August 26, 1999, both using the General Number Field Sieve.

☆ 6.3. Prove that there are no solutions $x, y, z \in \mathbb{N}$ to $x^4 + y^4 = z^2$. (*Hint: Use Exercise 6.2.*)

6.4. Suppose that an equation defining an elliptic curve $E = E(\mathbb{F}_p)$ over $\mathbb{F}_p$, p a prime is given by $y^2 = x^3 + ax + b$, where $a, b \in \mathbb{Z}$. Prove that the number of elements on E, counting the point at infinity is given by

$$p + 1 + \sum_{x \in \mathbb{F}_p} \left(\frac{x^3 + ax + b}{p} \right),$$

where the symbol in the sum is the Legendre Symbol.

6.5. The points $P = (-1, 2)$ and $Q = (1, 0)$ lie on the curve

$$y^2 = x^3 - 3x + 2.$$

Show that although $P \neq \pm Q$ and $P \neq \mathfrak{o}$, the straight line through P and Q does not intersect the curve at a third distinct point. Given the discussion on page 225, explain why there fails to be a third distinct point R in this case.

6.6. Suppose that E is an elliptic curve over a field $F \subset \mathbb{C}$. Prove that the set of points on E of order two form a subgroup that is isomorphic to one of

$$\mathbb{Z}/2\mathbb{Z} \times \mathbb{Z}/2\mathbb{Z}, \ \mathbb{Z}/2\mathbb{Z}, \text{ or } \{\mathfrak{o}\}.$$

6.7. Prove that the set of torsion points $E(F)_t$ forms a subgroup of the F-rational points $E(F)$ on an elliptic curve E.

6.8. Prove that

$$x^3 + y^3 = z^3$$

has solutions $x, y, z \in \mathbb{Z}$ with $xyz \neq 0$ if and only if

$$Y^2 = X^3 - 432$$

has solutions $X, Y \in \mathbb{Q}$ with $|Y| \neq 36$. Conclude that by the solution of FLT, the elliptic curve given by

$$y^2 = x^3 - 432$$

has only $(\pm 36, 12)$ and $\mathfrak{o}$ as torsion points, so $|E(\mathbb{Q})_t| = 3$.

6.9. Prove that the elliptic curve E given by

$$y^2 = x^3 - 2$$

has no nontrivial torsion points.

6.10. Prove that the set consisting of $\mathfrak{o}$ and all of the rational points on an elliptic curve E form a subgroup of $E(F)$. (*Hint: See the discussion on page 228.*)

6.11. A given $n \in \mathbb{N}$ is called a *congruent number* or simply *congruent* if it is the area of a right-angled triangle. Prove that the following are equivalent.

(1) $n = ab/2$ is congruent, where (a, b, c) is a Pythagorean triple (see Exercise 6.2).

(2) There exists an integer x such that x, $x - n$ and $x + n$ are all perfect squares of rational numbers.[6.18]

6.12. Let E be an elliptic curve over $\mathbb{Q}$ given by

$$y^2 = (x - \alpha_1)(x - \alpha_2)(x - \alpha_3),$$

where $\alpha_j \in \mathbb{Q}$ for $j = 1, 2, 3$. Assume that for a given point $(x_2, y_2) \neq o$ on E, there exists a point (x_1, y_1) on E such that $2(x_1, y_1) = (x_2, y_2)$. Prove that $x_2 - \alpha_j$ are squares of rational numbers for $j = 1, 2, 3$.

6.13. Let E be an elliptic curve over $\mathbb{Q}$ defined by

$$y^2 = x^3 - n^2 x$$

for some squarefree $n \in \mathbb{N}$.[6.19] Prove that the conditions in Exercise 6.11 are equivalent to E having a rational point other than $(\pm n, 0)$, $(0, 0)$ and o. In other words, n is congruent if and only if E has a rational point other than $(\pm n, 0)$, $(0, 0)$ and o. (See Exercises 6.11–6.12.)[6.20]

6.14. Let $n \in \mathbb{N}$ be squarefree. Prove that the following are equivalent.

(1) n is a congruent number.

(2) The simultaneous (homogeneous Diophantine)[6.21] equations

$$x^2 + ny^2 = z^2 \text{ and } x^2 - ny^2 = t^2$$

have a solution in integers x, y, z, t with $y \neq 0$.

(3) If E is the elliptic curve over $\mathbb{Q}$ defined by

$$y^2 = x^3 - n^2 x,$$

then the Mordell-Weil rank $r_{\mathbb{Q}}(E)$ is positive.

[6.18]Fermat used his method of infinite descent (see the solution to Exercise 6.3 on page 326) to prove that $n = 1, 2, 3$ are not congruent.

[6.19]Notice that there is no loss in generality by assuming that n is squarefree since $m^2 n$ is a congruent number if and only if n is one.

[6.20]It can be shown (see [102, Theorem 5.2, p. 134]) that when E is given by $y^2 = x^3 + Ax$ with $A \in \mathbb{Z}$ assumed to be fourth-power free, then $E(\mathbb{Q})_t = \mathbb{Z}/2\mathbb{Z} \oplus \mathbb{Z}/2\mathbb{Z}$ if $-A$ is a perfect square, $E(\mathbb{Q})_t = \mathbb{Z}/4\mathbb{Z}$ when $A = 4$, and $E(\mathbb{Q})_t = \mathbb{Z}/2\mathbb{Z}$ otherwise. Thus, for the case given in Exercise 6.13, n is congruent if and only if E has a point of infinite order. This is verified in Exercise 6.14 using the Mordell-Weil rank.

[6.21]A polynomial of degree d is said to be *homogeneous* if each term has degree d. For example, $x^3 + xyz = z^3$ is a homogeneous polynomial of degree $d = 3$ and $x + y = z$ is one of degree $d = 1$. A *Diophantine equation* is an equation in which only integer solutions are considered.

6.15. In Example 6.29, verify that the curve for which $a = 2$ also leads to an exhaustion of the divisors of M without achieving a factorization.

6.16. Perform the task in Exercise 6.15 for $a = 3$.

6.17. Use Lenstra's Elliptic Curve Algorithm to factor $n = 899$ using the (E, P) pair

$$(y^2 = x^3 + ax + 4, (0, 2))$$

and testing for each $a \in \mathbb{N}$ until a factorization occurs.

6.18. Given $n = 3551$, use Lenstra's Elliptic Curve Algorithm to find a factor with the (E, P) pair

$$(y^2 = x^3 + ax + 1, (0, 1))$$

and test for $a \in \mathbb{N}$.

6.19. Use the Elliptic Curve Primality Test to verify that $n = 1231$ is prime by verifying the hypothesis of Theorem 6.30 and setting up a table as in Example 6.31. Use the elliptic curve given by

$$y^2 = x^3 + 9x + 1$$

and the point $P = (0, 1)$.

6.20. Use the Elliptic Curve Primality Proving Method to test $n = 1117$ for primality using the curve given by

$$y^2 = x^3 + 3x + 4$$

and the point $(0, 2)$. Use the Elliptic Curve Primality Test to verify that n is prime by setting up a table as requested in Exercise 6.19.

6.21. Assuming that the ciphertext $c = ((11, 1), (1, 3))$ was encrypted using the ElGamal Cryptosystem over the elliptic curve E over $\mathbb{F}_{13}$ given by

$$y^2 = x^3 + 4x + 4,$$

recover the plaintext with the knowledge that $a = 2$ is the private key.

6.22. With the same curve and assumptions as in Exercise 6.21 concerning E, decipher $c = ((11, 1), (11, 12))$ where $a = 3$ is the private key in this case.

6.23. Given the same curve E and point P as in Example 6.37, decipher $c = ((12, 8), 2, 8)$ assuming that it was enciphered using the Menezes-Vanstone Elliptic Curve Cryptosystem with $k = 2$ and $a = 5$.

6.24. With the same instructions as in Exercise 6.23, decipher $c = ((0, 2), 4, 6)$ assuming that $k = 7$ and $a = 6$.

6.2   Zero-Knowledge

Motivation for the topics in this section may be derived from the age-old cryptographic problem of supplying mutually mistrustful entities with a means of interchanging messages, where each entity desires to reveal as little as possible about its own secrets. For instance, suppose that entity A wants to convince entity B that it has knowledge of the factorization of an RSA modulus $n = pq$. It is certainly simple to do this, since A can merely reveal p and q (an instance of what is called a *maximum disclosure proof*). However, A has cryptographically sound reasons for not wanting to do this. So how does A proceed to convince B without revealing the factorization? In other words, is it possible for A to prove to B the assertion that it has knowledge of the factorization of n without yielding anything beyond the validity of the assertion? If such a proof exists, it is called a *Zero-knowledge Proof*.[6.22] In order to formally define this and related phenomena in mathematical terms, we must first become acquainted with the following notion.

Definition 6.38 (Interactive (Minimum Disclosure) Proof Systems)
 An interactive proof system *is a game*[6.23] *involving an exchange of messages between an entity A, called a* prover *who executes a computationally unbounded algorithm (strategy), and an entity B, called a* verifier, *executing a probabilistic polynomial time strategy.*[6.24] *Here A's objective is to convince (or prove to) B the truth of some assertion, such as claimed knowledge of some secret S.*[6.25] *The interactive proof system is called* minimum disclosure, *also called a* proof of knowledge, *if the following two properties are satisfied.*

(1) *If the prover does not know the proof, then the chances of convincing the verifier of knowledge of the proof are negligible. In other words, with probability close to 1, B accepts A's proof whenever S is true. This is called the* completeness *property.*

[6.22]Zero-knowledge Protocols such as the ones we describe below are designed to address some concerns over Password Protocols such as the situation where entity A gives password to entity B who can thereafter impersonate A. Although it might be argued that Zero-knowledge Protocols are not in significant use, this should not detract from their interest and no book on cryptography should be without them. Their status is changing, as noted in [61, p. 485]: "Since their introduction, Zero-knowledge Proofs have proven to be very useful as a building block in the construction of cryptographic protocols... Due to their importance, the efficiency of Zero-knowledge Protocols has received considerable attention."

[6.23]A *game* is defined to be a conflict, involving gains and losses, between two or more opponents who follow formal rules. *Game theory* is that branch of mathematics involving logical analysis of games.

[6.24]This means that there are no bounds placed upon the computational time required by A, but B is computationally bounded by probabilistic polynomial time.

[6.25]"Proof" in this context is not a fixed notion, but rather is a randomized dynamic procedure of interaction between A and B. In other words, this is a *probabilistic proof*, correct only with a certain bounded error probability. However, the bounded probability can be made to be arbitrarily close to 0. This can be accomplished by repeating the verification procedure sufficiently many times until the error probability is made exponentially small. Thus, interactive proofs are often called *Proofs by Protocol*.

(2) *A dishonest prover C (impersonating A), who has a non-negligible probability of successfully executing the protocol must be able to compute S in polynomial time. In other words, any entity C, capable of successfully impersonating A, must essentially "know" S (be able to compute it in polynomial time). This is called the* soundness property.

The completeness property is the standard requirement for a protocol to function properly given honest entities, whereas the soundness property prevents a dishonest prover from convincing an honest verifier. Thus, the soundness property ensures that the only aspect of the proof that the verifier gets is that the prover knows the proof. In particular, the verifier cannot convey the proof to anyone else without actually reconstructing the proof from scratch. However, it should be noted that this does not, in itself, guarantee that the protocol is cryptographically secure. In other words, it does not guarantee that the probability of the protocol being defeated is negligible. In order to be cryptographically secure, the underlying mathematical problem faced by an enemy must be computationally hard (see page 52).

What Definition 6.38 says is that protocols which are minimum disclosure proofs yield one no insights into the proof of an an assertion simply from being convinced that the assertion is true. Whatever one learns from a minimum disclosure proof, one can learn from the assertion itself. Interactive proofs used for *identification* (meaning the corroboration of the identity of an entity) may be reformulated as *proofs of knowledge*. For instance, if entity A, who knows a secret S, wants to establish its identity to entity B, then A convinces B, via a Proof of Knowledge Protocol, of knowledge of S. If one is given that only A knows S, then A has been *identified* to B.

An example of an Interactive (minimum disclosure) Protocol is for A to use a one-way function to prove to B that it has a specific morsel of information, but the *proof* does not yield any way of determining what that morsel happens to be. If we require more of B, namely not only that B learns nothing from A's proof, but rather *learns nothing whatsoever*, then this is an instance of the following, first formalized in [90]–[91][6.26] and in [85]–[87].

Definition 6.39 (Zero-knowledge Proofs of Knowledge)

A Zero-knowledge Proof of Knowledge *is a proof of knowledge that satisfies the additional property:*

(3) *The verifier learns nothing from the prover that could not have been learned without the prover's participation — called the* Zero-knowledge Property.[6.27]

[6.26]For the reader with knowledge of the theory of languages (in the complexity theory sense), Goldwasser, Micali, and Rackoff formalized Zero-knowledge Proofs in the context of an interactive *proof of membership* of a string x in a language $\mathcal{L}$. They did this by showing that the languages of quadratic residues and quadratic nonresidues each have Zero-knowledge Interactive Proof Systems revealing only that $x \in \mathcal{L}$, called *revealing only a single bit of knowledge.*

[6.27]The reader is cautioned that the literature is far from consistent in the definition of "Zero-knowledge Proofs of Knowledge." For instance, see [134, pp. 421–424] and [14]. The definition

A protocol that is a Zero-knowledge Proof is called an Zero-knowledge Interactive Protocol. An example of such a protocol is the following: A convinces B of the settling of some theorem without revealing a single bit of information, not even if the theorem has been proved or a counterexample has been found.

Definition 6.39 tells us that a Zero-knowledge Interactive Protocol allows a proof of the truth of a statement (such as knowledge of a secret), without leaking any information about the statement aside from its validity. A cryptographically similar notion is an answer provided by an *oracle*, which may be an algorithm, or a subroutine in a program, or a trusted entity that can give an answer to any question that a presupposed algorithm is able to settle, without revealing any details of that algorithm.[6.28]

Another way of looking at property (3) in Definition 6.39 is that B is able to simulate the protocol as if A were actually participating although A, in fact, is not. Often in the literature, a protocol having the Zero-knowledge Property is defined as a protocol that is simulatable in the sense that there exists an expected polynomial time algorithm, called a *simulator* which takes the assertion to be proven as input, and produces, without interacting with the real prover, output which is indistinguishable from that which would occur from interacting with the real prover. The output, comprising a collection of messages, resulting from the execution of a protocol is often called a *transcript*. This brings us to another variant of Zero-knowledge Protocols.

Definition 6.40 (Computational and Perfect Zero-Knowledge)

A protocol is Computationally *Zero-knowledge if an observer (who witnesses an Interactive Zero-knowledge Proof), restricted to probabilistic polynomial-time tests, cannot distinguish simulated from real transcripts. A protocol is* Perfect[6.29] *Zero-knowledge if the probability distributions of the aforementioned transcripts are identical.*

Roughly speaking, Computational Zero-knowledge means that there does not exist an efficient method to distinguish the transcripts mentioned in Definition 6.40. With Perfect Zero-knowledge, transcripts are identically distributed rather than computationally indistinguishable.

---

given here more closely reflects the formal notions elucidated in [74]. Furthermore, by a "gain of knowledge" we will mean the following. Entity B will be said to have *gained knowledge* from entity A about a fact F if B's computational ability concerning F has increased. In other words, B can compute something about F after interaction with A that was not possible before that interaction. Lastly, for those with an understanding of some basic *information theory*, "knowledge" and "information" are not the same. Knowledge pertains primarily to publicly known facts, whereas *information* pertains to facts on which only partial data are publicly known. For instance, if A answers B's queries by telling B the outcome of a coin toss, then B gets from A information concerning the event (from the perspective of information theory). However, B gains *no knowledge* from A, since B can flip coins without A.

[6.28]An "oracle" may be defined generally (without reference to cryptography) as any entity that can provide an answer in negligible time. However, as noted in [134, p. 406], a Zero-knowledge Proof (in the sense given above) is similar to an answer given by a trusted oracle.

[6.29]There also exists a notion of *almost-perfect* or *Statistical* Zero-knowledge. For this notion, the probability distributions of the transcripts must be "statistically indistinguishable" (see [84], and [88]).

In the above discussion, we mentioned one application of Zero-knowledge Proofs, namely *identification*. The following was introduced in [77] by Fiat and Shamir in 1987 as an authentication and digital signature scheme. It was modified in [73]–[74] to a Zero-knowledge Proof of identity. (There is a very interesting story involving patents, the military, and this protocol (see [113]).) Also, it is based on the Goldwasser-Micali-Rackoff Zero-knowledge Proof System for quadratic residuacity (see [90]–[91] and Footnote 6.26). The following is probably the best-known Zero-knowledge Protocol for identity.[6.30]

◆ **Feige-Fiat-Shamir Identification Protocol — Simplified Version**

In the following protocol, a prover A who wants to prove knowledge of a secret s_A to a verifier B. Also, the acronym *TTP* will mean a *trusted third party*.[6.31]

setup stage:

(1) An RSA modulus $n = pq$ is chosen by the TTP and made public, but the p and q are kept secret. Also, a parameter $a \in \mathbb{N}$ is chosen.

(2) A and B select, respectively, a secret $s_A, s_B \in \mathbb{N}$ relatively prime to n with $s_A, s_B \leq n-1$. Then $t_A \equiv s_A^2 \pmod{n}$, and $t_B \equiv s_B^2 \pmod{n}$, are computed by A and B, respectively. They then register s_A and s_B, respectively, with the TTP as private keys, where t_A, t_B are their public keys.

protocol: Execute the following steps.

(1) A selects an $m \in \mathbb{N}$, called a *commitment* such that $m \leq n - 1$ and sends $w \equiv m^2 \pmod{n}$, called the *witness*, to B.

(2) B chooses $c \in \{0, 1\}$, called a *challenge*, and sends it to A.

(3) A computes $r \equiv ms_A^c \pmod{n}$, called the *response*, and sends it to B.

(4) B computes r^2 modulo n. If $r^2 \equiv wt_A^c \pmod{n}$, then reset a's value to $a-1$ and go to step (1) if $a > 0$. If $a = 0$, then terminate the protocol with B accepting the proof. If $r^2 \not\equiv wt_A^c \pmod{n}$, then terminate the protocol with B rejecting the proof.

[6.30]Zero-knowledge Proofs are ideally suited for identification of an owner A (who has a PIN number or password S) of a credit card, ID card, or computer account by allowing A to convince a merchant B of knowledge of S without revealing even a single bit of S.

[6.31]In many identification schemes including digital signatures (see Section Three of Chapter Three), a "judge" is needed to resolve disputes. This judge is sometimes called a "trusted third party" upon whom all parties agree in advance. Another name often used for this disinterested third party is an *arbitrator*, who is needed to complete a protocol. It is essential that this "disinterested" arbitrator has no allegiance to any entity involved in the protocol and has no particular reason to complete the protocol. In this way, all entities participating in the protocol can accept that what is done is not only correct, but also that their portion of the protocol will be complete. Among the roles of a TTP (in general) are *certification authority* (see Definition 4.56) who is responsible for authenticity of public keys, *key managers*, and *registration authority* responsible for authorizing entities, to name only a few.

Suppose that an enemy E tries to impersonate A. Then E could try to cheat by selecting any $m \leq n - 1$ and sending the witness $w \equiv m^2 t_A^{-1} \pmod{n}$ to B. If B then sends the challenge $c = 0$, E would send the correct answer $r = m$, which passes step (4). However, if $c = 1$ is sent as a challenge, then E , who does not know the square root s_A of t_A, cannot respond correctly. Hence, E has a probability $1/2$ of escaping detection in the first iteration, $1/4$ in the second, and so on. To diminish this kind of arbitrary cheating, to an acceptably small probability of 2^{-a}, one performs iterations of the protocol for a sufficiently high value of a. Thus, the Feige-Fiat-Shamir Protocol is a Zero-knowledge Proof of Knowledge of a modular square root of t_A, namely the secret s_A. Given that A is the only entity in possession of knowledge of the square root of t_A, then success in the above protocol is taken as verification that the prover is A. The Zero-knowledge Property ensures that interacting with A as delineated in the protocol does not leak information that can be used to impersonate A. The following is an illustration of the above protocol with small parameters.

Example 6.41 Let $a = 2$, and $n = pq = 101 \cdot 757 = 76457$. The prover A selects $s_A = 3$, so $t_A = 9$, and B selects $s_B = 7$, so $t_B = 49$. In the first round A chooses a commitment $m = 98$ and sends the witness $w = m^2 = 9604$ to B. B chooses the challenge $c = 0$, and A responds with $r = 98 = m \cdot t_A^c$. Then B computes $98^2 = 9604 = w \cdot t_A^c$, so B accepts, and sets $a = 1$. In the second round A selects $m = 747$ and sends $w \equiv m^2 \equiv 22810 \pmod{n}$ to B. B sends $c = 1$ to A, who computes $r \equiv 747 \cdot 3 \equiv 2241 \pmod{n}$, which gets sent to B. Lastly, B computes $r^2 \equiv 52376 \pmod{n}$, checks that $52376 \equiv w \cdot t_A^c \pmod{n}$, and sets $a = 0$. Hence, B accepts A's proof.

The Feige-Fiat-Shamir Protocol is representative of a general structure for a larger class of what are called *Three-move, Zero-knowledge Protocols*. In such protocols, A sends a witness, B replies with a challenge, and A replies with a response. One iteration of these "moves" is called a *round.* There is a hidden randomization factor built into the protocol when prover A selects a (private) commitment from some pre-determined set. Then A computes the associated (public) witness. This initial randomness guarantees that the iterations in the protocol are sequential and independent, namely that there is a variation from one protocol round to another. Also, this establishes a set of "questions," which A lays claim to being able to answer, so naturally A's subsequent responses are restricted. The given design of the protocol means that only the prover A with knowledge of the secret S is capable of answering all of the challenges with correct responses, none of which provide information about S. B's challenge selects a question from the set, A provides a response, and B checks that it is valid. The iteration of rounds necessarily restricts the probability of cheating to arbitrarily chosen limits specified by the bound given in the choice of the parameter a.

The process by which A determines the set of questions and B chooses the question, (1)–(2) in the Feige-Fiat-Shamir Protocol for example, is an instance of what is called a *Cut and Choose Protocol*. This terminology arises from the classic technique for dividing anything fairly between two people, namely A cuts the object (say a piece of fruit) in "half," and B chooses one of the halves, leaving the other half for A. Also, the process involved in (2)–(3) of the Feige-Fiat-Shamir Protocol, given a witness from (1), is an instance of *Challenge-response Protocols*. In this type of protocol it is essential, for security reasons, that A responds to at most one challenge for a given witness, and should never reuse a witness. Hence, Interactive Zero-knowledge Protocols combine the ideas of both Cut and Choose and Challenge-response Protocols. Other Zero-knowledge Interactive Proofs of Identity may be found in [29], [39], [149]–[150], and [175]. See also [134].

In 1978, before the aforementioned formalization by Goldwasser, Micali, and Rackoff of zero-knowledge proofs, Rabin [167] was using the Cut and Choose Protocol for cryptographic applications. Also, it is worth noting that at about the same time as the aforementioned formalization was being developed, Babai and Moran [9]–[10], were trying to formalize the notion of efficient provability. They did this by developing a theory of randomized interactive games, called *Arthur-Merlin Games* in an effort to provide a statistical foundation for their notion of efficient provability.

There is a rather informal and informative description of the Cut and Choose Protocol given by Quisquater, Guillou,[6.32] and Berson in [166] using an analogy of a cave.

◆ Cut and Choose Protocol — Cave Analogy

The following protocol is illustrated by Diagram 6.42 on the next page.

The setup is that A knows the secret magic words that can open the secret door deep in a cave. A seeks to prove knowledge of the secret to B, but does not want to reveal the magic words. This is how A proceeds. A given parameter $t \in \mathbb{N}$ is chosen and the following steps are executed.

(1) B stands at point P_0 out of sight of point P_1.

(2) A walks all the way into the cave, either to point P_2 or P_3.

(3) After A has disappeared into the cave, B walks to point P_1.

(4) B shouts to A with one of two instructions:

    (a) Leave out of passage number 1, or

    (b) Leave out of passage number 2.

(5) If A obeys, using the magic words to open the door if necessary, then go to step (6). Otherwise, terminate the protocol with B rejecting the proof.

(6) Reset the value of t to $t - 1$ and go to step (1) if $t > 0$. Otherwise, terminate the protocol with B accepting the proof.

[6.32]Guillou and Quisquater [165] also introduced a Zero-knowledge Identification Protocol based upon coin flipping into a well (see page 130).

Diagram 6.42 Schematic for The Cut and Choose Cave Analogy

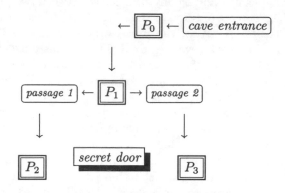

In the cave analogy, A cannot repeatedly guess which passage B will choose in step (4). If A did not know the secret words, then A would be forced to come out via the passage entered, but not the other passage since the secret door is the only connection between them. Thus, A has a one in two chance of guessing which passage B will choose in round one, and a one in four chance in round two. Hence, after t rounds, the chance of A fooling B is one in 2^t. Thus, for sufficiently large choice of the parameter t, A has a relatively insignificant chance of fooling B in all t rounds. For instance, if $t = 30$, then A has one chance in $1,073,741,824$ of fooling B for all thirty rounds. (A is more likely to win the lottery!) The attentive reader will note that we did not need to go through all the rounds since to convince B, all A has to do is let B witness entrance through one passage and return through the other. This just shows that the cave analogy is imperfect to describe the mathematical reasoning underlying it. Nevertheless, the protocol as illustrated by the constraints in the cave analogy are valid. The following application will give a more mathematical formulation.

First, we remind the reader that a *graph* is a set of points, called *vertices* or *nodes* together with a set of lines, called *edges* joining pairs of distinct vertices such that at most one edge joins any pair of vertices. Two vertices are called *adjacent* if there exists an edge joining them. Given vertices V_j for $1 \leq j \leq n \in \mathbb{N}$ on a graph, we say that $(V_1, \ldots, V_n)$ is a *path* from V_1 to V_n if V_k is adjacent to V_{k+1} for all $k = 1, 2, \ldots, n - 1$. A *circuit* is a path from a vertex back to itself. A *cycle* is a circuit that visits each node at most once, in other words a circuit without repeated nodes. A *Hamiltonian* Cycle[6.33] in a graph G is a cycle that contains every vertex of G, namely that visits each vertex of G exactly once. There does not exist an efficient method for determining when a given general

---

[6.33]These were named after the Irish mathematician, Sir William Rowan Hamilton (1805–1865), who invented a puzzle for finding such cycles along the edges of an icosahedron (a Platonic solid with twelve vertices, thirty edges, and twenty equilateral triangles). Hamilton was a child prodigy who knew Greek, Hebrew, and Latin by the age of five. He studied at Trinity College in Dublin, where he was appointed Professor of Astronomy at the age of twenty-two. He is best known for his discovery of the *quaternions* that bear his name. He spent the last two decades of his life working on applications of the quaternions. He died on September 2, 1865 from gout. *Elements of the Quaternions* was published posthumously.

graph has a Hamiltonian Cycle. Thus, for a sufficiently large graph, finding such a cycle in G is computationally infeasible.

◆ Zero-Knowledge Proofs Via Hamiltonian Cycles

Suppose that a prover A has knowledge of a Hamiltonian Cycle in a large graph G, and A wants to convince a verifier B of this knowledge without leaking knowledge of the cycle. Then A proceeds as follows. Let $t \in \mathbb{N}$ be a chosen parameter and execute the following steps.

(1) A randomly permutes the vertices and edges of G to create a graph H (which is isomorphic to G).[6.34]

(2) A gives a copy of H to B.

(3) B asks A to perform one of the following tasks:

   (a) Prove that G and H are isomorphic.

   (b) Provide a Hamiltonian Cycle in H.

(4) Either A responds by either performing one of (a) or (b):

   (a) Proving that G and H are isomorphic, but without providing any Hamiltonian Cycle in either G or H.

   (b) Providing a Hamiltonian Cycle in H, without proving that G and H are isomorphic.

   or A fails in which case B rejects the proof.

(5) Reset the value t to $t-1$ and go to step (1) if $t > 0$. Otherwise, terminate the protocol with B accepting the proof.

The above protocol is a Zero-knowledge Proof since B gets no information which would allow the determination of a Hamiltonian Cycle for G. If A shows, in step (4), that G and H are isomorphic, this does not assist B since the Hamiltonian Cycle is equally difficult to find in either graph. If A provides the Hamiltonian Cycle in H, this does not help B either since the problem of determining an isomorphism between the graphs is not much easier than finding the Hamiltonian Cycle. Given that, at each round, A generates a new H, then B gets no new information, no matter what the size of the parameter t happens to be.

Suppose that someone, not involved in the above interactive protocols, needs to be convinced of some fact. To do this, we need a *Noninteractive Protocol* or *mono-directional protocol.*

[6.34]Determining, for general graphs, an isomorphism between them is a problem for which there is no known polynomial-time algorithm to solve it. (However, it is not known to be **NP**-complete, whereas the problem of determining a Hamiltonian Cycle in a graph *is* known to be **NP**-complete.) Thus, if A did not construct the isomorphism, finding the Hamiltonian Cycle would be computationally infeasible. However, A can easily find this cycle in H since the isomorphism that maps the vertices and edges was A's construction.

◆ **Basic Zero-Knowledge Noninteractive Protocol**

Suppose that an entity A has a secret S which is the solution to some hard mathematical problem H (such as the Hamiltonian Cycle problem discussed above), and A must convince an entity C of this knowledge without interaction and without revealing any aspect of the secret. This is accomplished when entity C verifies the following steps.

(1) A uses S and a random number generator R to transform H into $n \in \mathbb{N}$ isomorphic problems P_j for $j = 1, 2, \ldots, n$. Using R and S, A solves P_j for each $j = 1, 2, \ldots, n$.

(2) A commits to the solutions S_j of the P_j. Call the commitments C_j, which are saved as bitstrings (see the discussion of bit commitment on page 130).

(3) A uses the C_j as inputs for a one-way hash function F (see Definition 3.7) and saves the first n bits B_j of the output of F for $j = 1, 2, \ldots, n$.

(4) For each $j = 1, 2, \ldots, n$, A takes B_j and:

   (a) If $B_j = 0$, then A proves that P_j is isomorphic to H.

   (b) If $B_j = 1$, then A demonstrates that S_j is a solution of P_j.

(5) A publishes the C_j from step (2) and the B_j from step 4(b).

The reason that the above protocol is a Noninteractive Zero-knowledge Proof is that A publishes data in step (5) that contains no information about S. Also, A cannot cheat since there is no way of knowing (or forcing) the outcome of the bits from F. In essence, F replaces B (from our Interactive Protocols) in step (3) of the above Noninteractive Protocol. The notion of Noninteractive Zero-knowledge Proofs was introduced in 1988 by Blum, Feldman, and Micali in [33].

We conclude this section with a proof of knowledge based upon discrete log.

◆ **Zero-knowledge Proof of Discrete Log**

Suppose that entity A has knowledge of x in $a^x \equiv c \pmod{n}$, where $a, c, n \in \mathbb{N}$ are public, $n = pq$ is an RSA modulus, and x is randomly chosen with $\gcd(x, \phi(n)) = \gcd(x, (p-1)(q-1)) = 1$. The following are the steps for convincing entity B of this knowledge without revealing x. Let $t \in \mathbb{N}$ be a chosen parameter and execute the following.

(1) A chooses $y \in \mathbb{N}$ at random such that $y < (p-1)(q-1)$ and computes $z \equiv a^y \pmod{n}$, then sends z to B.

(2) B sends a random bit b to A.

(3) A computes $w \equiv y + bx \pmod{(p-1)(q-1)}$ and sends w to B.

(4) If B confirms that $a^w \equiv zc^b \pmod{n}$, then go to step (5). Otherwise, terminate the protocol with B rejecting the proof.

(5) Reset the value of t to $t - 1$, and go to step (1) if $t > 0$. Otherwise, terminate the protocol with B accepting the proof.

We have barely touched upon the complex issues involved in the rapidly developing area of zero-knowledge. The reader may go to any number of the references given herein to expand horizons in this regard. There are numerous other aspects not mentioned herein such as *multi-prover proof systems* (see [27] for instance), and a variety of other deep and important connections that the reader may touch upon by further reading in this area. The next (concluding) section of this closing chapter of the book takes us to the realm of *quantum computing* and all the future possibilities that it suggests.

Exercises

6.25. Describe how one might use Blum integers $n = pq$, defined in Exercise 3.8 on page 136, to set up a Zero-knowledge Protocol for a prover A to convince a verifier B of knowledge of the factorization of n.

6.26. Develop a Zero-knowledge Interactive Proof based upon quadratic non-residues with an RSA modulus $n = pq$ (with secret factorization), where A, the prover, wants to prove that x is a quadratic residue, and wants to convince a verifier B, without revealing a square root of x.

6.27. Given $n = 919 \cdot 2087 = 1917953$, use the Feige-Fiat-Shamir Interactive Protocol to demonstrate how a prover A, given knowledge of a secret $s_A = 757$, could convince a verifier B of that knowledge. Do so by selecting $a = 2$. Then select $m = 3$ and $c = 1$ in round one and $m = 99$ and $c = 0$ in round two. Also, assume that B has chosen $s_B = 101$.

6.28. Given $n = pq = 2087 \cdot 487 = 1016369$, use the Feige-Fiat-Shamir Protocol to show how A can prove knowledge of a secret $s_A = 111$ to B, when $s_B = 90$. Assume that $a = 2$ and that on the first round $m = 21$ and $c = 0$, whereas on the second round $m = 12$ and $c = 1$.

☆ 6.29. Assume that p is a prime. Derive a bit commitment scheme (see page 131), where entity B commits to a bit b that is unconditionally concealing to A but binding to B only under the assumption of the infeasibility of the discrete log problem in $(\mathbb{Z}/p\mathbb{Z})^*$.

☆ 6.30. Let $n = pq$ be an RSA modulus and let m be a quadratic residue modulo ☆ n. Suppose that p, q and a square root of m are entrusted to a TTP. Devise a bit commitment scheme for B to commit to a bit b where the concealing property is unconditional but the binding property is based upon the infeasibility of computing the square root of m modulo n.

6.3 Quantum Cryptography

"Anyone who can contemplate quantum mechanics without getting dizzy doesn't understand it." — **Niels Bohr**[6.35]

"I think that it is safe to say that nobody understands quantum mechanics." — **Richard Feynman**[6.36]

In the late 1960s, a graduate student at Columbia University named Stephen Weisner came up with an idea for *quantum money* that would be impossible to counterfeit. His ideas were based on quantum theory. However, nobody (not even his supervisor) took him seriously. He even wrote a paper on the idea and it was rejected by four separate journals. However, Weisner had a friend named Charles Bennet from his undergraduate days at Brandeis University. He explained the quantum money idea to Bennet, who was the first to take him seriously. Later Bennet explained these ideas to a professor of computer science from the University of Montreal, named Gilles Brassard. Bennet and Brassard

---

[6.35] Niels Henrik David Bohr (1885–1962) was born in Copenhagen, Denmark on October 7, 1885. In 1911, he received his doctorate for a thesis on the electronic theory of metals, and by 1913, he had combined the concept of the nuclear atom with the quantum theory of Max Planck and Albert Einstein in a treatise that gave a quantum accounting for the lines observed in the spectrum of light emitted by atomic hydrogen. His explanation involved a radical departure from classical physics by postulating that the atom could neither emit nor absorb energy continuously, but only in finite steps, called *quantum jumps* or *leaps*. He was the first to apply quantum theory to the problem of atomic structure. By 1920, he was director of the newly created *Institute of Theoretical Physics* in Copenhagen, a position that he held until his death. In 1922, he was awarded the Nobel Prize for Physics, honoring his work on atomic structure. By 1925, quantum mechanics had arrived, embodied in the work of Werner Heisenberg, Erwin Schrödinger, Max Born, Wolfgang Pauli, and Paul Dirac, among others. Hence, Bohr's institute became the world centre for atomic physics. Bohr's concept of the atomic nucleus as a "liquid droplet" played a vital role in our understanding of nuclear fission — the splitting of a heavy atomic nucleus with the release of a tremendous amount of energy. Bohr took part in the projects for making a nuclear fission bomb, first in England, then at Los Alamos, New Mexico. By 1944, Bohr's concern for the impending cataclysm caused him to implore Churchill and Roosevelt to try to achieve international cooperation and by 1957, he was awarded the first *U. S. Atoms for Peace Award*. Bohr influenced generations of physicists both in his approach and through his vision. He died in his sleep on November 18, 1962.

[6.36] Richard Phillips Feynman (1918–1988) was born in New York City on May 11, 1918. He obtained his Ph.D. from Princeton University in 1942. He was a member of the team that developed the first atomic bomb at Los Alamos. By 1948, he had reconstructed a large part of quantum mechanics and electrodynamics. In 1950, he was appointed professor of theoretical physics at California Institute of Technology. In 1965, along with Julian Schwinger of Harvard and Shinichero Tomonaga of Japan, he was awarded the Nobel Prize in Physics for his work in correcting earlier inaccuracies in quantum electrodynamics — the theory that explains the interactions between radiation, electrons, and positrons (anti-electrons). In that year, he was also elected a foreign member of the Royal Society in London, England. After the *Challenger* disaster which saw the space shuttle explode 73 seconds after launch on January 28, 1986, he headed the inquiry and solved the mystery surrounding it. He was widely known for his self-effacing humour, often referring to his Nobel Prize a "pain in the neck", and he was a dedicated teacher always willing to share his insights with students. This was made abundantly clear when, even though he had been battling abdominal cancer for eight years, he taught until two weeks before his death. He died on February 15, 1988 in Los Angeles at the age of sixty-nine.

discussed the applications of the ideas to cryptography. The rest, as they say, is history. Bennet and Brassard, along with others, took the idea of quantum cryptography and developed it in an incredible series of papers [15]–[26], [34], [36]–[37], and [66] (see [23] or [210] for an overview). They even went on to build a prototype quantum cryptography machine (see [19]), which has been described as the first device whose capabilities exceed those of a Turing Machine (see [59]). However, as the quotes at the outset of this section indicate, the quantum theory behind all of this is not as concrete as what we have been studying thus far in the text. We are in for a wild ride, so get ready for the experience. The future of cryptography may prove to be a strange place indeed.[6.37]

Quantum cryptography is based on the following principle, which we must understand before proceeding

◆ Heisenberg's Uncertainty Principle

Heisenberg[6.38] proposed a new mathematical foundation for mechanics where physical objects are represented by matrices involving only *observable*, namely *measurable* objects. In 1927, he published his *uncertainty* (or *indeterminacy*) principle, which says that it is logically impossible to measure *simultaneously* every facet of a given object with perfect precision. Then Bohr and Heisenberg developed a philosophy of *complementarity* to account for these new *relative* physical variables — being dependent upon an appropriate measuring process. At its core, complementarity says that a person doing the measuring interacts with the observed object, resulting in its being a function of measurement, so the object is not revealed as itself.

Heisenberg translated his uncertainty principle into a simple statement:

[6.37]It turns out that the quantum money idea, however brilliant, is totally impractical. Even if there were a way to protect dollar bills using this idea of quantum money, it is estimated that it would cost a million dollars U. S. to protect each dollar bill! For an entertaining, informal, and easy-to-read overview of Weisner's idea (and a general history of "secret writing") see [188]. Weisner's original idea for quantum coding is described in Weisner's 1983 paper [201] (although the original manuscript probably pre-dates 1970).

[6.38]Werner Karl Heisenberg (1901–1976) was born on December 5, 1901 in Würzburg, Germany. He began studying physics at the University of Munich, obtaining his doctorate in 1923 on turbulence in fluid streams. In the fall of 1924, after a brief stint studying under Max Born at Göttingen, Heisenberg went to study at the Universitits Institut for Teoretisk Fysik in Copenhagen under Bohr, whose work on the atom had inspired him. By mid-1925, he had solved a major physical problem that laid the bedrock for the development of *quantum mechanics* — the science that explains the discrete energy states and other forms of quantized energy, such as in the light of atomic spectra and in the phenomenon of stability exhibited by macroscopic bits of matter. (By "quantization" is meant that observable quantities do not vary continuously, but rather are formed into discrete nuggets called *quanta*, which in the context of energy, means a discrete quantity of energy proportional in magnitude to the frequency of radiation it represents. This is the feature of quantum mechanics that makes computation, classical or quantum, possible.) From 1927 to 1941, Heisenberg was professor at the University of Leipzig, then for the next four years director of the Kaiser Wilhelm Institute for Physics in Berlin. In 1932, he was awarded the Nobel Prize for Physics. During World War II, he worked on the development of a nuclear reactor with Otto Hahn, one of the discoverers of nuclear fission. After the war, he organized and became director of the Max Planck Institute for Physics and Astrophysics at Göttingen. In 1958, he and the institute moved to Munich, where he died in 1976.

"Even in principle, we *cannot* know the present in all detail. For that reason, everything observed is a selection from a plenitude of possibilities and a limitation on what is possible in the future." In particular, this can be illustrated simply by the inescapable fact that one cannot measure both the velocity and the position of a particle at the same time. The very act of measuring one alters or influences the other, whether the other is measured or not. The reason is that if we want to measure the position of a particle, it is necessary to use light with very short wavelength, because in order to provide data on position we need wavelengths comparable to the object from which we seek to gather data. This is where the problem arises since short wavelength light transmits a big velocity boost to the electron when it bounces off it to provide position data. On the other hand, if we want to measure velocity, then we use very long wavelengths, which alters its position. This is the built-in fundamental uncertainty formulated by Heisenberg's Indeterminacy Principle. It is precisely this uncertainty that can be utilized to generate a secret key for a quantum cryptosystem. Now we describe how to do this.

A photon[6.39] travelling through space vibrates (or oscillates) as it does so. This vibration can be left-right (—) up-down (|) or at an angle (/) or (\). The angle of the vibration is known as the *polarization* of the photon. (Here we are discussing the simpler of the types of polarization called *linear* polarization, which means that as the photon travels, the electric field stays in the same plane.) We will maintain these symbols to denote the vibration of photons. For instance, the sun generates photons of all polarizations, including up-down, left-right, and *all* angles in between. Our notation is meant to simplify this situation by allowing only four possible polarizations, rather than the infinitely many possibilities.

If we place a polarizer in the path of photons, we can ensure that only some photons, those polarized in the same direction as the polarizer axis, get through. If the polarizer axis is vertically polarized (|), then only the photons polarized (|) should get through. Those that are polarized horizontally (—) have no chance and those polarized at an angle *should not* get through. However, in the quantum world, where uncertainty reigns, this is not always the case. Quantum theory dictates that if α is the angle that the plane of the electric field of the photon makes with the polarizer axis, then there is a probability of $\cos^2 \alpha$ that the photon will emerge with its polarization reset to that of the polarizer axis, and a probability of $1 - \cos^2 \alpha$ that it will be absorbed (to be re-emitted later as heat). Thus, if the polarizer axis is vertical, and if α is only slightly off vertical the photon has a high probability of passing through. If it is forty-five degrees, then it has a fifty percent chance of getting through, and this decreases to zero at the horizontally polarized photons.[6.40] In our simplified four position scenario,

---

[6.39] A photon is a quantum of electromagnetic radiation energy proportional to the frequency of radiation. Photons are *transverse electromagnetic waves*, meaning that their electric and magnetic fields are perpendicular to their direction of travel, and their electric and magnetic fields are perpendicular to each other.

[6.40] This is the basic principle upon which Polaroid sunglasses work, so the filter we have been discussing is often called a *Polaroid filter*. One can demonstrate this principle, using a pair

half of the diagonally polarized photons would pass through being vertically repolarized, all of the vertically polarized ones would pass through, and none of the horizontally polarized ones would pass through the filter. This is so since for $\alpha = 45$ or 135 degrees, $\cos^2 \alpha = 1/2$, $\cos^2 0 = 1$, and $\cos^2 90 = 0$.

◆ Quantum Key Generation

Suppose that entity A wishes to send entity B an enciphered message in binary. A has two schemes for doing this. One is called the *rectilinear scheme*, denoted by $+$, which uses a | for a 1 and a — for a 0. The other called the *diagonal scheme*, denoted by $\times$, which uses $/$ for 1 and $\setminus$ for 0. To send a message A can switch between these two schemes in a random fashion. Thus, for instance, A might send a photon burst consisting of | | $/$ $\setminus$ using the following combination of methods $++\times\times$ so the message is 1110.

(1) Entity A sends entity B a string of photon pulses with random settings on the rectilinear and diagonal schemes.[6.41]

(2) Entity B has a polarization detector with two settings. One is a $+$ detector that can measure — and |, with perfect accuracy, but misinterprets $/$ or $\setminus$ as one of | or — . The other, called a $\times$ detector, can measure $/$ and $\setminus$ but misinterprets | and — as one of the diagonally polarized ones. Both settings cannot be used at the same time due to Heisenberg's Uncertainty Principle. Thus, B sets the polarization detector at random settings. Sometimes the correct detector is picked and sometimes not.

(3) A telephones B over an unsecured line and discloses the polarization method used for each photon, but not how the photon was polarized. For instance, A might say that the first photon was sent using the $+$ method but not whether | or — was sent. Then B tells A which of the random polarizations in step (2) were correctly chosen. Lastly, A and B agree to ignore all photons for which B chose the wrong method. This generates a new shorter bitstring, consisting of B's correct measurements. This

---

of Polaroid sunglasses, by taking one lens out and place it in front of the fixed lens. There will be an orientation that is exactly the same for both lenses, so that the fixed lens has no effect on the loose lens. If the loose lens is now rotated ninety degrees, the effect will be complete blackness. This is because the polarization of the lenses are now perpendicular, so that photons that get through the one lens are blocked by the other. By rotating the loose lens forty-five degrees, one now gets an intermediate stage between complete blackness and no effect. This is because half of the photons that pass through the one lens succeed in getting through the other. The Polaroid Corporation was founded by Edwin Land in 1928, when he was still an undergraduate at Harvard College. Strangely enough, Land learned about the unusual properties of crystals that were formed when iodine was dropped into the urine of a dog that had been fed quinine. Those crystals were made of the material that Land later used for making polarizers.

[6.41] A device known as a *Pockels Cell* can be used to place a photon in a prescribed polarization state. A sequence of bits can be used to control the bias in the Pockels Cell and thereby determine the polarization orientations from the stream of photons emerging from the cell. This provides a mechanism for a sequence of bits to be transformed into a sequence of polarized photons, which can then be fed into an optical fiber, for instance. The Pockels Cell was invented in 1893 by the German physicist Fredrich Pockels.

bitstring is truly random since A's initial photon burst was random and B's choice of polarization methods was random. Hence, this agreed upon bitstring can be used for a one-time pad.

To see why the above key-generation scheme is unbreakable, suppose that a cryptanalyst C has also attempted to measure the initial photon burst from A. Then B and C are in exactly the same boat because both of them will choose the wrong detector half of the time (but not the same half). So even if C is eavesdropping on the telephone conversation between A and B, thereby gaining knowledge of the correct polarization settings, this does not help because C will have measured half of these incorrectly. Hence, this one-time pad is absolutely unbreakable, since C cannot intercept A's message without making errors. In other words, C cannot replicate A's original setting. The Heisenberg Uncertainty Principle guarantees it. Moreover, C's presence is detected by the very act of measuring. If A sends a $\diagdown$, for instance, and C uses the $+$ detector, then the incoming $\diagdown$ will emerge as one of $|$ or $-$, for the simple reason that this is the only way that photon can get through C's detector. If B measured the transformed photon with the $\times$ detector and $\diagup$ emerges, then a correct setting of the detector will result in an incorrect reading. In this case C has *twisted* the result. Of course, it might also occur that B's reading results in the correct $\diagdown$ emerging. Thus, C has a one in four chance of being detected for each bit checked. If n bits are checked, then the probability of detecting C is $1 - (3/4)^n$. Hence, for arbitrarily large n, we can make this as close to 1 as we desire. A and B can do an error check by telephone where they can determine correctness by reading off a subset of the bitstring and discarding those bits. Thus, we can add two more steps to the above algorithm for quantum key generation.

(4) A and B check the security and integrity of their one-time pad by testing some of the digits.

(5) If step (4) checks out, then they can use the one-time pad to send messages. If not, they know that a cryptanalyst has been tapping their lines, and they start over again at step (1).

The bitstream sent by A to B is unimportant. It only matters that they both agree on some subset of the bits as a secure key. Quantum cryptography allows key distribution between two entities (who share no prior keying material) that is provably secure against enemies with unlimited computing power, provided that the entities have access to a conventional channel, aside from the quantum channel. In [151] is a discussion of background on basic quantum channels for key distributions and experimental implementations.

The experimental prototype built in 1989 (see [19]) resulted in the first successful quantum cryptographic exchange between two computers in Bennett's lab at the IBM T. J. Watson Research Center. However, the computers were only thirty centimeters apart. Moreover, as later admitted by Brassard, the devices that were used to polarize the photons made *noticeably different* noises depending on the type of polarization chosen. Hence, security went out the window. However, more progress was made. In 1995, a group of researchers at the University of Geneva implemented a quantum cryptographic exchange over an optic fiber twenty-three kilometers long from Geneva to the town of Nyon. A quantum system has also been implemented over a distance of thirty kilometers (see [126]). However, what is needed is a means of using satellites to create a quantum cryptosystem. A research group at Los Alamos has been successful in transmitting a quantum key through the air over a distance of one kilometer. If we can find a way to do this over extended distances, then we could establish a global network cryptosystem that is absolutely unbreakable. The only way for that last statement to be false is if quantum mechanics itself is flawed, and if that is the case physicists around the world will have the task of starting over again at ground level — not very likely. In any case, the task is formidable since once a polarized photon is sent through the air and it interacts with air molecules, then it may change its polarization which is unacceptable.

In 1991, Ekert used a strikingly different approach to quantum cryptography. He devised a method using the property of *entanglement* [6.42] shared by pairs of correlated quantum systems (see [66], [169], [195]–[196]). There have been other alternative approaches to quantum cryptography presented such as the use of *phase modulation* (see [67]) introduced in 1992.

What about quantum computers? If a quantum computer could actually be built, then most conventional cryptosystems would be rendered useless. In 1994, Shor [184][6.43] discovered a polynomial time quantum algorithm for factoring large integers, thereby outperforming any classical computer. What this means is that *RSA*, for instance, could be broken! Hence, we would *need* quantum cryptography for security of sensitive data suddenly made vulnerable. Another powerful algorithm was devised in 1996 by Grover [93], who devised a quantum algorithm for finding a single item in an unsorted database in the square root of the time it would take on a classical computer. These algorithms show how quantum computers would outperform classical computers. Yet, even more is true. Quantum computers can do things that classical computers *cannot do*. A quantum computer, for example, can generate a truly random number! As we saw earlier, a classical computer is capable only of *pseudorandomness*. Also, we saw in the above discussion of quantum cryptography that a quantum computer would allow entities to communicate with messages virtually guaranteed to reveal the presence of an eavesdropper. Another important feature of quantum

---

[6.42] "Entangled states" or "entanglement" refers to the quantum fact that the properties of a composite system, even when the components are distant and non-interacting, cannot be fully expressed by descriptions of the properties of all the component systems. This is the feature of quantum mechanics that actually makes quantum cryptography possible.

[6.43] Shor used the quantum property of *interference* to show how a quantum computer could be used to factor a large composite number superefficiently. "Interference" is that feature of quantum mechanics which dictates that the outcome of a quantum process in general depends on all possible histories of that process. Hence, quantum interference can occur whenever there is more than one way to obtain a specific result. This is the feature that makes quantum computers qualitatively more powerful than classical ones.

computers in this regard is that they could simulate quantum physical forces. The work of Feynman [76] in 1982 showed that this is impossible on a classical computer. Hence, a quantum computer would be a means of testing aspects of quantum physics. If this is not fantastic enough, how about *teleportation*?

In common language, teleportation is understood to mean the transporting of a physical object from one location to the other by dissolution, transmission, and reconstruction *without* passing through the intervening distance —*Star Trek* stuff. The Heisenberg Uncertainty Principle would, in itself, seem to preclude any such possibility. The reason of course is that it is impossible to measure *all* attributes of an object at once, so how do we "scan" an object to ensure that we reconstruct it at the other end, given that our information is doomed to be incomplete? The answer came in 1993 when the authors of [21] showed how to exploit the notion of *entangled states*, and other phenomena to do an end-run around the limitations presented by the Heisenberg Uncertainty Principle. There is a catch. They could create an exact copy of the original if the original was *destroyed* in the process — *Don't* beam me up, Scotty! What they showed, in essence, was that there is a scheme to teleport the "quantum state" of an object, but not the object itself. For instance, a simple illustrative case would be the teleportation of an electron. Their scheme would not allow us to teleport that electron but rather just its "spin orientation" to another electron at a different location. Of what use is this?

To understand an important application of this scheme, we need the notion of a *qubit*, or *quantum bit*, which may be depicted as a small vector contained in a sphere where ↑ represents 0 and ↓ represents 1. Any other orientation such as ↖ or ↗ represents the angle with the vertical axis as a measure of the "0 to 1"-ness of the qubit. If we are able to teleport quantum states and we are able to encipher qubits using quantum states, then we can teleport a qubit from one location to another. Hence, we have a means of transmitting secret information from one quantum computer to another without ever having to pass through an unsecured channel! The mechanisms by which this can be done are delineated in [204], which also contains a chapter on quantum cryptography. The details, beyond the scope of this book, are described in an easy-to-read fashion therein. There is also a chapter on a quantum computer's outstanding error-correcting abilities and the potential for correctable quantum computations of *unlimited duration*.

There exist some fantastic developments in terms of the potential for some unusual quantum computer designs. For example, in 1988, the authors of [194] proposed a heteropolymer-based quantum computer, the basis of which would be the use of a linear array of atoms as memory cells. In 1995, Cirac and Zoller [49] proposed an ion trap-based quantum computer. In this scheme, the quantum memory register would be based on "trapped ions". An idea even more radically different than the above is that proposed in 1996 (see [53], [82]) using nuclear magnetic resonance techniques to perform the basic operations of a quantum computer. In this scheme, a liquid in a container (such as a test tube) would serve as memory with each molecule of the liquid performing the task of an independent memory register. Measuring the state of the memory in such a quantum computer is closer to the DNA-based computer posed by Adleman [1] in 1994 than the previous schemes for a quantum computer.

Adleman's idea for a DNA computer uses the toolbox of molecular biology to solve an instance of the Hamiltonian Cycle Problem discussed in Section Two of this chapter. The specific instance of the problem was encoded in the molecules of the DNA with the computation executed by using enzymes with standard protocols, described throughout this text. Adleman suggested that if we could build a DNA computer, it would execute a minimum of 10^{20} operations per second, whereas the fastest supercomputer today can execute 10^{12} operations per second. What the above demonstrates is that the building of either a DNA or a quantum computer would have devastating effects on public-key cryptography.

There exist tremendous technological stumbling blocks to clear before we can possibly use quantum teleportation for communication. The same may be said of course for the very construction of quantum computers themselves. However, given what we have seen in this section in terms of the tremendous strides already made, it would not be implausible to see these challenges overcome in a generation or two.

In this text, we have gone from the basic to the fantastic and the future will tell just how far we can go into that "outer limits" realm. At the turn of the twentieth century, automobiles, airplanes, and heart transplants were fantastic, if not implausible. Today, they are so commonplace as to not even raise an eyebrow. Will the turn of the twenty-second century see a similar attitude toward quantum computing, teleportation, and a global cryptosystem by satellite?

Appendix A_0: The Rijndael S-Box

The means by which Rijndael's invertible S-box, explicitly given below, was constructed consists of composing two functions. For each i, j with $1 \leq i \leq 32$, $1 \leq j \leq 8$, the following is executed. The first is to take the multiplicative inverse $\sum_{k=0}^{7} b_k 2^j$ of each nonzero $8i + j - 9$ in $\mathbb{F}_{2^8}$, with 0 getting mapped to 0, (see pages 289–292 in Appendix A). Thus, $a_{i,j} = 8i + j - 9$ gets mapped to $\sum_{k=0}^{7} b_k 2^j$ (where we have suppressed any reference to the i, j in the coefficients b_k for convenience of presentation in what follows). Then the following Affine function is applied:

$$
\begin{pmatrix}
1 & 1 & 1 & 1 & 1 & 0 & 0 & 0 \\
0 & 1 & 1 & 1 & 1 & 1 & 0 & 0 \\
0 & 0 & 1 & 1 & 1 & 1 & 1 & 0 \\
0 & 0 & 0 & 1 & 1 & 1 & 1 & 1 \\
1 & 0 & 0 & 0 & 1 & 1 & 1 & 1 \\
1 & 1 & 0 & 0 & 0 & 1 & 1 & 1 \\
1 & 1 & 1 & 0 & 0 & 0 & 1 & 1 \\
1 & 1 & 1 & 1 & 0 & 0 & 0 & 1
\end{pmatrix}
\begin{pmatrix}
b_7 \\ b_6 \\ b_5 \\ b_4 \\ b_3 \\ b_2 \\ b_1 \\ b_0
\end{pmatrix}
+
\begin{pmatrix}
0 \\ 1 \\ 1 \\ 0 \\ 0 \\ 0 \\ 1 \\ 1
\end{pmatrix}
=
\begin{pmatrix}
s_7 \\ s_6 \\ s_5 \\ s_4 \\ s_3 \\ s_2 \\ s_1 \\ s_0
\end{pmatrix}.
$$

Thus, $s_7 s_6 s_5 s_4 s_3 s_2 s_1 s_0$ is the binary equivalent of the decimal digit appearing the the S-box at position (i, j).

To illustrate how this S-box was constructed, we first observe that the column matrix, added on the left of the equality, is binary for the decimal digit 99 (or equivalently, the hexadecimal digit 63). So for instance, $a_{0,0} = 0$ gets mapped to 0, so each $b_k = 0$ for $1 \leq k \leq 7$. Adding the zero transformation to the column matrix yields 99, which is the first entry in the S-box. A less trivial illustration is the entry $a_{11,3} = 82$ in the position matrix. It has representation as the binary polynomial $x^6 + x^4 + x$ in $\mathbb{F}_{2^8} \cong \mathbb{F}_2[x]/(m(x))$, where $m(x) = x^8 + x^4 + x^3 + x + 1$ is the irreducible Rijndael polynomial (see Example A.63 on page 296 in Appendix A). The multiplicative inverse of 82 in $\mathbb{F}_{2^8}$ is given by $x^2 + 1$, so $(b_7, b_6, b_5, b_4, b_3, b_2, b_1, b_0) = (0, 0, 0, 0, 0, 1, 0, 1)$. Thus:

$$
\begin{pmatrix}
1 & 1 & 1 & 1 & 1 & 0 & 0 & 0 \\
0 & 1 & 1 & 1 & 1 & 1 & 0 & 0 \\
0 & 0 & 1 & 1 & 1 & 1 & 1 & 0 \\
0 & 0 & 0 & 1 & 1 & 1 & 1 & 1 \\
1 & 0 & 0 & 0 & 1 & 1 & 1 & 1 \\
1 & 1 & 0 & 0 & 0 & 1 & 1 & 1 \\
1 & 1 & 1 & 0 & 0 & 0 & 1 & 1 \\
1 & 1 & 1 & 1 & 0 & 0 & 0 & 1
\end{pmatrix}
\begin{pmatrix} 0 \\ 0 \\ 0 \\ 0 \\ 0 \\ 1 \\ 0 \\ 1 \end{pmatrix}
+
\begin{pmatrix} 0 \\ 1 \\ 1 \\ 0 \\ 0 \\ 0 \\ 1 \\ 1 \end{pmatrix}
=
\begin{pmatrix} 0 \\ 0 \\ 0 \\ 0 \\ 0 \\ 0 \\ 0 \\ 0 \end{pmatrix},
$$

and 0 is the decimal entry in position $(11, 3)$ of the S-box:

99	124	119	123	242	107	111	197
48	1	103	43	254	215	171	118
202	130	201	125	250	89	71	240
173	212	162	175	156	164	114	192
183	253	147	38	54	63	247	204
52	165	229	241	113	216	49	21
4	199	35	195	24	150	5	154
7	18	128	226	235	39	178	117
9	131	44	26	27	110	90	160
82	59	214	179	41	227	47	132
83	209	0	237	32	252	177	91
106	203	190	57	74	76	88	207
208	239	170	251	67	77	51	133
69	249	2	127	80	60	59	168
81	163	64	143	146	157	56	245
188	182	218	33	16	255	243	210
205	12	19	236	95	151	68	23
196	167	126	61	100	93	25	115
96	129	79	220	34	42	144	136
70	238	184	20	222	94	11	219
224	50	58	10	73	6	36	92
194	211	172	98	145	149	228	121
231	200	55	109	141	213	78	169
108	86	244	234	101	122	174	8
186	120	37	46	28	166	180	198
232	221	116	31	75	189	139	138
112	62	181	102	72	3	246	14
97	53	87	185	134	193	29	158
225	248	152	17	105	217	142	148
155	30	135	233	206	85	40	223
140	161	137	13	191	230	66	104
65	153	45	15	176	84	187	22

Appendix A: Fundamental Facts

In this appendix, we set down some fundamental facts beginning with the fundamental notion of a set. Proofs may be found in standard introductory texts on the subject matter.

◆ Sets

Definition A.1 *A* set *is a* well-defined *collection of* distinct *objects. The terms* set, collection, *and* aggregate *are synonymous. The objects in the set are called* elements *or* members. *We write* $a \in S$ *to denote membership of an element a in a set S, and if a is not in S, then we write $a \notin S$.*[A.1]

This definition *avoids* the problems of the contradictions that arise in such discussions as the Russell Antinomy.

Set notation is given by putting elements between two braces. For instance, an important set is the set of *natural numbers*:

$$\mathbb{N} = \{1, 2, 3, 4, \ldots\}.$$

In general, we may specify a set by properties. For instance,

$$\{x \in \mathbb{N} : x > 3\}$$

specifies those natural numbers that satisfy the property of being bigger than 3, which is the same as $\{x \in \mathbb{N} : x \neq 1, 2, 3\}$.

Definition A.2 (Subsets and Equality)

A set T is called a subset *of a set S, denoted by $T \subseteq S$, if every element of T is in S. On the other hand, if there is an element $t \in T$ such that $t \notin S$, then we write $T \not\subseteq S$, and say that T is* not *a subset of S. We say that two sets S and T are* equal, *denoted by $T = S$ provided that $t \in T$ if and only if $t \in S$, namely both $T \subseteq S$, and $S \subseteq T$. If $T \subseteq S$, but $T \neq S$, then we write $T \subset S$, and call T a proper* subset *of S. All sets contain the* empty set, *denoted by $\varnothing$, or $\{\}$, consisting of no elements. The set of all* subsets *of a given set S is called its* power set.

[A.1] The reader unfamiliar with the notion of *well-definedness* should view the concept in the following manner. A set of objects is well-defined provided that it is always possible to determine whether or not a particular element belongs to the set. The classical example of a collection that is *not* well defined is described as follows. Suppose that there is a library with many books, and each of these books may be divided into two categories, those that list themselves in their own index, and those that don't. The chief librarian decides to set up a Master Directory which will keep track of those books that don't list themselves. Now, the question arises: Does the Master Directory list itself? If it does not, then it *should* since it only lists those that do not list themselves. If it does, then it *should not* for the same reason—a paradox! This is called the *Russell Paradox* or *Russell Antinomy*. The problem illustrated by the Russell Paradox is with *self-referential* collections of objects. We see that Russell's collection is not *well-defined*, so it is not a set. Russell's example may be symbolized as $S = \{x : x \notin S\}$. The term "unset" is often used to describe such a situation.

Definition A.3 (Complement, Intersection, and Union)

The intersection *of two sets* S *and* T *is the set of all elements common to both, denoted by* $S \cap T$, *namely*

$$S \cap T = \{a : a \in S \text{ and } a \in T\}.$$

The union *of the two sets consists of all elements that are in* S *or in* T *(possibly both), denoted by* $S \cup T$, *namely*

$$S \cup T = \{a : a \in S \text{ or } a \in T\}.$$

If $T \subseteq S$, *then the complement of* T *in* S, *denoted by* $S \smallsetminus T$ *is the set of all those elements of* S *that are not in* T, *namely*

$$S \smallsetminus T = \{s : s \in S \text{ and } s \notin T\}.$$

Two sets S *and* T *are called* disjoint *if* $S \cap T = \varnothing$.

For instance, if $S = \mathbb{N}$, and $T = \{1, 2, 3\}$, then $S \cap T = T = \{1, 2, 3\}$, and $S \cup T = \mathbb{N}$. Also, $S \smallsetminus T = \{x \in \mathbb{N} : x > 3\}$.

Definition A.4 (Set Partitions)

Let S *be a set , and let* $\mathfrak{G} = \{S_1, S_2, \ldots\}$ *be a set of nonempty subsets of* S. *Then* $\mathfrak{G}$ *is called a* partition *of* S *provided both of the following are satisfied.*

(a) $S_j \cap S_k = \varnothing$ for all $j \neq k$.

(b) $S = S_1 \cup S_2 \cup \cdots \cup S_j \cdots$, namely $s \in S$ if and only if $s \in S_j$ for some j.

For an example of partitioning, see the notion of *congruence* on page 57.

Definition A.5 (Binary Relations and Operations)

Let s_1, s_2 *be elements of a set* S. *Then we call* (s_1, s_2) *an* ordered pair, *where* s_1 *is called the* first component, *and* s_2 *is called the* second component. *If* T *is another set, then the* Cartesian product *of* S *with* T, *denoted by* $S \times T$ *is given by the set of ordered pairs:*

$$S \times T = \{(s, t) : s \in S, t \in T\}.$$

A relation R *on* $S \times T$ *is a subset of* $S \times T$ *where* $(s, t) \in R$ *is denoted by* sRt. *A relation on* $S \times S$ *is called a* binary relation. *A relation* R *on* $(S \times S) \times S$ *is called a* binary operation *on* S *if* R *associates with each* $(s_1, s_2) \in S \times S$, *a unique element* $s_3 \in S$. *In other words, if* $(s_1, s_2)Rs_3$ *and* $(s_1, s_2)Rs_4$, *then* $s_3 = s_4$.

For example, a relation on $S \times T = \{1, 2, 3\} \times \{1, 2\}$ is $\{(1, 1), (1, 2)\}$. Notice that there does not exist a unique second element for 1 in this relation. We cannot discuss a binary operation here since $S \neq T$. The next section provides us with an important notion of a binary operation.

◆ **Functions**

Definition A.6 (Functions)

A function f (also called a mapping *or* map) *from a set S to a set T is a relation on S × S, denoted by f : S → T, which assigns each S ∈ S a unique t ∈ T, called the* image *of s under f, denoted by f(s) = t. The set S is called the* domain *of f and T is called the* range *of f. If S₁ ⊆ S, then the* image *of S₁ under f, denoted by f(S₁), is the set {t ∈ T : t = f(s) for some s ∈ S₁}. If S = S₁, then f(S) is called the* image *of f, denoted by* img(S). *If T₁ ⊆ T, the* inverse image *of T₁ under f, denoted by f⁻¹(T₁), is the set {s ∈ S : f(s) ∈ T₁}..*

A function f : S → T is called injective (*also called* one-to-one) *if and only if for each s₁, s₂ ∈ S, f(s₁) = f(s₂) implies that s₁ = s₂. A function f is* surjective (*also called* onto) *if f(S) = T, namely if for each t ∈ T, t = f(s) for some s ∈ S. A function f is called* bijective (*or a* bijection) *if it is both injective and surjective. Two sets are said to be in a* one-to-one correspondence *if there exists a bijection between them.*

Each of the following may be verified for a given function $f : S \to T$.

A.6. If $S_1 \subseteq S$, then $S_1 \subseteq f^{-1}(f(S_1))$.

A.6. If $T_1 \subseteq T$, then $f(f^{-1}(T_1)) \subseteq T_1$.

A.6. The identity map, $1_S : S \to S$, given by $1_S(s) = s$ for all $s \in S$, is a bijection.

A.6. f is injective if and only if there exists a function $g : T \to S$ such that $gf = 1_S$, and g is called a *left inverse of f*.

A.6. f is surjective if and only if there exists a function $h : T \to S$ such that $fh = 1_T$, and h is called a *right inverse for f*.

A.6. If f has both a left inverse g and a right inverse h, then $g = h$ is a unique map called the *two-sided inverse of f*.

A.6. f is bijective if and only if f has a two-sided inverse.

Notice that in Definition A.5 a binary operation on S is just a function on S × S. The number of elements in a set is of central importance.

Definition A.7 (Cardinality)

If S and T are sets, and there exists a one-to-one mapping from S to T, then S and T are said to have the same cardinality. *A set S is* finite *if either it is empty or there is an n ∈ N and a bijection f : {1, 2, ..., n} ↦ S. The number of elements in a finite set S is sometimes called its* cardinality, *or* order, *denoted by |S|. A set is said to be* countably infinite *if there is a bijection between the set and N. If there is no such bijection and the set is infinite, then the set is said to be* uncountably infinite.

Example A.8 If $n \in \mathbb{N}$ is arbitrary and $n_0 \in \mathbb{N}$ is arbitrary but *fixed*, then the map $f : \mathbb{N} \mapsto n_0 \mathbb{N}$ via $f(n) = n_0 n$ is bijective, so the multiples of $n_0 \in \mathbb{N}$ can be identified with $\mathbb{N}$. For instance, the case where $n_0 = 2$ shows that the even natural numbers may be identified with the natural numbers themselves.

Definition A.9 (Indexing Sets and Set Operations)
   Let I be a set, which may be finite, or infinite (possibly uncountably infinite), and let $\mathfrak{U}$ be a universal set, *which means a set that has the property of containing all sets under consideration. We define*

$$\cup_{j \in I} S_j = \{s \in \mathfrak{U} : s \in S_j \text{ for some } j \in I\},$$

and

$$\cap_{j \in I} S_j = \{s \in \mathfrak{U} : s \in S_j \text{ for all } j \in I\}.$$

Here, I is called the indexing set, *$\cup_{j \in I} S_j$ is called a* generalized set-theoretic union, *and $\cap_{j \in I} S_j$ is called a* generalized set-theoretic intersection.

Example A.10 The reader may verify both of the following properties about generalized unions and intersections. In what follows, $T, S_j \subseteq \mathfrak{U}$.

(a) $T \cup (\cap_{j \in I} S_j) = \cap_{j \in I} (T \cup S_j)$.

(b) $T \cap (\cup_{j \in I} S_j) = \cup_{j \in I} (T \cap S_j)$.

   We now provide the *Fundamental Laws of Arithmetic* as a fingertip reference for the convenience of the reader.

   The Laws of Arithmetic:
   ◆ **The Laws of Closure.** If $a, b \in \mathbb{R}$, then $a + b \in \mathbb{R}$ and $ab \in \mathbb{R}$.
   ◆ **The Commutative Laws.** If $a, b \in \mathbb{R}$, then $a + b = b + a$, and $ab = ba$.
   ◆ **The Associative Laws.** If $a, b, c \in \mathbb{R}$, then $(a + b) + c = a + (b + c)$, and $(ab)c = a(bc)$.
   ◆ **The Distributive Law.** If $a, b, c \in \mathbb{R}$, then $a(b + c) = ab + ac$.
   ◆ **The Cancellation Law** Let $a, b, c \in \mathbb{R}$. If $a + c = b + c$, then $a = b$ for any $c \in \mathbb{R}$. Also, if $ac = bc$, then $a = b$ for any $c \in \mathbb{R}$, with $c \neq 0$.

   Note that as a result of the distributive law, we may view $-a$ for any $a \in \mathbb{R}$ as $(-1) \cdot a$, or -1 *times a*.
   We now look at inverses under multiplication.

   ◆ **The Multiplicative Inverse**
   If $z \in \mathbb{R}$ with $z \neq 0$, then the *multiplicative inverse* of z is that number $1/z = z^{-1}$ (since $z \cdot \frac{1}{z} = 1$, the multiplicative identity). In fact, division may be considered the inverse of multiplication.

For instance, if $z \neq \pm 1$, then $z^{-1} \notin \mathbb{Z}$. For this we need a larger set. The following are called the *rational numbers*.

$$\mathbb{Q} = \{a/b : a, b \in \mathbb{Z}, \text{ and } b \neq 0\}.$$

Rational numbers have *periodic* decimal expansions. In other words, they have patterns that repeat *ad infinitum*. For instance, $1/2 = 0.5000\ldots$ and $1/3 = 0.333\ldots$. However, there are numbers whose decimal expansions have *no* repeated pattern, such as

$$\sqrt{2} = 1.41421356237\ldots,$$

so it is not a quotient of integers. These numbers, having decimal expansions that are not periodic, are called irrational numbers, denoted by $\mathfrak{J}$. It is possible that a sequence of rational numbers may *converge* to an irrational one. For instance, define

$$q_0 = 2, \text{ and } q_{j+1} = 1 + \frac{1}{q_j} \text{ for } j \geq 0.$$

Then

$$\lim_{j \to \infty} q_j = \frac{1 + \sqrt{5}}{2},$$

called the *Golden Ratio*, denoted by $\mathfrak{g}$ which we will study in this appendix. The reader familiar with Fibonacci Numbers (see page 20) will have recognized that for $j \geq 0$,

$$q_{j+1} = 1 + \frac{1}{q_j} = \frac{q_j + 1}{q_j} = \frac{F_{j+3}}{F_{j+2}},$$

so

$$\lim_{j \to \infty} \frac{F_{j+3}}{F_{j+2}} = \mathfrak{g}.$$

The *real numbers* consist of the set-theoretic union:

$$\mathbb{R} = \mathbb{Q} \cup \mathfrak{J}.$$

To complete the hierarchy of numbers (at least for our purposes), the *complex numbers* employ $\sqrt{-1}$, as follows:

$$\mathbb{C} = \{a + b\sqrt{-1} : a, b \in \mathbb{R}\}.$$

If $a < 0$, then $\sqrt{a} \notin \mathbb{R}$. For instance, $\sqrt{-1} \notin \mathbb{R}$ and $\sqrt{-5} \notin \mathbb{R}$. Consider, $\sqrt{25} = 5 \in \mathbb{R}$. A common error is to say that $\sqrt{25} = \pm 5$, but this is **false**. The error usually arises from the confusion of the solutions to $x^2 = 25$ with the solutions to $\sqrt{5^2} = x$. Solutions to $x^2 = 25$ are certainly $x = \pm 5$, but the **only** solution to $\sqrt{5^2} = x$ is $x = 5$, the unique *positive* integer such that $x^2 = 25$. A valid way of avoiding confusion with $\sqrt{x^2}$ is the following development.

We may define *exponentiation* by observing that for any $x \in \mathbb{R}$, $n \in \mathbb{N}$,

$$x^n = x \cdot x \cdots x,$$

multiplied n times. Note that by convention $x^0 = 1$ for any nonzero real number x (and 0^0 is undefined). In what follows, the notation $\mathbb{R}^+$ means all of the *positive* real numbers. For rational exponents, we have the following.

Definition A.11 (Rational Exponents)
Let $n \in \mathbb{N}$. If n is even and $a \in \mathbb{R}^+$, then $\sqrt[n]{a} = b$ means that unique value of $b \in \mathbb{R}^+$ such that $b^n = a$. If n is even and $a \in \mathbb{R}$ with a negative, then $\sqrt[n]{a}$ is undefined. If n is odd, then $\sqrt[n]{a} = b$ is that unique value of $b \in \mathbb{R}$ such that $b^n = a$. In each case, a is called the base *for the exponent.*

Based upon Definition A.11. the symbol $a^{\frac{m}{n}}$ for $a \in \mathbb{R}^+$ and $m, n \in \mathbb{N}$ is given by

$$a^{\frac{m}{n}} = \left(a^{\frac{1}{n}}\right)^m.$$

Definition A.12 (Absolute Value)
If $x \in \mathbb{R}$, then
$$|x| = \begin{cases} x & \text{if } x \geq 0, \\ -x & \text{if } x < 0, \end{cases}$$
called the absolute value *of x.*

With Definition A.12 in mind, we see that if $x > 0$, then

$$\sqrt{x^2} = (x^2)^{1/2} = (x)^{2 \cdot 1/2} = x^1 = x = |x|,$$

and if $x < 0$, then

$$\sqrt{x^2} = \sqrt{(-x)^2} = (-x)^{2 \cdot 1/2} = (-x)^1 = -x = |x|.$$

Hence,
$$\sqrt{x^2} = |x|.$$

Also,
$$a^{-\frac{m}{n}} = \frac{1}{a^{\frac{m}{n}}}.$$

In general, we have the following laws.

Theorem A.13 (Laws for Exponents)
Let $a, b \in \mathbb{R}^+$, and $n, m \in \mathbb{N}$.

(a) $a^n b^n = (ab)^n.$

(b) $a^m a^n = a^{m+n}.$

(c) $(a^m)^n = a^{mn}.$

(d) $(a^m)^{\frac{1}{n}} = \sqrt[n]{a^m} = a^{\frac{m}{n}} = (a^{\frac{1}{n}})^m.$

Proof. See [144, Proposition 1.4.1, p. 46].   □

Corollary A.14 *Let $a, n \in \mathbb{N}$. Then $\sqrt[n]{a} \in \mathbb{Q}$ if and only if $\sqrt[n]{a} \in \mathbb{Z}$.*

Note that we cannot have a *negative* base in Theorem A.13. The reason for this assertion is given in the following discussion. If we were to allow $-5 = \sqrt{25}$, then by Theorem A.13,

$$-5 = \sqrt{25} = 25^{1/2} = (5^2)^{1/2} = 5^{2 \cdot 1/2} = 5^1 = 5,$$

which is a contradiction. From another perspective, suppose that we allowed for negative bases in Theorem A.13. Then

$$5 = \sqrt{25} = \sqrt{(-5)^2} = ((-5)^2)^{1/2} = (-5)^{2 \cdot 1/2} = (-5)^1 = -5,$$

again a contradiction. Hence, only positive bases are allowed for the laws in Theorem A.13 to hold.

In the above, we have used the symbol $>$ (greater than) and $<$ (less than). We now formalize this notion of ordering as follows.

Definition A.15 (Ordering) *If $a, b \in \mathbb{R}$, then we write $a < b$ if $a - b$ is negative, and say that a is strictly less than b. Equivalently, $b > a$ means that b is strictly bigger than a. (Thus, to say that $b - a$ is positive, is equivalent to saying that $b - a > 0$.) We also write $a \leq b$ to mean that $a - b$ is not positive, namely $a - b = 0$ or $a - b < 0$. Equivalently, $b \geq a$ means that $b - a$ is nonnegative, namely $b - a = 0$ or $b - a > 0$.*

Now we state the principle governing order.

◆ **The Law of Order.** If $a, b \in \mathbb{R}$, then *exactly one* of the following must hold: $a < b$, $a = b$, or $a > b$.

A basic rule, which follows from the Law of Order, is the following.

◆ **The Transitive Law.** Let $a, b, c \in \mathbb{R}$. If $a < b$ and $b < c$, then $a < c$.

What now follows easily from this is the connection between order and the operations of addition and multiplication, namely if $a < b$, then $a + c < b + c$ for any $c \in \mathbb{R}$, and $ac < bc$ for any $c \in \mathbb{R}^+$. However, if $c < 0$, then $ac > bc$.

Since we have the operations of addition and multiplication, it would be useful to have a notation that would simplify calculations.

◆ **The Sigma Notation**

We can write $n = 1 + 1 + \cdots + 1$ for the sum of n copies of 1. We use the Greek letter upper case *sigma* to denote *summation*. For instance, $\sum_{i=1}^{n} 1 = n$ would be a simpler way of stating the above. Also, instead of writing the sum of the first one hundred natural numbers as $1 + 2 + \cdots + 100$, we may write it as $\sum_{i=1}^{100} i$. In general, if we have numbers $a_m, a_{m+1}, \cdots, a_n$ $(m \leq n)$, we may write their sum as

$$\sum_{i=m}^{n} a_i = a_m + a_{m+1} + \cdots a_n,$$

and by convention

$$\sum_{i=m}^{n} a_i = 0 \text{ if } m > n.$$

The letter i is the *index of summation* (and any letter may be used here) n is *the upper limit of summation*, m is *the lower limit of summation*, and a_i is a *summand*. In the previous example, $\sum_{i=1}^{n} 1$, there is no i in the summand since we are adding the *same* number n times. The upper limit of summation tells us how many times that is. Similarly, we can write, $\sum_{j=1}^{4} 3 = 3 + 3 + 3 + 3 = 12$. This is the simplest application of the sigma notation. Another example is $\sum_{i=1}^{10} i = 55$.

Theorem A.16 (Properties of the Summation (Sigma) Notation)

Let $h, k, m, n \in \mathbb{Z}$ with $m \leq n$ and $h \leq k$. If R is a ring, then:

(a) If $a_i, c \in R$, then $\sum_{i=m}^{n} ca_i = c \sum_{i=m}^{n} a_i$.

(b) If $a_i, b_i \in R$, then $\sum_{i=m}^{n} (a_i + b_i) = \sum_{i=m}^{n} a_i + \sum_{i=m}^{n} b_i$.

(c) If $a_i, b_j \in R$, then

$$\sum_{i=m}^{n} \sum_{j=h}^{k} a_i b_j = \left(\sum_{i=m}^{n} a_i \right) \left(\sum_{j=h}^{k} b_j \right) = \sum_{j=h}^{k} \sum_{i=m}^{n} a_i b_j = \left(\sum_{j=h}^{k} b_j \right) \left(\sum_{i=m}^{n} a_i \right).$$

A close cousin of the summation symbol is the product symbol defined as follows.

◆ The Product Symbol

The multiplicative analogue of the summation notation is the *product symbol* denoted by Π, upper case Greek *pi*. Given $a_m, a_{m+1}, \ldots, a_n \in R$, where R is a given ring and $m \leq n$, their product is denoted by:

$$\prod_{i=m}^{n} a_i = a_m a_{m+1} \cdots a_n,$$

and by convention $\prod_{i=m}^{n} a_i = 1$ if $m > n$.

The letter i is the *product index*, m is the *lower product limit*, n is the *upper product limit*, and a_i is a *multiplicand* or *factor*.

For example, if $x \in \mathbb{R}^+$, then

$$\prod_{j=0}^{n} x^j = x^{\sum_{j=0}^{n} j} = x^{n(n+1)/2},$$

(see Theorem A.29 below). Now we return to a discussion of the integers with a study of divisibility.

◆ **Divisibility.** If $a, b \in \mathbb{Z}$, then to say that b *divides* a, denoted by $b \mid a$, means that $a = bx$ for a *unique* $x \in \mathbb{Z}$, denoted by $x = a/b$. Note that the existence and uniqueness of x implies that b cannot be 0. We also say that a is *divisible* by b. If b does *not* divide a, then we write $b \nmid a$, and say that a is *not divisible* by b. Note that x does not exist for $b = 0$, $a \neq 0$ and x is not unique for $a = b = 0$. We say that *division by zero is undefined.*[A.2]

We may classify integers according to whether they are divisible by 2, as follows.

Definition A.17 (Parity)
If $a \in \mathbb{Z}$, and $a/2 \in \mathbb{Z}$, then we say that a is an even *integer. In other words, an even integer is one which is divisible by 2. If $a/2 \notin \mathbb{Z}$, then we say that a is an* odd *integer. In other words, an odd integer is one which is not divisible by 2. If two integers are either both even or both odd, then they are said to have the same* parity. *Otherwise they are said to have* opposite *or* different *parity.*

Of particular importance for divisibility is the following algorithm.

Theorem A.18 (The Division Algorithm)
If $a \in \mathbb{N}$ and $b \in \mathbb{Z}$, then there exist unique integers $q, r \in \mathbb{Z}$ with $0 \leq r < a$, and $b = aq + r$.

Now we look more closely at our terminology. To say that b divides a is to say that a is a *multiple* of b, and that b is a *divisor* of a. Also, note that b dividing a is equivalent to the remainder upon dividing a by b is zero. Any divisor $b \neq a$ of a is called a *proper divisor* of a. If we have two integers a and b, then a *common divisor* of a and b is a natural number n which is a divisor of *both* a and b. There is a special kind of common divisor which deserves special recognition.

Definition A.19 (The Greatest Common Divisor)
If $a, b \in \mathbb{Z}$ are not both zero, then the[A.3] *greatest common divisor or gcd of a and b is the natural number g such that $g \mid a$, $g \mid b$, and g is divisible by any common divisor of a and b, denoted by $g = \gcd(a, b)$.*

We have a special term for the case where the gcd is 1.

[A.2] There are mathematical structures $\mathcal{S}$ that have what are called *zero divisors*. These are elements $s, t \in \mathcal{S}$ such that both s and t are nonzero, but $st = 0$. For instance, in the *ring* $\mathcal{S} = \mathbb{Z}/6\mathbb{Z}$ (see Definition 2.8), $2 \cdot 3 = 0$ in $\mathbb{Z}/6\mathbb{Z}$, so $\mathbb{Z}/6\mathbb{Z}$ has zero divisors, but $\mathbb{Z}$ has no zero divisors.

[A.3] The word "the" is valid here since g is indeed unique. See part (a) of Theorem A.24 below.

Definition A.20 (Relative Primality)
  If $a, b \in \mathbb{Z}$, and $\gcd(a, b) = 1$, then a and b are said to be relatively prime *or* coprime. *Sometimes the phrase a* is prime to *b is also used.*

  By applying the Division Algorithm, we get the following due to Euclid.

Theorem A.21 (The Euclidean Algorithm)
  Let $a, b \in \mathbb{Z}$ $(a \geq b > 0)$, and set $a = r_{-1}, b = r_0$. *By repeatedly applying the Division Algorithm, we get* $r_{j-1} = r_j q_{j+1} + r_{j+1}$ *with* $0 < r_{j+1} < r_j$ *for all* $0 \leq j < n$, *where* n *is the least nonnegative number such that* $r_{n+1} = 0$, *in which case* $\gcd(a, b) = r_n$.

  Proof. See [144, Theorem 1.3.3, p. 37]. □

  It is easily seen that any common divisor of $a, b \in \mathbb{Z}$ is also a common divisor of an expression of the form $ax + by$ for $x, y \in \mathbb{Z}$. Such an expression is called a *linear combination* of a and b. The greatest common divisor is a special kind of linear combination, which can be computed using a more general form of Theorem A.21, as follows.

Theorem A.22 (The Extended Euclidean Algorithm)
  Let $a, b \in \mathbb{N}$, and let q_i for $i = 1, 2, \ldots, n+1$ *be the quotients obtained from the application of the Euclidean Algorithm to find* $g = \gcd(a, b)$, *where* n *is the least nonnegative integer such that* $r_{n+1} = 0$. *If* $s_{-1} = 1$, $s_0 = 0$, *and*

$$s_i = s_{i-2} - q_{n-i+2} s_{i-1},$$

for $i = 1, 2, \ldots, n+1$, *then*

$$g = s_{n+1} a + s_n b.$$

  Proof. See [144, Theorem 1.3.4, p. 38]. □

Corollary A.23 *If* $c \mid a$ *and* $c \mid b$, *then* $c \mid (ax + by)$ *for any* $x, y \in \mathbb{Z}$. *In particular, the least positive value of* $ax + by$ *is* g.

  There are also numerous properties of the gcd that we will need in the main text. We gather them here as a convenient reference.

Theorem A.24 (Properties of the gcd)
  Let $a, b \in \mathbb{Z}$, and $c \in \mathbb{N}$, *with* $g = \gcd(a, b)$.

 (a) g is unique.

 (b) If $c \mid a$ and $c \mid b$, then $c \mid g$, so $c \leq g$.

 (c) $\gcd(a, b) = a$ if and only if $a \mid b$.

(d) For any $m \in \mathbb{N}$, $mg = \gcd(ma, mb)$.

(e) If c is a common divisor of a and b, then $\gcd(a/c, b/c) = g/c$.

(f) $\gcd(a/g, b/g) = 1$.

(g) $\gcd(a, b) = \gcd(b, a) = \gcd(-a, b) = \gcd(a, -b) = \gcd(-a, -b)$.

Proof. See [144, Theorem 1.3.2, p. 35]. □

In the main text, we will also need a concept, closely related to the gcd, as follows.

Definition A.25 (The Least Common Multiple)
If $a, b \in \mathbb{Z}$, then the[A.4] smallest natural number which is a multiple of both a and b is the least common multiple of a and b, denoted by $\mathrm{lcm}(a, b)$.

For instance, if $a = 22$ and $b = 14$, then $\gcd(a, b) = 2$, and $\mathrm{lcm}(a, b) = 154$.

We will find it convenient to have a list of some properties of the least common multiple.

Theorem A.26 (Properties of the Least Common Multiple)
Let $\ell = \mathrm{lcm}(a, b)$ for $a, b \in \mathbb{Z}$.

(a) $\ell = b$ if and only if $a \mid b$.

(b) $n\ell = \mathrm{lcm}(an, bn)$ for any $n \in \mathbb{N}$.

(c) $\ell \leq ab$.

(d) If $c \mid a$ and $c \mid b$, then $\mathrm{lcm}(a/c, b/c) = \ell/c$.

Proof. See [144, Theorem 1.3.5, p. 39]. □

We will also need the following comparative properties of the lcm and the gcd.

Theorem A.27 (Relative Properties of the gcd and lcm)
Let $a, b \in \mathbb{N}, \ell = \mathrm{lcm}(a, b)$, and $g = \gcd(a, b)$.

(a) $\ell g = ab$.

(b) If $g = 1$, then $\ell = ab$.

[A.4] Here the uniqueness of the lcm follows from the uniqueness of the gcd via part (a) of Theorem A.27.

Proof. See [144, Theorem 1.3.6, pp. 39–40].   □

In Section Two of Chapter One, we introduced *primes*, the building blocks or atoms of arithmetic. In the study of the integers in terms of factorization into primes, the following is sometimes called the *Unique Factorization Theorem* for integers.

Theorem A.28 (The Fundamental Theorem of Arithmetic)

Let $n \in \mathbb{N}, n > 1$. Then n has a factorization into a product of prime powers (existence). Moreover, $n = \prod_{i=1}^{r} p_i = \prod_{i=1}^{s} q_i$, where the p_i and q_i are primes, then $r = s$, and the factors are the same if their order is ignored (uniqueness).

Proof. See [144, pp. 44–45].   □

For example, $617,400 = 2^3 \cdot 3^2 \cdot 5^2 \cdot 7^3$. Before leaving the discussion of primes it is worthy of note that one of the most elegant proofs to remain from antiquity is Euclid's proof of the infinitude of primes. Suppose that $p_1, p_2, \ldots, p_n$ for $n \in \mathbb{N}$ are all of the primes. Then set $N = \prod_{j=1}^{n} p_j$. Since $N + 1 > p_j$ for any natural number $j \leq n$, then N must be composite. Hence, $p_j \mid (N + 1)$ for some such j by the Fundamental Theorem of Arithmetic. Since $p_j \mid N$, then $p_j \mid N + 1 - N = 1$, a contradiction.

In the proof of the Division Algorithm, the following concept on order is used to establish the existence of a quotient and a remainder.

◆ **The Well-Ordering Principle.** Every nonempty subset of $\mathbb{N}$ contains a least element.

We say that $\mathbb{N}$ is *well-ordered*. Any nonempty set, denoted by $\mathcal{S} \neq \varnothing$, with $\mathcal{S} \subseteq \mathbb{Z}$, having a least element is said to be well-ordered. The Well-Ordering Principle is sometimes called the *Principle of the Least Element*.

It is known that the Well-Ordering Principle is equivalent to the following important principle (see [144, Exercise 1.2.42, p. 30]).

◆ **The Principle of Mathematical Induction.**
Suppose that $\mathcal{S} \subseteq \mathbb{N}$. If

(a) $1 \in \mathcal{S}$, and

(b) If $n > 1$ and $n - 1 \in \mathcal{S}$, then $n \in \mathcal{S}$,

then $\mathcal{S} = \mathbb{N}$.

In other words, the Principle of Mathematical Induction says that any subset of the natural numbers that contains 1, and can be shown to contain $n > 1$ whenever it contains $n - 1$ must *be* $\mathbb{N}$. Part (a) is called the *induction step*, and the assumption that $n \in \mathcal{S}$ is called the *induction hypothesis*. Typically, one establishes the induction step, then assumes the induction hypothesis and proves the conclusion, that $n \in \mathcal{S}$. Then we simply say that *by induction, $n \in \mathcal{S}$* for all $n \in \mathbb{N}$. This principle is illustrated in the following two results.

Theorem A.29 (A Summation Formula)

$$\sum_{j=1}^{n} j = \frac{n(n+1)}{2}.$$

Proof. If $n = 1$, then $\sum_{j=1}^{n} j = 1 = n(n+1)/2$, and the induction step is secured. Assume that

$$\sum_{j=1}^{n-1} j = (n-1)n/2,$$

the induction hypothesis. Now consider

$$\sum_{j=1}^{n} j = n + \sum_{j=1}^{n-1} j = n + (n-1)n/2,$$

by the induction hypothesis. Hence,

$$\sum_{j=1}^{n} j = [2n + (n-1)n]/2 = (n^2 + n)/2 = n(n+1)/2,$$

as required. Hence, by induction, this must hold for all $n \in \mathbb{N}$. □

Theorem A.30 (A Geometric Formula) *If $a, r \in \mathbb{R}$, $r \neq 1$, $n \in \mathbb{N}$, then*

$$\sum_{j=0}^{n} ar^j = \frac{a(r^{n+1} - 1)}{r - 1}.$$

Proof. If $n = 1$, then

$$\sum_{j=0}^{n} ar^j = a + ar = a(1 + r) = a(1 + r)(r - 1)/(r - 1) = a(r^2 - 1)/(r - 1) =$$

$$a(r^{n+1} - 1)/(r - 1),$$

which is the induction step. By the induction hypothesis, we get,

$$\sum_{j=0}^{n+1} ar^j = ar^{n+1} + \sum_{j=0}^{n} ar^j = ar^{n+1} + a(r^{n+1} - 1)/(r - 1) = a(r^{n+2} - 1)/(r - 1),$$

as required. □

The sum in Theorem A.30 is called a *geometric sum* where a is the *initial term* and r is called the *ratio*.

In the text, we will need another form of induction that is equivalent to the above (see [144, Theorem 1.2.5, p. 23] for the proof of the equivalence). This version is as follows.

♦ **The Principle of Mathematical Induction (Second Form)**
Suppose that $\mathcal{S} \subseteq \mathbb{Z}$, and $m \in \mathbb{Z}$ with

(a) $m \in S$, and

(b) If $m < n$ and $\{m, m+1, \ldots, n-1\} \subseteq S$, then $n \in S$.

Then $k \in S$ for all $k \in \mathbb{Z}$ such that $k \geq m$.

An illustration of the use of this form of induction is as follows. For a definition of Fibonacci Numbers, see page 20 of the main text. In what follows,

$$\mathfrak{g} = \frac{1 + \sqrt{5}}{2},$$

called the golden ratio.

Theorem A.31 *For any* $n \in \mathbb{N}$, $F_n \geq \mathfrak{g}^{n-2}$.

Proof. We use the Principle of Induction in its second form. We need to handle $n = 1, 2$ separately since $F_n = F_{n-1} + F_{n-2}$ only holds for $n \geq 3$. If $n = 1$, then $F_n = 1 > \frac{1}{\mathfrak{g}} = \mathfrak{g}^{n-2} = \frac{2}{1+\sqrt{5}}$. Also, if $n = 2$, then $F_2 = 1 = \mathfrak{g}^0 = \mathfrak{g}^{n-2}$. This establishes the induction step. Now assume that $F_m \geq \mathfrak{g}^{m-2}$ for all $m \in \mathbb{N}$ with $m \leq n$, the induction hypothesis. By the induction hypothesis

$$F_{n+1} = F_n + F_{n-1} \geq \mathfrak{g}^{n-2} + \mathfrak{g}^{n-3} = \mathfrak{g}^{n-3}(\mathfrak{g} + 1).$$

By Exercise 1.32 on page 31, $(\mathfrak{g} + 1) = \mathfrak{g}^2$, so $F_{n+1} \geq \mathfrak{g}^{n-3}\mathfrak{g}^2 = \mathfrak{g}^{n-1}$. By the Principle of Induction (second form) we have proved that this holds for all $n \in \mathbb{N}$. $\square$

Above, we defined the product notation. For instance, $\prod_{i=1}^{7} i = 1 \cdot 2 \cdot 3 \cdot 4 \cdot 5 \cdot 6 \cdot 7 = 5,040$. This is an illustration of the following concept.

Definition A.32 (Factorial Notation!)

If $n \in \mathbb{N}$, then $n!$ (read "enn factorial"), is the product of the first n natural numbers. In other words,

$$n! = \prod_{i=1}^{n} i.$$

We agree, by convention, that $0! = 1$. In other words, multiplication of no factors yields the identity.

The factorial notation gives us the number of distinct ways of arranging n objects. For instance, if you have 10 books on your bookshelf, then you can arrange them in $10! = 3,628,800$ distinct ways.

Now that we have the factorial notation under our belts, we may introduce another symbol, based upon it, which is valuable in number theory.[A.5]

[A.5]Certain counting arguments rely upon a simple idea as follows. If n sets contain $n + 1$ distinct elements in total, then at least one set must contain two or more elements. This is *the Pigeonhole Principle*, from the application of $n + 1$ pigeons flying into n holes. This principle is equivalent to the *Dirichlet Box Principle*, which says: if more than $m \in \mathbb{N}$ objects are placed in m boxes, then at least one of the boxes contains at least two elements.

Definition A.33 (Binomial Coefficients)

If $k, n \in \mathbb{Z}$ with $0 \leq k \leq n$, then the symbol $\binom{n}{k}$, (read "n choose k") is given by

$$\binom{n}{k} = \frac{n!}{k!(n-k)!},$$

the binomial coefficient.[A.6]

Proposition A.34 (Properties of the Binomial Coefficient)

If $n, k \in \mathbb{Z}$ and $0 \leq k \leq n$, then

(a) $\binom{n}{n-k} = \binom{n}{k}$.   (**Symmetry Property**)

(b) $\binom{n+1}{k+1} = \binom{n}{k+1} + \binom{n}{k}$.   (**Pascal's Identity**)[A.7]

(c) $\sum_{i=0}^{n} (-1)^i \binom{n}{i} = 0$.   (**Null Summation Property**)

(d) $\sum_{i=0}^{n} \binom{n}{i} = 2^n$.   (**Full Summation Property**)

Proof. See [144, Proposition 1.2.1, pp. 18–19].   □

An important fundamental result involving binomial coefficients that we will need in the text is the following.

Theorem A.35 (The Binomial Theorem)

Let $x, y \in \mathbb{R}$, and $n \in \mathbb{N}$. Then

$$(x+y)^n = \sum_{i=0}^{n} \binom{n}{i} x^{n-i} y^i.$$

[A.6] It is used in the theory of probability as the number of different combinations of n objects taken k at a time. For instance, the number of ways of choosing two objects from a set of five objects, *without regard for order*, is $\binom{5}{2} = 5!/(2!3!) = 10$ distinct ways.

[A.7] Blaise Pascal (1623-1662) with his contemporaries René Descartes (1596-1650) and Pierre de Fermat (see Footnote 1.40) among others, made France the center of mathematics in the second third of the seventeenth century. When Pascal was only sixteen years old, he published a paper, which was only one page long, and has become known as *Pascal's Theorem*, which says that opposite sides of a hexagon, inscribed in a conic, intersect in three collinear points. Pascal is most remembered for his connections between the study of probability and the arithmetic triangle. Although this triangle had been around for centuries before, Pascal made new and fascinating discoveries about it. Therefore, it is now called *Pascal's Triangle* (see [144, pp. 18–21]). On November 23, 1654 Pascal had an intense religious experience which caused him to abandon mathematics. However, one night in 1658, he was kept awake by a toothache, and began to distract himself by thinking about the properties of the *cycloid*. Suddenly, the toothache disappeared, which he took as divine intervention, and returned to mathematics. He died on August 19, 1662.

Proof. See [144, Theorem 1.2.3, p. 19].   □

Note that the full and null summation properties in Proposition A.34 are just special cases of the binomial theorem (with $x = y = 1$ and $x = 1 = -y$, respectively.)

In the text, we will be in need of some elementary facts concerning matrix theory. We now list these facts, without proof, for the convenience of the reader. The proofs, background, and details may be found in any text on elementary linear algebra.

✦ Basic Matrix Theory

If $m, n \in \mathbb{N}$, then an $m \times n$ matrix (read "m by n matrix") is a rectangular array of entries with m rows and n columns. We will assume that the entries come from a commutative ring with identity R (see page 60). If A is such a matrix, and $a_{i,j}$ denotes the entry in the ith row and j^{th} column, then

$$A = (a_{i,j}) = \begin{pmatrix} a_{1,1} & a_{1,2} & \cdots a_{1,n} \\ a_{2,1} & a_{2,2} & \cdots a_{2,n} \\ \vdots & \vdots & \vdots \\ a_{m,1} & a_{m,2} & \cdots a_{m,n} \end{pmatrix}.$$

Two $m \times n$ matrices $A = (a_{i,j})$, and $B = (b_{i,j})$ are equal if and only if $a_{i,j} = b_{i,j}$ for all i and j. The matrix $(a_{j,i})$ is called the *transpose* of A, denoted by

$$A^t = (a_{j,i}).$$

Addition of two $m \times n$ matrices A and B is done in the natural way.

$$A + B = (a_{i,j}) + (b_{i,j}) = (a_{i,j} + b_{i,j}),$$

and if $r \in R$, then $rA = r(a_{i,j}) = (ra_{i,j})$, called *scalar multiplication*, which is used most often in practice for $R = \mathbb{R}$.

Under the above definition of addition and scalar multiplication, the set of all $m \times n$ matrices with entries from R, a commutative ring with identity, form a set, denoted by $\mathcal{M}_{m \times n}(R)$.[A.8] When $m = n$, this set is in fact a ring given by the following.

If $A = (a_{i,j})$ is an $m \times n$ matrix and $B = (b_{j,k})$ is an $n \times r$ matrix, then the *product* of A and B is defined as the $m \times r$ matrix:

$$AB = (a_{i,j})(b_{j,k}) = (c_{i,k}),$$

where

$$c_{i,k} = \sum_{\ell=1}^{n} a_{i,\ell} b_{\ell,k}.$$

[A.8]For the reader with some knowledge of module theory, this set is in fact an R-module.

Multiplication, if defined, is associative, and distributive over addition. If $m = n$, then $\mathcal{M}_{n \times n}(R)$ is a ring, with identity given by the $n \times n$ matrix:

$$I_n = \begin{pmatrix} 1_R & 0 & \cdots & 0 \\ 0 & 1_R & \cdots & 0 \\ \vdots & \vdots & \vdots & \vdots \\ 0 & 0 & \cdots & 1_R \end{pmatrix},$$

called *the $n \times n$ identity matrix*, where 1_R is the identity of R.

Another important aspect of matrices that we will need throughout the text is motivated by the following. We maintain the assumption that R is a commutative ring with identity. Let $(a, b), (c, d) \in \mathcal{M}_{1 \times 2}(R)$. If we set up these row vectors into a single 2×2 matrix

$$A = \begin{pmatrix} a & b \\ c & d \end{pmatrix},$$

then $ad - bc$ is called the *determinant* of A, denoted by $\det(A)$. More generally, we may define the determinant of any $n \times n$ matrix in $\mathcal{M}_{n \times n}(R)$ for any $n \in \mathbb{N}$. The determinant of any $r \in \mathcal{M}_{1 \times 1}(R)$ is just $\det(r) = r$. Thus, we have the definitions for $n = 1, 2$, and we may now give the general definition inductively. The definition of the determinant of a 3×3 matrix

$$A = \begin{pmatrix} a_{1,1} & a_{1,2} & a_{1,3} \\ a_{2,1} & a_{2,2} & a_{2,3} \\ a_{3,1} & a_{3,2} & a_{3,3} \end{pmatrix}$$

is defined in terms of the above definition of the determinant of a 2×2 matrix, namely $\det(A)$ is given by

$$a_{1,1} \det \begin{pmatrix} a_{2,2} & a_{2,3} \\ a_{3,2} & a_{3,3} \end{pmatrix} - a_{1,2} \det \begin{pmatrix} a_{2,1} & a_{2,3} \\ a_{3,1} & a_{3,3} \end{pmatrix} + a_{1,3} \det \begin{pmatrix} a_{2,1} & a_{2,2} \\ a_{3,1} & a_{3,2} \end{pmatrix}.$$

Therefore, we may inductively define the determinant of any $n \times n$ matrix in this fashion. Assume that we have defined the determinant of an $n \times n$ matrix. Then we define the determinant of an $(n + 1) \times (n + 1)$ matrix $A = (a_{i,j})$ as follows. First, we let $A_{i,j}$ denote the $n \times n$ matrix obtained from A by deleting the ith row and j^{th} column. Then we define the *minor* of $A_{i,j}$ at position (i, j) to be $\det(A_{i,j})$. The *cofactor* of $A_{i,j}$ is defined to be

$$\text{cof}(A_{i,j}) = (-1)^{i+j} \det(A_{i,j}).$$

We may now define the determinant of A by

$$\det(A) = a_{i,1}\text{cof}(A_{i,1}) + a_{i,2}\text{cof}(A_{i,2}) + \cdots + a_{i,n+1}\text{cof}(A_{i,n+1}). \quad \text{(A.36)}$$

This is called the *expansion of a determinant by cofactors* along the ith row of A. Similarly, we may expand along a column of A.

$$\det(A) = a_{1,j}\text{cof}(A_{1,j}) + a_{2,j}\text{cof}(A_{2,j}) + \cdots + a_{n+1,j}\text{cof}(A_{n+1,j}),$$

called the *cofactor expansion along the jth column of A*. Both expansions can be shown to be identical. Hence, a determinant may be viewed as a function that assigns a real number to an $n \times n$ matrix, and the above gives a method for finding that number. Other useful properties of determinants that we will have occasion to use in the text are given in the following.

Theorem A.37 (Properties of Determinants)
    Let R be a commutative ring with identity and let $A = (a_{i,j})$, $B = (b_{i,j}) \in \mathcal{M}_{n \times n}(R)$. Then each of the following hold.

(a) $\det(A) = \det(a_{i,j}) = \det(a_{j,i}) = \det(A^t)$.

(b) $\det(AB) = \det(A) \det(B)$.

(c) *If matrix A is achieved from matrix B by interchanging two rows (or two columns), then $\det(A) = -\det(B)$.*

(d) *If $\mathcal{S}_n$ is the symmetric group on n symbols, then*

$$det(A) = \sum_{\sigma \in \mathcal{S}_n} (\text{sgn}(\sigma)) a_{1,\sigma(1)} a_{2,\sigma(2)} \cdots a_{n,\sigma(n)},$$

where $\text{sgn}(\sigma)$, is 1 or -1 according as σ is even or odd (see Exercise 2.40 on page 100).

If $A \in \mathcal{M}_{n \times n}(R)$, then A is said to be *invertible*, or *nonsingular* if there is a unique matrix denoted by
$$A^{-1} \in M_{n \times n}(R)$$
such that
$$AA^{-1} = I_n = A^{-1}A.$$

Here are some properties of invertible matrices.

Theorem A.38 (Properties of Invertible Matrices)
    Let R be a commutative ring with identity, $n \in \mathbb{N}$, and A invertible in $\mathcal{M}_{n \times n}(R)$. Then each of the following holds.

(a) $(A^{-1})^{-1} = A$.

(b) $(A^t)^{-1} = (A^{-1})^t$, *where "t" denotes the transpose.*

(c) $(AB)^{-1} = B^{-1}A^{-1}$

In order to provide a formula for the inverse of a given matrix, we need the following concept.

Definition A.39 (Adjoint)

Let R be a commutative ring with identity. If $A = (a_{i,j}) \in \mathcal{M}_{n \times n}(R)$, then the matrix

$$A^a = (b_{i,j})$$

given by

$$b_{i,j} = (-1)^{i+j} \det(A_{j,i}) = \mathrm{cof}(A_{j,i}) = \left[(-1)^{i+j} \det(A_{i,j})\right]^t$$

is called the adjoint *of A.*

Some properties of adjoints related to inverses, including a formula for the inverse, are as follows.

Theorem A.40 (Properties of Adjoints)

If R is a commutative ring with identity and $A \in \mathcal{M}_{n \times n}(R)$, then each of the following holds.

(a) $AA^a = \det(A)I_n = A^a A$.

(b) *A is invertible in $\mathcal{M}_{n \times n}(R)$ if and only if $\det(A)$ is a unit[A.9] in R, in which case*

$$A^{-1} = \frac{A^a}{\det(A)}.$$

Example A.41 If $n = 2$, then the inverse of a nonsingular matrix

$$A = \begin{pmatrix} a & b \\ c & d \end{pmatrix}$$

is given by

$$A^{-1} = \begin{pmatrix} \frac{d}{\det(A)} & \frac{-b}{\det(A)} \\ \frac{-c}{\det(A)} & \frac{a}{\det(A)} \end{pmatrix}$$

✦ Polynomials and Polynomial Rings

If R is a ring, then a *polynomial $f(x)$* in an *indeterminant x* with *coefficients* in R is an infinite formal sum

$$f(x) = \sum_{j=0}^{\infty} a_j x^j = a_0 + a_1 x + \cdots + a_n x^n + \cdots,$$

[A.9] A *unit* or *invertible element* u in a commutative ring with identity R is an element for which there exists a multiplicative inverse. In other words, an element $u \in R$ is a unit if there exists an element $u^{-1} \in R$ such that $uu^{-1} = 1_R$.

where the *coefficients* a_j are in R for $j \geq 0$ and $a_j = 0$ for all but a finite number of those values of j. The set of all such polynomials is denoted by $R[x]$. If $a_n \neq 0$, and $a_j = 0$ for $j > n$, then a_n is called the *leading coefficient* of $f(x)$. If the leading coefficient $a_n = 1_R$, in the case where R is a commutative ring with identity 1_R, then $f(x)$ is said to be *monic*.

We may add two polynomials from $R[x]$, $f(x) = \sum_{j=0}^{\infty} a_j x^j$ and $g(x) = \sum_{j=0}^{\infty} b_j x^j$, by

$$f(x) + g(x) = \sum_{j=0}^{\infty} (a_j + b_j) x^j \in R[x],$$

and multiply them by

$$f(x)g(x) = \sum_{j=0}^{\infty} c_j x^j,$$

where

$$c_j = \sum_{i=0}^{j} a_i b_{j-i}.$$

Also, $f(x) = g(x)$ if and only if $a_j = b_j$ for all $j = 0, 1, \ldots$. Under the above operations $R[x]$ is a ring, called the *polynomial ring over R in the indeterminant* x. Furthermore, if R is commutative, then so is $R[x]$, and if R has identity 1_R, then 1_R is the identity for $R[x]$. Notice that with these conventions, we may write $f(x) = \sum_{j=0}^{n} a_j x^j$, for some $n \in \mathbb{N}$, where a_n is the leading coefficient since we have tacitly agreed to "ignore" zero terms.

If $\alpha \in R$, we write $f(\alpha)$ to represent the element $\sum_{j=0}^{n} a_j \alpha^j \in R$, called the *substitution* of α for x. When $f(\alpha) = 0$, then α is called a *root* of $f(x)$. The substitution gives rise to a mapping $\overline{f} : R \mapsto R$ given by $\overline{f} : \alpha \mapsto f(\alpha)$, which is determined by $f(x)$. Thus, $\overline{f}$ is called a *polynomial function* over R.

Definition A.42 (Degrees of Polynomials)

If $f(x) \in R[x]$, with $f(x) = \sum_{j=0}^{d} a_j x^j$, and $a_d \neq 0$, then $d \geq 0$ is called the degree of $f(x)$ over R, denoted by $\deg_R(f)$. If no such d exists, we write $\deg_R(f) = -\infty$, in which case $f(x)$ is the zero polynomial in $R[x]$ (for instance, see Example A.48 below). If F is a field of characteristic zero, then

$$\deg_{\mathbb{Q}}(f) = \deg_F(f)$$

for any $f(x) \in \mathbb{Q}[x]$.[A.10] If F has characteristic p, and $f(x) \in \mathbb{F}_p[x]$, then

$$\deg_{\mathbb{F}_p}(f) = \deg_F(f).$$

In either case, we write $\deg(f)$ for $\deg_F(f)$, without loss of generality, and call this the degree of $f(x)$.

[A.10]The characteristic of a ring R is the smallest $n \in \mathbb{N}$ (if there is one) such that $n \cdot r = 0$ for all $r \in R$. If there is no such n, then R is said to have characteristic 0. Any field containing $\mathbb{Q}$ has characteristic zero, while any field containing the finite field $\mathbb{F}_p$ for a prime p has characteristic p (see the discussion following Definition A.45 below) .

With respect to roots of polynomials, the following is important.

Definition A.43 (Discriminant of Polynomials)

Let $f(x) = a \prod_{j=1}^{n} (x - \alpha_j) \in F[x]$, $\deg(f) = n > 1$, $a \in F$ a field in $\mathbb{C}$, where $\alpha_j \in \mathbb{C}$ are all the roots of $f(x) = 0$ for $j = 1, 2, \ldots, n$. Then the discriminant of f is given by

$$\mathrm{disc}(f) = a^{2n-2} \prod_{1 \leq i < j \leq n} (\alpha_j - \alpha_i)^2.$$

From Definition A.43, we see that f has a multiple root in $\mathbb{C}$ (namely for some $i \neq j$ we have $\alpha_i = \alpha_j$, also called a *repeated root*) if and only if $\mathrm{disc}(f) = 0$.

Example A.44 If $f(x) = ax^2 + bx + c$ where $a, b, c \in \mathbb{Z}$, then $\mathrm{disc}(f) = b^2 - 4ac$ and if $f(x) = x^3 - c$, then $\mathrm{disc}(f) = -27c^2$.

Definition A.45 (Division of Polynomials)

We say that a polynomial $g(x) \in R[x]$ divides $f(x) \in R[x]$, if there exists an $h(x) \in R[x]$ such that $f(x) = g(x)h(x)$. We also say that $g(x)$ is a factor of $f(x)$.

Definition A.46 (Irreducible Polynomials over Rings)

A polynomial $f(x) \in R[x]$ is called irreducible (over R), if $f(x)$ is not a unit in R and any factorization $f(x) = g(x)h(x)$, with $g(x), h(x) \in R[x]$ satisfies the property that one of $g(x)$ or $h(x)$ is in R, called a constant polynomial. In other words, $f(x)$ cannot be the product of two nonconstant polynomials. If $f(x)$ is not irreducible, then it is said to be reducible.[A.11]

In general, it is important to make the distinction between degrees of a polynomial over various rings, since the base ring under consideration may alter the makeup of the polynomial.

For the following example, recall that a *finite field* is a field with a finite number of elements $n \in \mathbb{N}$, denoted by $\mathbb{F}_n$. In general, if K is a finite field, then $K = \mathbb{F}_{p^m}$ for some prime p and $m \in \mathbb{N}$, also called *Galois fields*. The field $\mathbb{F}_p$ is called the *prime subfield* of K. In general, a prime subfield is a field having no proper subfields, so $\mathbb{Q}$ is the prime subfield of any field of characteristic 0 and $\mathbb{Z}/p\mathbb{Z} = \mathbb{F}_p$ is the prime field of any field $K = \mathbb{F}_{p^m}$. Also, we have the following result.

[A.11]Note that it is possible that a reducible polynomial $f(x)$ could be a product of two polynomials of the *same degree* as that of f. For instance, $f(x) = (1 - x) = (2x + 1)(3x + 1)$ in $R = \mathbb{Z}/6\mathbb{Z}$.

Theorem A.47 (Multiplicative Subgroups of Fields)
If F is any field and F^* is a finite subgroup of the multiplicative subgroup of nonzero elements of F, then F^* is cyclic.[A.12] *In particular, if $F = \mathbb{F}_{p^n}$ is a finite field, then F^* is a finite cyclic group.*

Example A.48 The polynomial $f(x) = 2x^2 + 2x + 2$ is of degree two over $\mathbb{Q}$. However, over $\mathbb{F}_2$, $\deg_{\mathbb{F}_2}(f) = -\infty$, since f is the zero polynomial in $\mathbb{F}_2[x]$.

Some facts concerning irreducible polynomials will be needed in the text as follows.

Theorem A.49 (Irreducible Polynomials Over Finite Fields)
The product of all monic irreducible polynomials over a finite field $\mathbb{F}_q$ whose degrees divide a given $n \in \mathbb{N}$ is equal to $x^{q^n} - x$.

Based upon Theorem A.49, the following may be used as an algorithm for testing polynomials for irreducibility over prime fields and thereby generate irreducible polynomials.

Corollary A.50 *The following are equivalent.*

(a) f is irreducible over $\mathbb{F}_p$, where p is prime, and $\deg_{\mathbb{F}_p}(f) = n$.

(b) $\gcd(f(x), x^{p^i} - x) = 1$ for all natural numbers $i \leq \lfloor n/2 \rfloor$.

The following is also a general result concerning irreducible polynomials over any field.

Theorem A.51 (Irreducible Polynomials Over Arbitrary Fields)
Let F be a field and $f(x) \in F[x]$. Denote by $(f(x))$ the principal ideal in $F[x]$ generated by $f(x)$ (see Definition A.59 below). Then the following are equivalent.

(a) f is irreducible over F.

(b) $F[x]/(f(x))$ is a field.

Another useful result is the following.

[A.12]Recall that a multiplicative abelian group is *cyclic* whenever the group generated by some $g \in G$ coincides with G. Note that any group of prime order is cyclic and the product of two cyclic groups of relatively prime order is also a cyclic group. Also, if $\mathcal{S}$ is a nonempty subset of a group G, then the intersection of all subgroups of G containing $\mathcal{S}$ is called the subgroup *generated* by $\mathcal{S}$.

Theorem A.52 (Polynomials, Traces, and Norms)

Suppose that $f(x) \in R[x]$ is a monic, irreducible polynomial (over R where R is an integral domain), $\deg(f) = d \in \mathbb{N}$, and α_j for $j = 1, 2, \ldots, d$ are all of the roots of $f(x)$ in $\mathbb{C}$. Then

$$f(x) = x^d - Tx^{d-1} + \cdots \pm N,$$

where

$$T = \sum_{j=1}^{d} \alpha_j \text{ and } N = \prod_{j=1}^{d} \alpha_j,$$

where T is called the trace *and N is called the* norm *(of any of the roots of $f(x)$).*

Now that we have the notion of irreducibility for polynomials, we may state a unique factorization result for polynomials over fields.

Theorem A.53 (Unique Factorization for Polynomials)

If F is a field, then every nonconstant polynomial $f(x) \in F[x]$ can be factored in $F[x]$ into a product of irreducible polynomials $p(x)$, each of which is unique up to order and units (nonzero constant polynomials) in F.

The Euclidean Algorithm applies to polynomials in a way that allows us to talk about common divisors of polynomials in a fashion similar to that for integers.

Definition A.54 (The GCD of Polynomials)

If $f_i(x) \in F[x]$ for $i = 1, 2$, where F is a field, then the greatest common divisor *of $f_1(x)$ and $f_2(x)$ is the unique monic polynomial $g(x) \in F[x]$ satisfying both:*

(a) *For $i = 1, 2$, $g(x) | f_i(x)$.*

(b) *If there is a $g_1(x) \in F[x]$ such that*

$$g_1(x) | f_i(x)$$

for $i = 1, 2$, then

$$g_1(x) | g(x).$$

If $g(x) = 1$, we say that $f_1(x)$ and $f_2(x)$ are relatively prime, *or* coprime *denoted by*

$$\gcd(f_1(x), f_2(x)) = 1.$$

There is also a Euclidean result for polynomials over a field.

Theorem A.55 (Euclidean Algorithm for Polynomials)

If $f(x), g(x) \in F[x]$, where F is a field, and $g(x) \neq 0$, there exist unique $q(x), r(x) \in F[x]$ such that

$$f(x) = q(x)g(x) + r(x),$$

where $\deg(r) < \deg(g)$. (Note that if $r(x) = 0$, the zero polynomial, then $\deg(r) = -\infty$.)

Finally, if $f(x)$ and $g(x)$ are relatively prime, there exist $s(x), t(x) \in F[x]$ such that

$$1 = s(x)f(x) + t(x)g(x).$$

We will also need the following important result of Lagrange concerning polynomial solutions in modular arithmetic (see Section One of Chapter Two).

Theorem A.56 (Lagrange's Theorem)

Suppose that p is a prime and $f(x) \in \mathbb{Z}[x]$ is polynomial of degree $d \geq 1$ modulo p. Then

$$f(x) \equiv 0 \,(\mathrm{mod}\, p)$$

has at most d incongruent solutions.

Proof. See [144, Theorem 2.51, p. 104].   □

We will need the following important polynomial in the main text.

Definition A.57 (Cyclotomic Polynomials)

If $n \in \mathbb{N}$, then the nth cyclotomic polynomial is given by

$$\Phi_n(x) = \prod_{\substack{\gcd(n,j)=1 \\ 1 \leq j < n}} (x - \zeta_n^j).$$

Also, the degree of $\Phi_n(x)$ is $\phi(n)$, the Euler Totient (see Definition 2.19).

Note that despite the form of the cyclotomic polynomial given in Definition A.57, it can be shown that $\Phi_n(x) \in \mathbb{Z}[x]$. The reader may think of the term *cyclotomic* as "circle dividing," since the nth roots of unity divide the unit circle into n equal arcs. Also, the ζ_n^j are sometimes called *De Moivre*[A.13] *Numbers* (see [202, p. 388]).

[A.13] Abraham De Moivre (1667–1754) was a French-born Huguenot, who left for England when Louis XIV revoked the Edict of Nantes in 1685. He was one of the pioneers of the theory of probability in the early eighteenth century. He became acquainted with Newton and Halley when he went to England. However, as a Frenchman, he was unable to secure a university position there, and remained mostly self-supporting through fees for tutorial services. Yet he produced a considerable amount of research, perhaps the most famous of which is his *Doctrine of Chances* first published in 1718. This and subsequent editions had more than fifty problems on probability. Perhaps the most famous theorem with de Moivre's name attached to it is the one that says: For a, b coordinates in the complex plane, r the radius and ϕ the angle that the radius vector makes with the real axis, $(a + bi)^n = r^n(\cos(n\phi) + i\sin(n\phi))$.

The following section is of importance for us in the main text as a tool for the description of numerous cryptographic devices (see page 61).

✦ Action on Rings

Definition A.58 (Morphisms of Rings)

If R and S are two rings and $f : R \to S$ is a function such that $f(ab) = f(a)f(b)$, and $f(a + b) = f(a) + f(b)$ for all $a, b \in R$, then f is called a ring homomorphism. *If, in addition, $f : R \to S$ is an injection as a map of sets, then f is called a* ring monomorphism. *If a ring homomorphism f is a surjection as a map of sets, then f is called a* ring epimorphism. *If a ring homomorphism f is a bijection as a map of sets, then f is called a* ring isomorphism, *and R is said to be* isomorphic *to S, denoted by $R \cong S$. Lastly, $\ker(f) = \{s \in S : f(s) = 0\}$ is called the* kernel *of f. Also, f is injective if and only if $\ker(f) = \{0\}$.*

There is a fundamental result that we will need in the text. In order to describe it, we need the following notion.

Definition A.59 (Ideal, Cosets, and Quotient Rings) *An* ideal *I in a commutative ring R with identity is a subring of R satisfying the additional property that $rI \subseteq I$ for all $r \in R$. If I is an ideal in R then a* coset *of I in R is a set of the form $r + I = \{r + \alpha : \alpha \in I\}$ where $r \in R$. The set $R/I = \{r + I : r \in R\}$ becomes a ring under multiplication and addition of cosets given by,*

$$(r + I)(s + I) = rs + I, \text{ and } (r + I) + (s + I) = (r + s) + I,$$

for any $r, s \in R$ (and this can be shown to be independent of the representatives r and s). R/I is called the quotient ring *of R by I, or the* factor ring *of R by I, or the* residue class ring modulo I. *The cosets are called the* residue classes modulo I. *A mapping*

$$f : R \mapsto R/I,$$

which takes elements of R to their coset representatives in R/I is called the natural map *of R to R/I, and it is easily seen to be an epimorphism. The cardinality of R/I is denoted by $|R : I|$.*

Example A.60 Consider the ring of integers modulo $n \in \mathbb{N}$, $\mathbb{Z}/n\mathbb{Z}$ introduced in Definition 2.8. Then $n\mathbb{Z}$ is an ideal in $\mathbb{Z}$, and the quotient ring is the residue class ring modulo n.

Remark A.61 *Since rings are also groups, then the above concept of cosets and quotients specializes to groups. In particular, we have the following. Note that an index of a subgroup H in a group G can be defined similarly to the above situation for rings as follows. The index of H in G, denoted by $|G : H|$, is the*

cardinality of the set of distinct right (respectively left) cosets of H in G. Our principal interest is when this cardinality is finite (so this allows us to access the definition of cardinality given earlier). Then Lagrange's Theorem for groups says that

$$|G| = |G : H| \cdot |H|,$$

so if G is a finite group, then $|H| \mid |G|$. In particular, a finite abelian group G has subgroups of all orders dividing $|G|$.

Now we are in a position to state the important result for rings. The reader unfamiliar with the notation "img" of a function should consult Definition A.6 for the description.

Theorem A.62 (Fundamental Isomorphism Theorem for Rings)

If R and S are commutative rings with identity, and

$$\phi : R \to S$$

is a homomorphism of rings, then

$$\frac{R}{\ker(\phi)} \cong \text{img}(\phi).$$

Example A.63 If $\mathbb{F}_q$ is a finite field where $q = p^n$ (p prime) and $f(x) \in \mathbb{F}_p[x]$ is an irreducible polynomial of degree n (see page 289), then

$$\mathbb{F}_q \cong \frac{\mathbb{F}_p[x]}{(f(x))}.$$

The situation in Example A.63 is related to the following definition and theorem.

Definition A.64 (Maximal and Proper Ideals)

Let R be a commutative ring with identity. An ideal $I \neq R$ is called maximal *if whenever $I \subseteq J$, where J is an ideal in R, then $I = J$ or $I = R$. (An ideal $I \neq R$ is called a* proper *ideal.)*

Theorem A.65 (Rings Modulo Maximal Ideals).

If R is a commutative ring with identity then M is a maximal ideal in R if and only if R/M is a field.

Example A.66 If F is a field and $r \in F$ is a fixed nonzero element, then

$$I = \{f(x) \in F[x] : f(r) = 0\}$$

is a maximal ideal and

$$F \cong F[x]/I.$$

Another aspect of rings that we will need in the text is the following. If $\mathcal{S} = \{R_j : j = 1, 2, \ldots, n\}$ is a set of rings, then let R be the set of n-tuples $(r_1, r_2, \ldots, r_n)$ with $r_j \in R_j$ for $j = 1, 2, \ldots n$, with the *zero element* of R being the n-tuple, $(0, 0, \ldots, 0)$. Define addition in R by

$$(r_1, r_2, \ldots, r_n) + (r'_1, r'_2, \ldots, r'_n) = (r_1 + r'_1, r_2 + r'_2, \ldots, r_n + r'_n),$$

for all $r_j, r'_j \in R_j$ with $j = 1, 2, \ldots, n$, and multiplication by

$$(r_1, r_2, \ldots, r_n)(r'_1, r'_2, \ldots, r'_n) = (r_1 r'_1, r_2 r'_2, \ldots, r_n r'_n),$$

This defines a structure on R called the *direct sum* of the rings R_j, $j = 1, 2, \ldots, n$, denoted by

$$\oplus_{j=1}^{n} R_j = R_1 \oplus \cdots \oplus R_n, \tag{A.67}$$

which is easily seen to be a ring. Similarly, when the R_j are groups, then this is a direct sum of groups, which is again a group.

In the text, we will have occasion to refer to such items as vector spaces, so we remind the reader of the definition. The reader is referred to pages 60–62, where we discussed the axioms for algebraic objects such as groups, rings, and fields. In particular, for the sake of completeness, note that any set satisfying all of the axioms of Theorem 2.7, except (g), is called a *division ring*.

✦ Vector Spaces

A *vector space* consists of an additive abelian group V and a field F together with an operation called *scalar multiplication* of each element of V by each element of F on the left, such that for each $r, s \in F$ and each $\alpha, \beta \in V$ the following conditions are satisfied:

A.68. $r\alpha \in V$.

A.69. $r(s\alpha) = (rs)\alpha$.

A.70. $(r + s)\alpha = (r\alpha) + (s\alpha)$.

A.71. $r(\alpha + \beta) = (r\alpha) + (r\beta)$.

A.72. $1_F \alpha = \alpha$.

The set of elements of V are called *vectors* and the elements of F are called *scalars*. The generally accepted abuse of language is to say that V is a *vector space over F*. If V_1 is a subset of a vector space V that is a vector space in its own right, then V_1 is called a *subspace of V*.

Example A.73 For a given prime p, $m, n \in \mathbb{N}$, the finite field $\mathbb{F}_{p^n}$ is an n-dimensional vector space over $\mathbb{F}_{p^m}$ with p^{mn} elements.

Definition A.74 (Bases, Dependence, and Finite Generation)
   If S is a subset of a vector space V, then the intersection of all subspaces of V containing S is called the subspace generated by S, *or* spanned by S. *If there is a finite set S, and S generates V, then V is said to be* finitely generated. *If $S = \varnothing$, then S generates the zero vector space. If $S = \{m\}$, a singleton set, then the subspace generated by S is said to be the* cyclic subspace generated by m.
   A subset S of a vector space V is said to be linearly independent *provided that for distinct $s_1, s_2, \ldots, s_n \in S$, and $r_j \in V$ for $j = 1, 2, \ldots, n$,*

$$\sum_{j=1}^{n} r_j s_j = 0 \text{ implies that } r_j = 0 \text{ for } j = 1, 2, \ldots, n.$$

If S is not linearly independent, then it is called linearly dependent. *A linearly independent subset of a vector space that spans V is called a* basis *for V.*

   In the text, we will have need of the following notion, especially as it pertains to the infinite binary case.

✦ Sequences

Definition A.75 (Sequences)
   A sequence *is a function whose domain is $\mathbb{N}$, with images denoted by a_n, called the nth term of the sequence. The entire sequence is denoted by $\{a_n\}_{n=1}^{\infty}$, or simply $\{a_n\}$, called an* infinite sequence *or simply a* sequence. *If $\{a_n\}$ is a sequence, and $L \in \mathbb{R}$ such that*

$$\lim_{n \to \infty} a_n = L,$$

then the sequence is said to converge *(namely when the limit exists) whereas sequences that have no such limit are said to* diverge. *If the terms of the sequence are nondecreasing, $a_n \leq a_{n+1}$ for all $n \in \mathbb{N}$, or nonincreasing, $a_n \geq a_{n+1}$ for all $n \in \mathbb{N}$, then $\{a_n\}$ is said to be* monotonic. *A sequence $\{a_n\}$ is called* bounded above *if there exists an $M \in \mathbb{R}$ such that $a_n \leq M$ for all $n \in \mathbb{N}$. The value M is called an* upper bound *for the sequence. A sequence $\{a_n\}$ is called* bounded below *if there is a $B \in \mathbb{R}$ such that $B \leq a_n$ for all $n \in \mathbb{N}$, and B is called a* lower bound *for the sequence. A sequence $\{a_n\}$ is called* bounded *if it bounded above and bounded below.*

Some fundamental facts concerning sequences are contained in the following.

Theorem A.76 (Properties of Sequences) *Let $\{a_n\}$ and $\{b_n\}$ be sequences. Then*

(a) *If $\{a_n\}$ is bounded and monotonic, then it converges.*

(b) *If $\lim_{n\to\infty} a_n = \lim_{n\to\infty} b_n = L \in \mathbb{R}$, and $\{c_n\}$ is a sequence such that there exists a natural number N with $a_n \le c_n \le b_n$ for all $n > N$, then $\lim_{n\to\infty} c_n = L$.*

(c) *If $\lim_{n\to\infty} |a_n| = 0$, then $\lim_{n\to\infty} a_n = 0$.*

◆ **Continued Fractions**

Proofs for the results in this section can be found in [144, pp.221–272].

Definition A.77 (Continued Fractions)
If $q_j \in \mathbb{R}$ where $j \in \mathbb{Z}$ is nonnegative and $q_j \in \mathbb{R}^+$ for $j > 0$, then an expression of the form

$$\alpha = q_0 + \cfrac{1}{q_1 + \cfrac{1}{q_2+}}$$

$$+\cfrac{1}{q_k + \cfrac{1}{q_{k+1}}}$$

is called a continued fraction. If $q_k \in \mathbb{Z}$ for all $k \ge 0$, then it is called a simple continued fraction,[A.14] denoted by $\langle q_0; q_1, \ldots, q_k, q_{k+1}, \ldots \rangle$. If there exists a nonnegative integer n such that $q_k = 0$ for all $k \ge n$, then the continued fraction is called finite. *If no such n exists, then it is called* infinite.

Definition A.78 (Convergents) *Let $n \in \mathbb{N}$ and let α have continued fraction expansion $\langle q_0; q_1, \ldots, q_n, \ldots \rangle$ for $q_j \in \mathbb{R}^+$ when $j > 0$. Then*

$$C_k = \langle q_0; q_1, \ldots, q_k \rangle$$

is the kth convergent of α for any nonnegative integer k.

[A.14]Note that the classical definition of a *simple* continued fraction is a continued fraction that arises from the reciprocals as in the Euclidean Algorithm, so the "numerators" are all 1 and the denominators all integers. This is to distinguish from more general continued fractions in which the numerators and denominators can be functions of a complex variable, for instance. Simple continued fractions are also called *regular* continued fractions in the literature.

Theorem A.79 (Finite Simple Continued Fractions are Rational)
   Let $\alpha \in \mathbb{R}$. Then $\alpha \in \mathbb{Q}$ if and only if α can be written as a finite simple continued fraction.

Theorem A.80 (Representation of Convergents)
   Let $\alpha = \langle q_0; q_1, \ldots, q_n, \ldots \rangle$ for $n \in \mathbb{N}$ be a continued fraction expansion. Define two sequences for $k \in \mathbb{Z}$ nonnegative:

$$A_{-2} = 0, A_{-1} = 1, A_k = q_k A_{k-1} + A_{k-2},$$

and

$$B_{-2} = 1, B_{-1} = 0, B_k = q_k B_{k-1} + B_{k-2}.$$

Then

$$C_k = A_k/B_k = \frac{q_k A_{k-1} + A_{k-2}}{q_k B_{k-1} + B_{k-2}},$$

is the kth convergent of α for any nonnegative integer $k \le n$.

Theorem A.81 (Irrationals Are Infinite Simple Continued Fractions)
   Let $\alpha \in \mathbb{R}$. Then α is irrational if and only if α has a unique infinite simple continued fraction expansion $\alpha = \alpha_0 = \langle q_0; q_1, \ldots \rangle = \lim_{k \to \infty} C_k$, where $q_{k-1} = \lfloor \alpha_{k-1} \rfloor$ with $\alpha_k = 1/(\alpha_{k-1} - q_{k-1})$ and $C_k = A_k/B_k$ for $k \in \mathbb{N}$.

Theorem A.82 (Convergents of Surds[A.15])
   Suppose that $D > 0$ is not a perfect square, $n \in \mathbb{Z}$, and $|n| < \sqrt{D}$. If (x, y) is a positive solution of $x^2 - Dy^2 = n$, namely $x, y \in \mathbb{N}$, then x/y is a convergent in the simple continued fraction expansion of $\sqrt{D}$.

Definition A.83 (Periodic Continued Fractions)
   An infinite simple continued fraction $\alpha = \langle q_0; q_1, q_2, \ldots \rangle$ is called periodic *if there exists an integer $k \ge 0$ and $\ell \in \mathbb{N}$ such that $q_n = q_{n+\ell}$ for all integers $n \ge k$. We use the notation*

$$\alpha = \langle q_0; q_1, \ldots, q_{k-1}, \overline{q_k, q_{k+1}, \ldots, q_{\ell+k-1}} \rangle,$$

as a convenient abbreviation.[A.16] The smallest such natural number $\ell = \ell(\alpha)$ is called the period length *of α, and $q_0, q_1, \ldots, q_{k-1}$ is called the* pre-period *of α. If k is the* least *nonnegative integer such that $q_n = q_{n+\ell}$ for all $n \ge k$, then $q_k, q_{k+1}, \ldots, q_{k+\ell-1}$ is called the* fundamental period *of α. If $k = 0$ is the least such value, then α is said to be* purely periodic, *namely $\alpha = \langle \overline{q_0; q_1, \ldots, q_{\ell-1}} \rangle$.*

[A.15] A *surd* is the square root of an integer that is not a perfect square. The term *surd* is actually an archaic term for *square root*. The term *quadratic surd* also refers to the objects introduced in Definition A.85.
[A.16] Some texts call this *eventually* periodic.

In order to introduce the next concept, we need the following notion.

Definition A.84 (Discriminants) *Let $D_0 \neq 1$ be a square-free integer, and set*

$$\Delta_0 = \begin{cases} D_0 & \text{if } D_0 \equiv 1 \,(\text{mod } 4), \\ 4D_0 & \text{otherwise}. \end{cases}$$

Then Δ_0 is called a fundamental discriminant *with associated* fundamental radicand D_0. *Let $f_\Delta \in \mathbb{N}$, and set $\Delta = f_\Delta^2 \Delta_0$. Then*

$$\Delta = \begin{cases} D & \text{if } D \equiv 1 \,(\text{mod } 4) \text{ and } f_\Delta \text{ is odd}, \\ 4D & \text{otherwise}. \end{cases}$$

is a discriminant *with* conductor f_Δ,[A.17] *and associated* radicand

$$D = \begin{cases} f_\Delta^2 D_0 & \text{if } D_0 \not\equiv 1 \,(\text{mod } 4) \text{ or } f_\Delta \text{ is odd}, \\ (f_\Delta/2)^2 D_0 & \text{otherwise}, \end{cases}$$

having underlying fundamental discriminant Δ_0 *with associated* fundamental radicand D_0.

Definition A.85 (Quadratic Irrationals)
Suppose that Δ is a discriminant with underlying radicand $D > 1$. A quadratic irrational, *with underlying discriminant Δ, is a number of the form*

$$\alpha = \frac{P + \sqrt{D}}{Q}, \quad (P, Q \in \mathbb{Z})$$

where $Q \neq 0$ and $P^2 \equiv D \,(\text{mod } Q)$.

Theorem A.86 (Lagrange: Quadratic Irrationals are Periodic)
Let $\alpha \in \mathbb{R}$. Then α has a periodic infinite simple continued fraction expansion if and only if α is a quadratic irrational.

Theorem A.87 (Pure Periodicity Equals Reduction)
Let $\alpha = \langle q_0; q_1, \ldots \rangle$ be an infinite simple continued fraction, with $\ell(\alpha) = \ell \in \mathbb{N}$. Then α is purely periodic if and only if $\alpha > 1$ and $-1 < \alpha' < 0$, where α' is the algebraic conjugate of α. Any quadratic irrational which satisfies these two conditions is called reduced.

[A.17]We use the letter f here for conductor since the German word for it is *Führer*. The origins of the mathematical term *conductor* are rooted in the German language.

Corollary A.88 *If $D > 1$ is not a perfect square, then*

$$\sqrt{D} = \langle q_0; \overline{q_1, \ldots, q_{\ell-1}, 2q_0} \rangle,$$

where $q_j = q_{\ell-j}$ for $j = 1, 2, \ldots, \ell - 1$ and $q_0 = \lfloor \sqrt{D} \rfloor$.

Of crucial importance in the text involving the continued fraction factoring algorithm in Chapter Three is the following material.

Theorem A.89 (Continued Fractions and Recursion)
Let D be a positive integer that is not a perfect square, and let

$$\alpha_0 = (P_0 + \sqrt{D})/Q_0$$

be a quadratic irrational. Recursively define the following for $k \geq 0$:

$$\alpha_k = (P_k + \sqrt{D})/Q_k, \tag{A.90}$$

$$q_k = \lfloor \alpha_k \rfloor, \tag{A.91}$$

$$P_{k+1} = q_k Q_k - P_k, \tag{A.92}$$

and

$$Q_{k+1} = (D - P_{k+1}^2)/Q_k. \tag{A.93}$$

Then $P_k, Q_k \in \mathbb{Z}$ and $Q_k \neq 0$ for $k \geq 0$, and $\alpha_k = \langle q_k; q_{k+1}, \ldots \rangle$.

Theorem A.94 (Continued Fractions and Quadratic Irrationals)
Let $\alpha = (P + \sqrt{D})/Q$ be a quadratic irrational and set

$$G_{k-1} = Q_0 A_{k-1} - P_0 B_{k-1} \quad (k \geq -1),$$

where A_{k-1}, B_{k-1} are given in Theorem A.80 on page 300. Then

$$G_{k-1}^2 - B_{k-1}^2 D = (-1)^k Q_k Q_0 \quad (k \geq 1). \tag{A.95}$$

Corollary A.96 *If $\alpha = \sqrt{D}$, then Equation (A.95) becomes*

$$A_{k-1}^2 - B_{k-1}^2 D = (-1)^k Q_k. \tag{A.97}$$

Theorem A.98 (Reduction and Periodicity)
Let α be a reduced quadratic irrational with $\ell(\alpha) = \ell$. Then both of the following hold.

(a) $P_0 = P_{k\ell}$ *and* $Q_0 = Q_{k\ell}$ *for all $k \geq 0$.*

(b) *If $\beta = \sqrt{D}$ and $\ell(\beta) = \ell$, then $P_1 = P_{k\ell}$ for all $k \geq 1$ and $Q_0 = Q_{k\ell} = 1$ for all $k \geq 0$.*

Solutions to Odd-Numbered Exercises

Section 1.1

1.1 KNOWLEDGE IS POWER

1.3 WAR IS IMMINENT

1.5 One merely reverses the rules to decipher. For instance, the first pair BP of ciphertext letters occurs on the same row. so we choose the letters to their left, PR. The second set DV occurs on a diagonal with AC as the opposite ends (respectively) of the other diagonal. Then KT occurs in the same column, so we choose the letters immediately above them, TI, and so on to get:

PRACTICES ZEALOUSLY PURSUED PASS INTO HABITS,

where the last letter Z is ignored as the filler of the digraph.

1.7 First, put the letters by column, in numeric order, according to the *GERMAN* rectangle as follows.

G	E	R	M	A	N
3	2	6	4	1	5
F	A	X	X	A	F
F	G	A	G	F	G
X	F	A	G	A	X
A	A	F	V	X	F
A	X	X	X	X	A

Now unravel the rows of this rectangle as follows.

FAXXAF FGAGFG XFAGAX
AAFVXF AXXXXA

Now by going to Table 1.3 and determining each letter by the coordinates, such as C for FA, the first pair, we get the following plaintext message.

COME RETRIBUTION

(*It is more than mildly ironic, given what was to ensue, that this was actually used by Jefferson Davis as a Vigenère Key to encipher the very last official cryptogram of the Confederacy, a couple of weeks after Lee's surrender, which ended the American Civil War.*)

1.9 The translation is:

A GOOD GLASS IN THE BISHOP'S HOSTEL IN THE DEVIL'S SEAT TWENTY ONE DEGREES AND THIRTEEN MINUTES NORTHEAST AND BY NORTH

This and the completion of the text in Exercise 1.10 is due to Edgar Allan Poe, who was a cryptanalyst as well as a writer of fiction. A complete explanation is given in the solution of Exercise 1.10 in the Solutions Manual.

Section 1.2

1.11 Let R_n be the number of pairs of rabbits at month n. Let M_n denote the number of pairs of mature rabbits, and I_n the number of immature rabbits. Then

$$R_n = M_n + I_n.$$

For any $n \geq 3$, $M_n = R_{n-1}$, and $I_n = M_{n-1}$, since every newborn pair at time n is the product of a mature pair at time $n-1$. Thus,

$$R_n = R_{n-1} + M_{n-1}.$$

Moreover, $M_{n-1} = R_{n-2}$. Thus, we have

$$R_n = R_{n-1} + R_{n-2}.$$

1.13 By Exercise 1.12,

$$-1 < x - \lfloor x \rfloor - 1 = x + n - 1 - \lfloor x \rfloor - n \leq \lfloor x + n \rfloor - \lfloor x \rfloor - n \leq$$

$$x + n - \lfloor x \rfloor - n = x - \lfloor x \rfloor < 1,$$

so

$$\lfloor x + n \rfloor - \lfloor x \rfloor - n = 0.$$

1.15 Let $n = \lfloor x \rfloor$ and $m = \lfloor y \rfloor$, then $x = n + z$ and $y = m + w$ where $0 \leq z, w < 1$. Therefore,

$$\lfloor x \rfloor + \lfloor -x \rfloor = n + \lfloor -n - z \rfloor = n + \lfloor -n - 1 + 1 - z \rfloor$$

and by Exercise 1.13,

$$= n - n - 1 + \lfloor 1 - z \rfloor = \begin{cases} 0 & \text{if } z = 0, \\ -1 & \text{if } z > 0, \end{cases}$$

1.17 The stronger version of the second result has already been established in the proof of Corollary 1.8 since the last line of the proof shows that $(b^m - 1) \mid (b^t - 1)$.

1.19 (a) Since $p \mid (a^n + 1)$, then there exists an $\ell \in \mathbb{N}$ such that $a^n = p\ell - 1$. Also, by the Division Algorithm, there exist $q, r \in \mathbb{N}$ such that $p = 2nq + r$ with $1 \leq r < 2n$. If $r = 1$, then we have the first conclusion of part (a), so we assume that $r > 1$ and deduce the second conclusion. By Theorem 1.7, $p \mid (a^{p-1} - 1)$, so

$$p \mid (a^{2nq+r-1} - 1) = (a^n)^{2q} a^{r-1} - 1 = (p\ell - 1)^{2q} a^{r-1} - 1,$$

and by the Binomial Theorem this equals $(pt + 1)a^{r-1} - 1$ for some $t \in \mathbb{Z}$. We have shown that

$$p \mid (a^{r-1} - 1),$$

so we may let $a^{r-1} = 1 + pz$ for some $z \in \mathbb{N}$. Set $g = \gcd(r - 1, n)$. Then $r - 1 = gw$ and $n = gk$ with $\gcd(w, k) = 1$ (see Theorem A.24). Also, by Theorem A.22, there exist $u, v \in \mathbb{Z}$ such that $uk - vw = 1$. Thus, by the Binomial Theorem,

$$a^{nu} = (p\ell - 1)^u = (-1)^u + pm,$$

for some $m \in \mathbb{Z}$. Hence,

$$p \mid (a^{nu} - (-1)^u) = a^{gku} - (-1)^u = a^{g(1+vw)} - (-1)^u = a^{gvw}a^g - (-1)^u =$$

$$(a^{gw})^v a^g - (-1)^u = (a^{r-1})^v a^g - (-1)^u = (1+pz)^v a^g - (-1)^u,$$

and by the Binomial Theorem, this equals $(1+py)a^g - (-1)^u$ for some $y \in \mathbb{Z}$. We have shown that

$$p \mid (a^g - (-1)^u).$$

Since $(a^g - (-1)^u) \mid (a^{gk} - (-1)^{uk})$, and $p \mid (a^n + 1) = a^{gk} + 1$, then uk must
· be odd, so both u and k are odd. Therefore, in total, we have shown that

$$p \mid (a^g + 1) = a^{n/k} + 1$$

for an odd divisor k of n.

(b) This is done in the same fashion as in the proof of part (a), and is simpler since we do not have to be concerned about parity issues. Since $p \mid (a^n - 1)$, then $a^n = p\ell + 1$. By the Division Algorithm, there exist $q, r \in \mathbb{Z}$ such that $p = nq + r$ with $1 \leq r < n$. If $r = 1$, then we are done, so we assume that $r > 1$. Since $p \mid (a^{p-1} - 1)$, then as in the proof of part (a), $p \mid (a^{r-1} - 1)$. Also, if $g = \gcd(n, r-1)$, then as in the proof of part (a), $p \mid (a^g - 1)$. Since g is a divisor of n, we are done.

1.21 (a)

$$V_m V_n + \Delta U_n U_m =$$

$$(\alpha^m + \beta^m)(\alpha^n + \beta^n) + (\alpha - \beta)^2 \frac{(\alpha^n - \beta^n)(\alpha^m - \beta^m)}{(\alpha - \beta)^2} =$$

$$\alpha^{n+m} + \alpha^m \beta^n + \alpha^n \beta^m + \beta^{m+n} + \alpha^{n+m} - \alpha^n \beta^m - \alpha^m \beta^n + \beta^{m+n} =$$

$$2(\alpha^{n+m} + \beta^{n+m}) = 2V_{n+m}.$$

(b)–(c) We use induction on n.

Induction Step: For $n = 1$,

$$2^{n-1} U_n = \sum_{k=1}^{\lfloor (n+1)/2 \rfloor} \binom{n}{2k-1} V_1^{n-2k+1} \Delta^{k-1},$$

and

$$2^{n-1} V_n = \sum_{k=0}^{\lfloor n/2 \rfloor} \binom{n}{2k} V_1^{n-2k} \Delta^k.$$

Induction hypothesis:

$$2^{n-2} U_{n-1} = \sum_{k=1}^{\lfloor n/2 \rfloor} \binom{n-1}{2k-1} V_1^{n-2k} \Delta^{k-1},$$

and

$$2^{n-2} V_{n-1} = \sum_{k=0}^{\lfloor (n-1)/2 \rfloor} \binom{n-1}{2k} V_1^{n-2k-1} \Delta^k.$$

We may assume that n is even since the other case is similar. By part (a), $2V_n = V_1 V_{n-1} + \Delta U_{n-1} U_1$, and by the induction hypothesis,

$$2^{n-1} V_n = \sum_{k=0}^{n/2-1} \binom{n-1}{2k} V_1^{n-2k} \Delta^k + \sum_{k=1}^{n/2} \binom{n-1}{2k-1} V_1^{n-2k} \Delta^k =$$

$$V_1^n + \Delta^{n/2} + \sum_{k=1}^{n/2-1} \left(\binom{n-1}{2k} + \binom{n-1}{2k-1} \right) V_1^{n-2k} \Delta^k =$$

$$\sum_{k=0}^{n/2} \binom{n}{2k} V_1^{n-2k} \Delta^k,$$

where the last equality follows from *Pascal's Identity* (see part (b) of Proposition A.34). Now we turn to the proof for U_n.

By part (a), $2V_{n+1} = V_1 V_n + \Delta U_n U_1$. Thus, from what we have just proved we get,

$$U_n = \frac{1}{\Delta}(2V_{n+1} - V_1 V_n) =$$

$$\frac{2}{2^n \Delta} \sum_{k=0}^{n/2} \binom{n+1}{2k} V_1^{n+1-2k} \Delta^k - \frac{V_1}{2^{n-1}\Delta} \sum_{k=0}^{n/2} \binom{n}{2k} V_1^{n-2k} \Delta^k.$$

Therefore,

$$2^{n-1} U_n = \sum_{k=0}^{n/2} \left(\binom{n+1}{2k} - \binom{n}{2k} \right) V_1^{n-2k+1} \Delta^{k-1},$$

and again by Pascal's Identity this equals,

$$\sum_{k=1}^{n/2} \binom{n}{2k-1} V_1^{n-2k+1} \Delta^{k-1},$$

as required.

1.23 We use induction on m. If $m = 0$, then the result is clear. Assume that

$$2Q^{m-j} V_{n-m+j} = V_n V_{m-j} - \Delta U_n U_{m-j},$$

for $1 \le j < m$. Then by Exercise 1.22,

$$V_n V_m - \Delta U_n U_m = V_n(\sqrt{R} V_{m-1} - Q V_{m-2}) - \Delta U_n(\sqrt{R} U_{m-1} - Q U_{m-2}) =$$

$$\sqrt{R}(V_n V_{m-1} - \Delta U_n U_{m-1}) - Q(V_n V_{m-2} - \Delta U_n U_{m-2}) =$$

$$\sqrt{R}(2Q^{m-1} V_{n-m+1}) - Q(2Q^{m-2} V_{n-m+2}),$$

where the last equality is from the induction hypothesis, and this equals

$$2Q^{m-1}(\sqrt{R} V_{n-m+1} - V_{n-m+2}) = 2Q^m V_{n-m},$$

where the last equality is from Exercise 1.22.

1.25 We have

$$U_n V_m + V_n U_m = \frac{(\alpha^n - \beta^n)(\alpha^m + \beta^m)}{\alpha - \beta} + \frac{(\alpha^n + \beta^n)(\alpha^m - \beta^m)}{\alpha - \beta} =$$

$$\frac{\alpha^{n+m} + \alpha^n \beta^m - \alpha^m \beta^n - \beta^{m+n} + \alpha^{n+m} - \alpha^n \beta^m + \alpha^m \beta^n - \beta^{m+n}}{\alpha - \beta} =$$

$$2\frac{\alpha^{n+m} - \beta^{n+m}}{\alpha - \beta} = 2U_{n+m}.$$

1.27 We have

$$U_n V_m - V_n U_m = \frac{(\alpha^n - \beta^n)(\alpha^m + \beta^m)}{(\alpha - \beta)} - \frac{(\alpha^n + \beta^n)(\alpha^m - \beta^m)}{(\alpha - \beta)} =$$

$$\frac{\alpha^{n+m} + \alpha^n \beta^m - \alpha^m \beta^n - \beta^{m+n} - \alpha^{n+m} + \alpha^n \beta^m - \alpha^m \beta^n + \beta^{m+n}}{\alpha - \beta} =$$

$$2\frac{(\alpha^n \beta^m - \alpha^m \beta^n)}{(\alpha - \beta)} = 2(\alpha\beta)^m \frac{(\alpha^{n-m} - \beta^{n-m})}{(\alpha - \beta)} = 2Q^m U_{n-m},$$

where the last equality follows from Exercises 1.20–1.21.

1.29 Let $n = mm_1$. Then

$$U_n/U_m = (\alpha^n - \beta^n)/(\alpha^m - \beta^m) = (\alpha^{mm_1} - \beta^{mm_1})/(\alpha^m - \beta^m) =$$

$$\alpha^{m(m_1-1)} + \alpha^{m(m_1-2)}\beta^m + \alpha^{m(m_1-3)}\beta^{2m} + \cdots + \alpha^m \beta^{m(m_1-2)} + \beta^{m(m_1-1)} =$$

$$V_{m(m_1-1)} + V_{m(m_1-3)}Q^m + V_{m(m_1-5)}Q^{2m} + \cdots + T,$$

where $T = Q^{m(m_1-2)/2} V_m$ if m_1 is even, and $T = Q^{m(m_1-1)/2}$ if m_1 is odd. In either case, U_n/U_m is an integral multiple of $\sqrt{R}$. Hence, $U_m \mid U_n$.

1.31 We use induction. If $n = 1$, then

$$g_{n+2} = g_3 = g_{n+1} + g_n = a + b = aF_1 + bF_2 = aF_n + bF_{n+1}.$$

This is the induction step. Assume the induction hypothesis, namely: $g_{m+1} = aF_{m-1} + bF_m$ for all $m \le n$. Consider

$$g_{n+2} = g_{n+1} + g_n = aF_{n-1} + bF_n + aF_{n-2} + bF_{n-1},$$

by the induction hypothesis. The latter equals $a(F_{n-1} + F_{n-2}) + b(F_n + F_{n-1}) = aF_n + bF_{n+1}$.

1.33 Let

$$N = \sqrt{1 + \sqrt{1 + \sqrt{1 + \cdots}}} = \sqrt{1 + N},$$

so

$$N + 1 = N^2,$$

and from Exercise 1.32, we know that $N = \mathfrak{g}$ is the only solution, since $N > 0$.

1.35 We use induction. If $n = 1$, then

$$\frac{1}{\sqrt{5}}\left[\mathfrak{g}^n - \mathfrak{g}'^n\right] = \frac{1}{\sqrt{5}}\left[\frac{1+\sqrt{5}}{2} - \frac{1-\sqrt{5}}{2}\right] = \frac{\sqrt{5}}{\sqrt{5}} = 1 = F_n.$$

Assume that $F_n = \frac{1}{\sqrt{5}}[\mathfrak{g}^n - \mathfrak{g}'^n]$. By the induction hypothesis we have,

$$F_{n+1} = F_n + F_{n-1} = \frac{1}{\sqrt{5}}\left[\mathfrak{g}^n - \mathfrak{g}'^n\right] + \frac{1}{\sqrt{5}}\left[\mathfrak{g}^{n-1} - \mathfrak{g}'^{n-1}\right],$$

and by factoring out appropriate powers, this is equal to

$$\frac{1}{\sqrt{5}}\left[\mathfrak{g}^{n-1}(1 + \mathfrak{g}) - \mathfrak{g}'^{n-1}(1 + \mathfrak{g}')\right].$$

By Exercise 1.32, $1+\mathfrak{g} = \mathfrak{g}^2$, and the reader may similarly verify that $1+\mathfrak{g}' = \mathfrak{g}'^2$. Hence, $F_{n+1} = \frac{1}{\sqrt{5}}[\mathfrak{g}^{n+1} - \mathfrak{g}'^{n+1}]$.

Section 1.3

1.37 Since $R_{(b,n+1)} = bR_{(b,n)} + 1$ is a recursive definition for a repunit, then an induction yields the result. For $n = 1$, $R_{(b,1)} = 1 = (b^1 - 1)/(b - 1) = 1$. If $R_{(b,n)} = (b^n - 1)/(b - 1)$, then $R_{(b,n+1)} = bR_{(b,n)} + 1 = b(b^n - 1)/(b - 1) + 1 = (b^{n+1} - 1)/(b - 1)$

1.39 (a) $(1110101)_2$ (b) $(1101010)_2$ (c) $(0011110)_2$ (d) $(0100001)_2$

1.41 (a) $8,765,445,679$ (b) $8,989,898,990$ (c) 1 (d) $4,999,888,889$

1.43 (a) $1,111,111,111$ (b) $8,989,898,989$ (c) $0,000,000,009$ (d) $9,999,999,999$

1.45 (a) $1101 0\bar{1}$ (b) $\bar{1}0\bar{1}\bar{1}10$ (c) 11010 (d) 0

1.47 $(10000111111)_2$

1.49 $(999)_{10} = (1111100111)_2$ and $(88)_{10} = (1011000)_2$. For the result of addition, see Exercise 1.47.

1.51 $(1BDDC)_{16}$

1.53 $\bar{1}1\bar{1}0$

1.55 $(1111111030)_6$

1.57 $(10000110010000100001010)_2$

1.59 $(235)_9$

1.61 This is just Example 1.14 since we just negate $\bar{1}110$ to get $1\bar{1}\,\bar{1}0$ and *add* it to $1\bar{1}\,\bar{1}1$ to get 1011.

1.63 $(2222)_3 = (88)_9$, so $(8888)_9 - (88)_9 = (8800)_9$.

1.65 $(17)_{10} = (10001)_2$, $(19)_{10} = (10011)_2$, $(23)_{10} = (10111)_2$, $(29)_{10} = (11101)_2$, $(31)_{10} = (11111)_2$, $(37)_{10} = (100101)_2$.

1.67 By the Division Algorithm, there are $q_0, a_0 \in \mathbb{Z}$ with $n = q_0|b| + a_0$ where $0 \le a_0 < |b|$. Thus, $-q_0 b = n - a_0$, so $|q_0 b| \le |n| + a_0 \le |n| + |b| - 1$. Therefore, $|q_0| \le (|n| + |b| - 1)/|b|$.
If $(|n| + |b| - 1)/|b| \ge |n|$, then $|n| + |b| - 1 \ge |nb|$, so $|b| - 1 \ge |n|(|b| - 1)$. Since $|b| > 1$, then $n \in \{0, \pm 1\}$. Since $n \ne 0$ and the unique representations for 1 and

-1 are given by $1 = (1)_b$ and $-1 = (1, |b| - 1)_b = (1, -b - 1)_b$, then we may assume that $(|n| + |b| - 1)/|b| < |n|$.

In this case, $|n| > |q_0|$. We now use the Division Algorithm to get $-q_0 = q_1|b| + a_1$, where $0 \leq a_1 < |b|$, and we proceed as in the above argument for n. Eventually this leads to the required representation since

$$n = -q_0 b + a_0 = (-q_1 b + a_1)b + a_0 = -q_1 b^2 + a_1 b + a_0 = \cdots,$$

so $n = (a_{t_n} \ldots a_1 a_0)_b$. Uniqueness is now proved in essentially the same fashion as in the proof of Theorem 1.11.

1.69 (a) $-1 = (11)_{-2}$ (b) $-10 = (1010)_{-2}$ (c) $-17 = (110011)_{-2}$ (d) $-100 = (11101100)_{-2}$

1.71 If such an N exists, then either $N = 1$ in which case $r = \lfloor r \rfloor = \lfloor r \rfloor / b^0$, or $N > 1$ in which case we have

$$r = \lfloor r \rfloor + \sum_{j=1}^{N-1} a_j b^{-j} = \frac{\lfloor r \rfloor b^{N-1} + \sum_{j=1}^{N-1} a_j b^{N-1-j}}{b^{N-1}},$$

and $n = \lfloor r \rfloor b^{N-1} + \sum_{j=1}^{N-1} a_j b^{N-1-j} \in \mathbb{Z}$, as required.

Conversely, if $r = n/b^m$ for some $n, m \in \mathbb{Z}$ with $m \geq 0$, then by Exercise 1.70, for all $j \geq m$, we have

$$a_j = \lfloor b^j r \rfloor - b \lfloor b^{j-1} r \rfloor = \lfloor b^{j-m} n \rfloor - b \lfloor b^{j-1-m} n \rfloor = b^{j-m} n - b^{j-m} n = 0,$$

so the representation terminates.

1.73 $(1000101100100)_2$

1.75 $(FEDCAB)_{16}$

1.77 $(276)_8 = (167)_8 (1)_8 + (107)_8$.

1.79 $(101010)_2 = (10)_2 (1111)_2 + (1100)_2$.

1.81 First we set:

$$a = (a_{2n-1} \ldots a_1 a_0)_2 \text{ and } b = (b_{2n-1} \ldots b_1 b_0)_2.$$

Then we write

$$a = 2^n A_1 + A_0 \text{ with } A_1 = (a_{2n-1} \ldots a_n)_2 \text{ and } A_0 = (a_{n-1} \ldots a_0)_2,$$

and

$$b = 2^n B_1 + B_0 \text{ with } B_1 = (b_{2n-1} \ldots b_n)_2 \text{ and } B_0 = (b_{n-1} \ldots b_0)_2.$$

Then

$$ab = (2^{2n} + 2^n) A_1 B_1 + 2^n (A_1 - A_0)(B_1 - B_0) + (2^n + 1) A_0 B_0.$$

This recursive algorithm reduces the problem of multiplying $2n$-bit numbers to three multiplications of the n-bit numbers:

$$A_1 B_1, \ (A_1 - A_0)(B_1 - B_0), \text{ and } A_0 B_0,$$

together with some shifting and adding operations.

Section 1.4

1.83 As we saw in the description of the algorithm in Section Three, as well as in Example 1.18, we require that every digit of one number be multiplied by every digit of the other. In other words, there are mn such bit operations. This is one interpretation. However, as noted in Section Four, there is not a single "correct answer" in these estimations. For instance, suppose that $n \geq m$. Then in the multiplication process (with n on top in the multiplication diagram) we get at most m rows, each having at most $m + n$ bits. Thus, if we obtain the answer to the multiplication by adding the jth row to the $(j-1)^{st}$ row for each $j \geq 1$ until completion, then we need at most m additions of at most $m + n$-bit integers. Since each addition takes $m + n$ bit operations, the total number of bit operations to get our answer is at most $m(m + n)$ bit operations. Given that $m \leq n$, then the upper bound is $2mn$ bit operations. Either one of the above answers is accurate. The first estimate may be justified by observing that, because of the increasing number of zeros on the right in each row, each addition involves only n nontrivial bit operations. Nevertheless, we must always take the "worst-case" scenario as our answer, since this gives us the simplest and "safest" estimate. In any case, these are both $O(mn)$.

1.85 This is essentially the same as the solution to Exercise 1.83. In the division algorithm given on page 41, each subtraction requires at most $m + n$ bit operations and we have at most m subtractions, so there are at most $m(m + n)$ bit operations. Hence, we have $O(mn)$ as required.

1.87 By part (a) of Theorem 1.24, $O(a_j x^j) = O(x^j)$ for each j. Thus, by part (b) of the theorem, $O(f) = O(x^n)$.

1.89 $O(n \ln^2 n)$

Section 2.1

2.1 (a) If $n \in \mathbb{N}$, and $a, b, c \in \mathbb{Z}$, then $a - a = 0$, which implies that

$$a - a \equiv 0 \,(\text{mod } n).$$

In other words,

$$a \equiv a \,(\text{mod } n),$$

which establishes the reflexive property.

(b) Let $n \in \mathbb{N}$, $a, b, c \in \mathbb{Z}$, and $a \equiv b \,(\text{mod } n)$. Then $a - b = kn$ for some $k \in \mathbb{Z}$, so $b - a = (-k)n$. Therefore, $b \equiv a \,(\text{mod } n)$, which establishes the symmetric property.

To prove part (c), we use the definition of congruence. Since $a \equiv b \,(\text{mod } n)$, and $b \equiv c \,(\text{mod } n)$, then $n \mid (a - b)$ and $n \mid (b - c)$. Therefore, $n \mid (a - b) + (b - c) = (a - c)$, that is $a \equiv c \,(\text{mod } n)$.

2.3 Suppose that $a \equiv b \,(\text{mod } n)$. Then $a - b = kn$ for some integer k. Multiplying the equation by m, we get $m(a - b) = knm$. In other words, $am \equiv bm \,(\text{mod } n)$.

2.5 Since $a = b + kn$ for some $k \in \mathbb{Z}$ and $n = \ell m$ for some $\ell \in \mathbb{N}$, then $a = b + k\ell m$, so $a - b = (k\ell)m$, whence $a \equiv b \,(\text{mod } m)$.

2.7 Let $n \equiv 3 \,(\text{mod } 4)$, and suppose that $x^2 \equiv -1 \,(\text{mod } n)$, for some integer x. If all primes dividing n are of the form $p \equiv 1 \,(\text{mod } 4)$, then it follows that $n \equiv 1$ $(\text{mod } 4)$. Hence, there is a prime $p | n$ such that $p \equiv 3 \,(\text{mod } 4)$. Therefore, $x^2 \equiv -1$ $(\text{mod } p)$. By Fermat's Little Theorem, $x^{p-1} \equiv 1 \,(\text{mod } p)$. However,

$$x^{p-1} = (x^2)^{(p-1)/2} \equiv (-1)^{(p-1)/2} \equiv -1 \,(\text{mod } p)$$

since $(p-1)/2$ is odd. Therefore, we have that $-1 \equiv 1 \,(\text{mod } p)$, forcing $p = 2$, a contradiction.

2.9 A counterexample for (a) is obtained by letting $u = 4$ and $n = 9$. In that case, $a \equiv 1 \,(\text{mod } 3)$, but $a^2 \not\equiv 1 \,(\text{mod } 9)$. A proof of (b) is obtained by merely noting if $a^2 \equiv 1 \,(\text{mod } n)$, then any prime $p \,\big|\, n$ must satisfy $p \,\big|\, a^2 - 1 = (a-1)(a+1)$, so $p \,\big|\, (a-1)$ or $p \,\big|\, (a+1)$, namely $a \equiv \pm 1 \,(\text{mod } p)$. A counterexample for (c) is given by the one for (a).

2.11 This follows from the fact that

$$\binom{p}{j} = \frac{p!}{(p-j)!j!} \in \mathbb{N},$$

and $(p-j) \nmid p$, $j \nmid p$ for all natural numbers $j \le p - 1$.

2.13 By Exercise 2.12, we may take a solution x_0 to be in the least residue system modulo n/g. Thus, distinct solutions of $ax \equiv b \,(\text{mod } n)$ are precisely the values $x - x_0 + mn/g$ where $m = 0, 1, \ldots, g-1$. Hence, x_0 is the unique solution given by $m = 0$.

2.15 By Fermat's Little Theorem,

$$\sum_{j=1}^{p-1} j^{p-1} \equiv \sum_{j=1}^{p-1} 1 \equiv p - 1 \equiv -1 \,(\text{mod } p).$$

2.17 They are 5, 13, and 563.

2.19 By Fermat's Little Theorem, $2^{p-1} \equiv 1 \,(\text{mod } p)$ for all primes p dividing $\mathfrak{F}_n$. By the Division Algorithm, there exist $q, r \in \mathbb{Z}$ such that $p - 1 = 2^{n+1}q + r$ where $0 \le r < 2^{n+1}$. Therefore,

$$1 \equiv 2^{p-1} \equiv 2^{2^{n+1}q+r} \equiv (2^{2^{n+1}})^q 2^r \equiv 2^r \,(\text{mod } p)$$

by Exercise 1.18, so $r = 0$ and $2^{n+1} \,\big|\, (p-1)$. In other words, $p = 2^{n+1}m + 1$ for some $m \in \mathbb{N}$.

2.21 By Euler's Theorem, $m^{\phi(n)} \equiv 1 \,(\text{mod } n)$, so $m \cdot m^{\phi(n)-1} \equiv 1 \,(\text{mod } n)$. In other words, $m^{\phi(n)-1}$ is a multiplicative inverse of m modulo n.

2.23 This follows from a recursive use of Claim 2.23 in the proof of Theorem 2.22.

2.25 If $n = \prod_{j=1}^{k} p_j^{a_j}$ is the canonical prime factorization of n, then $d = \prod_{j=1}^{k} p_j^{b_j}$ for some $0 \le b_j \le a_j$. Therefore, by Theorem 2.22,

$$\phi(d) = \prod_{j=1}^{k} p_j^{b_j-1}(p_j - 1) \,\Big|\, \prod_{j=1}^{k} p_j^{a_j-1}(p_j - 1) = \phi(n).$$

2.27 A simple solution is $n = -4$ since $-4 = 5(-1) + 1$ is a division of five heaps with the monkey getting one coconut and the sailor getting -1 coconuts. Thus, since during the process we have to divide by 5 six times, the smallest positive solution is $n = -4 + 5^6 = 15,621$.

2.29 $x = 23$

2.31 In the notation of Theorem 2.29, $n_1 = 3$, $n_2 = 6$, $n_3 = 10$, $r_1 = 1$, $r_2 = 4$, and $r_3 = 1$. Since $\gcd(n_2, n_3) = 2 \nmid (r_2 - r_3) = 3$, then there can be no solution.

2.33 At the jth step for each $j \in \mathbb{N}$ with $j \leq k$, there exist at most two multiplications of numbers in size less than n^2, so each step takes $O(\ln^2(n^2))$ bit operations.

2.35 Define a map $\psi_j : \mathbb{Z} \mapsto \mathbb{Z}/n_j/\mathbb{Z}$ defined by $\psi_j(m) = \overline{m}$, the residue class of m modulo n. Then let

$$\psi : \mathbb{Z} \mapsto \mathbb{Z}/n_1\mathbb{Z} \oplus \cdots \oplus \mathbb{Z}/n_\ell\mathbb{Z}$$

be defined by

$$\psi(m) = (\psi_1(m), \psi_2(m), \ldots, \psi_\ell(m)).$$

This is a ring homomorphism. By Theorem 2.27, there exists an $m \in \mathbb{Z}$ such that $\psi(m) = (m_1, m_2, \ldots, m_\ell)$ if and only if $\psi_j(m) = m_j$ for each $j = 1, 2, \ldots, \ell$. Thus, ψ is onto. Since $\psi(m) = 0$ if and only if $m \equiv 0 \pmod{n_j}$ for each j and this holds in turn if and only if $m \equiv 0 \pmod{\prod_{j=1}^{\ell} n_j}$, then $\ker(\psi)$ is the ideal $\prod_{j=1}^{\ell} n_j\mathbb{Z}$ in $\mathbb{Z}$. By Theorem A.62 on page 296, we have the desired isomorphism.

Section 2.2

2.37 Since $a^{-1}(am + b) - a^{-1}b = m$, then by choosing $f^{-1}(x) = a^{-1}x - a^{-1}b = a^{-1}(x - b)$, we get the inverse function.

2.39 If H is a subgroup of G, then (a)–(c) follow from the definition of a subgroup. Conversely, if (a)–(c) hold, then by (a), part (b) of Theorem 2.7 holds. By (b), part (i) of Theorem 2.7 holds. By part (a) and the fact that associativity holds for G, we must have associativity for H, so part (h) of Theorem 2.7 is satisfied. By (c) multiplicative inverses exist, so H is a group in its own right by the definition given on page 62. In other words, it is a subgroup of G.

2.41 First, we convert the letters, including the key, to their numerical equivalents via Table 2.37 to get:

$$(CJUJ) = (2, 9, 20, 9), \quad (LAFY) = (11, 0, 5, 24), \quad (TCTU) = (19, 2, 19, 20),$$

$$(LTKG) = (11, 19, 10, 6), \text{ and } \quad (EASY) = (4, 0, 18, 24).$$

Employing the decryption/encryption function, and converting the numbers to letters via Table 2.37 we get,

$$e(2, 9, 20, 9) = (4 - 2, 0 - 9, 18 - 20, 24 - 9) = (2, 17, 24, 15) = (CRYP),$$

$$e(11, 0, 5, 24) = (4 - 11, 0 - 0, 18 - 5, 24 - 24) = (19, 0, 13, 0) = (TANA),$$

$$e(19, 2, 19, 20) = (4 - 19, 0 - 2, 18 - 19, 24 - 20) = (11, 24, 25, 4) = (LYZE),$$

and $\quad e(11, 19, 10, 6) = (4 - 11, 0 - 19, 18 - 10, 24 - 6) = (19, 7, 8, 18) = (THIS).$

The end result is:

CRYPTANALYZE   THIS

2.43 First we calculate that

$$e^{-1} = \begin{pmatrix} 1 & 24 \\ 25 & 3 \end{pmatrix}.$$

Then we decipher each pair of ciphertext numerical equivalents determined from Table 2.37 as follows.

$$(12, 6)e^{-1} = (6, 20),$$
$$(5, 18)e^{-1} = (13, 18),$$
$$(3, 24)e^{-1} = (5, 14),$$
$$(6, 15)e^{-1} = (17, 7),$$
$$(15, 7)e^{-1} = (8, 17),$$

and

$$(11, 7)e^{-1} = (4, 25).$$

Now we use Table 2.37 to get the plaintext equivalents of our number pairs:

$$G \; U \; N \; S \quad F \; O \; R \quad H \; I \; R \; E$$

where there is an extra Z at the end, which had been enciphered to complete the last plaintext pair, so we discard it.

2.45 Since $(01101) = 13, (01110) = 14, (10010) = 18, (10100) = 20, (00010) = 2, (00111) = 7, (00000) = 0, (00110) = 6, (00100) = 4, (01101) = 13, (00010) = 2$, and $(11000) = 24$, then via Table 2.37, we get the following plaintext.

NO SUCH AGENCY

The NSA is often said to be an acronym for this phrase. Another one is:
NEVER SAY ANYTHING

Section 2.3

2.47 The weak keys are:

$$(c_0, d_0) = (0^{28}, 0^{28}), (1^{28}, 1^{28}), (0^{28}, 1^{28}), (1^{28}, 0^{28}) \in \mathbb{Z}^{28} \times \mathbb{Z}^{28}.$$

2.49 Since B can reconstruct $y_1, \ldots, y_n$ from $x_1, \ldots, x_n$ using k, then the enciphered MAC y_n can be compared to the received MAC y_n to verify authenticity. (*Note that any change of the x_j by a cryptanalyst (who does not have k) will result in a change to the MAC, so that it will not be accepted by B.*)

Section 2.4

2.51

$$k + c =$$

$$(1100101000110011110001010111000101111111010101010001)+$$
$$(1011100101111010111100010100000111110000010101010010) =$$
$$(0111001101001001001101000011000010001110000000011) =$$

$$m.$$

2.53 Converting **LPXEHGM** to numerical equivalents, we have

$$11, 15, 23, 4, 7, 6, 12.$$

Thus, we compute the following. $m_1 = c_1 - k_1 = 11 - 7 = 4$, $m_2 = c_2 - k_2 = 15 - 2 = 13$, $m_3 = c_3 - m_1 = 23 - 4 = 19$, $m_4 = c_4 - m_2 = 4 - 13 \equiv 17 \pmod{26}$, $m_5 = c_5 - m_3 = 7 - 19 \equiv 14 \pmod{26}$, $m_6 = c_6 - m_4 = 6 - 17 \equiv 15 \pmod{26}$, and $m_7 = c_7 - m_5 = 12 - 14 \equiv 24 \pmod{26}$. Via Table 2.37, the letter equivalents give us:

ENTROPY

2.55 The output table is given by:

n	k_{4n+7}	k_{4n+6}	k_{4n+5}	k_{4n+4}
-1	1	0	0	0
0	1	1	0	0
1	1	1	1	0
2	1	1	1	1
3	0	1	1	1
4	1	0	1	1
5	0	1	0	1
6	1	0	1	0
7	1	1	0	1
8	0	1	1	0
9	0	0	1	1
10	1	0	0	1
11	0	1	0	0
12	0	0	1	0
13	0	0	0	1
14	1	0	0	0

Thus, the pn-sequence is given by

$$b = (100110101111000) = (k_{(m-1)\ell} k_{(m-2)\ell} \cdots, k_\ell k_0) = (k_{56} k_{52} \ldots k_4 k_0).$$

2.57 The answer is $\lambda(n)$, the *Carmichael Function*, defined as follows. If $n = 2^a \prod_{j=1}^{k} p_j^{a_j}$ is the canonical prime factorization of $n \in \mathbb{N}$, namely $2 < p_1 < p_2 < \cdots < p_k$, then

$$\lambda(n) = \begin{cases} \phi(n) & \text{if } n = 2^a, \text{ and } 1 \le a \le 2, \\ 2^{a-2} = \phi(n)/2 & \text{if } n = 2^a, a > 2, \\ \text{lcm}(\lambda(2^a), \phi(p_1^{a_1}), \ldots, \phi(p_k^{a_k})) & \text{if } k \ge 1. \end{cases}$$

(See [144] for more information on this function. The Carmichael Function was first discussed by Cauchy [45] in 1841.)

Section 3.1

3.1 First, we compute the multiplicative inverse of $e = 5$ modulo 3048, which is $d = 1829$. Then we may decipher 76 by $76^{1829} \equiv 5 \pmod{3049}$. Similarly the remaining ciphertext digits are deciphered to give the plaintext:

$$(5, 14, 8, 11, 4, 3, 0, 6, 0, 8, 13).$$

Then using table 2.37, we get the English equivalents: **FOILED AGAIN**.

3.3 Assume that $2^b \le e < 2^{b+1}$, and that we use the repeated squaring method given on page 71, to compute m^{2^j} modulo n for $j = 0, 1, \ldots, b$. This requires $\lfloor \log_2 e \rfloor$ squarings, and each squaring requires $O((\log_2 n)^2)$ bit operations. In total, this is $O((\log_2 n)^2 \log_2 e)$ bit operations. We must also multiply the residues, and reduce modulo n at each stage. This takes $O((\log_2 n)^2 \log_2 e)$ bit operations, as well since there are at most $\log_2 e$ multiplications, each taking $O((\log_2 n)^2)$ bit operations. Hence, the total must be $O((\log_2 n)^2 \log_2 e)$ bit operations.

3.5 Let α and β be two generators of $(\mathbb{Z}/p\mathbb{Z})^*$ and let $\gamma \in (\mathbb{Z}/p\mathbb{Z})^*$. Set $x = \log_\alpha \gamma$, $y = \log_\beta \gamma$, and $z = \log_\alpha \beta$. Then $\alpha^x = \gamma = \beta^y = (\alpha^z)^y$. Thus,

$$x \equiv zy \,(\mathrm{mod}\; p - 1),$$

by Fermat's Little Theorem. Hence,

$$\log_\alpha \gamma \equiv \log_\beta \gamma \log_\alpha \beta \,(\mathrm{mod}\; p),$$

or equivalently,

$$\log_\alpha \gamma \log_\alpha^{-1} \beta \equiv \log_\beta \gamma \,(\mathrm{mod}\; p).$$

In other words, any algorithm that can compute logarithms to base α can be used to compute logarithms to any other base β that generates $(\mathbb{Z}/p\mathbb{Z})^*$.

3.7 (1) A must not be able to invert B's output to recover R.

(2) B must not be able to find two different inputs of opposite parity that lead to the same output.

3.9 First, note that we have for each $k = 0, 1, \ldots, a_j - 1$,

$$x_k = \sum_{i=k}^{a_j - 1} b_i^{(j)} p_j^i \tag{S1}$$

and

$$\beta_k = \alpha^{x_k},$$

so

$$\beta_k^{(p-1)/p_j^{k+1}} \equiv \alpha^{(p-1)x_k/p_j^{k+1}} \,(\mathrm{mod}\; p).$$

Thus, it suffices to prove:

$$\alpha^{(p-1)x_k/p_j^{k+1}} \equiv \alpha^{(p-1)b_k^{(j)}/p_j} \,(\mathrm{mod}\; p),$$

which holds precisely when

$$\frac{(p-1)x_k}{p_j^{k+1}} \equiv \frac{(p-1)b_k^{(j)}}{p_j} \,(\mathrm{mod}\; p - 1)$$

by Fermat's Little Theorem. From (S1),

$$\frac{(p-1)x_k}{p_j^{k+1}} - \frac{(p-1)b_k^{(j)}}{p_j} = \frac{(p-1)(x_k - p_j^k b_k^{(j)})}{p_j^{k+1}} =$$

$$\frac{p-1}{p_j^{k+1}} \left(\sum_{i=k}^{a_j-1} b_i^{(j)} p_j^i - b_k^{(j)} p_j^k \right) = (p-1) \left(\sum_{i=k+1}^{a_j-1} b_i^{(j)} p_j^{i-k-1} \right),$$

which is congruent to 0 modulo $p - 1$.

Section 3.2

3.11 Since $ed \equiv 1 \,(\mathrm{mod}\ \phi(n))$, there exists a $g \in \mathbb{Z}$ such that $ed = 1 + g\phi(n)$. If $p \nmid m$, then by Fermat's Little Theorem, $m^{p-1} \equiv 1 \,(\mathrm{mod}\ p)$. Hence,

$$m^{ed} = m^{1+g(p-1)(q-1)} \equiv m(m^{g(q-1)})^{p-1} \equiv m \,(\mathrm{mod}\ p). \qquad (S2)$$

If $p \mid m$, then (S2) holds again since $m \equiv 0 \,(\mathrm{mod}\ p)$. Hence, $m^{ed} \equiv m \,(\mathrm{mod}\ p)$ for any m. Similarly, $m^{ed} \equiv m \,(\mathrm{mod}\ q)$. Since $p \neq q$, $m^{ed} \equiv m \,(\mathrm{mod}\ n)$. Thus,

$$c^d \equiv (m^e)^d \equiv m \,(\mathrm{mod}\ n).$$

3.13 We find $\phi(n) = \phi(16199) = 15936$. Then use the extended Euclidean Algorithm to solve $1 = 797d + 15936x$ to get $d = 11957$ (and $x = -598$). Decipher each $c \in \mathcal{C}$ via $c^{11957} \equiv m \,(\mathrm{mod}\ 16199)$. For instance, $14238^{11957} \equiv 13 \,(\mathrm{mod}\ 16199)$. Similarly, we get the rest.

$$\mathcal{M} = \{13, 14, 18, 20, 02, 07, 00, 06, 04, 13, 02, 24\}.$$

The English plaintext via Table 3.12 is

NO SUCH AGENCY.

(See the Solution to Exercise 2.45.)

3.15 Since

$$\overline{m} \equiv \overline{c}^d \equiv c^d(x^e)^d \equiv mx \,(\mathrm{mod}\ n),$$

then C merely computes $m \equiv \overline{m}x^{-1} \,(\mathrm{mod}\ n)$.

3.17 A computes

$$\beta = 6^{701} \equiv 1664 \,(\mathrm{mod}\ 1777) \text{ and } \gamma \equiv 1341 \cdot 1729^{701} \equiv 1103 \,(\mathrm{mod}\ 1777).$$

Thus , $(\beta, \gamma) = (1664, 1103)$ gets sent to B.

3.19 Since $n = pq$, a factorization of n can be achieved via

$$p + q = n - \phi(n) + 1 \text{ and } p - q = \sqrt{(p+q)^2 - 4n}.$$

Hence, that the two are computationally equivalent follows, so they have the same complexity.

Section 3.3

3.21 B first computes $m' = s^e = 49000^5 \equiv 2014 \,(\mathrm{mod}\ n)$. Since $m' \in \mathrm{img}(R)$, then B indeed accepts s and recovers the message $R^{-1}(m') = m' + 1 = 2015 = m$.

3.23 B computes $s^2 = 12535^2 \equiv 153673 \,(\mathrm{mod}\ 594593)$. Since $153673 \in \mathrm{img}(R)$, then B accepts m' and recovers $m = R^{-1}(m') = m + 1 = 153674$.

3.25 B verifies that $\beta = 32 < p - 1 = 3676$, and accepts it. Then B computes

$$x \equiv \alpha^{a\beta}\beta^s \equiv 1867^{32} \cdot 32^{3397} \equiv 1679 \,(\mathrm{mod}\ 3677),$$

and $h(m) = h(3011) = 3011 - 982 = 2029$ to get,

$$z = \alpha^{h(m)} = 2^{2029} \equiv 1679 \,(\mathrm{mod}\ 3677).$$

Since $x = z$, then B does accept m as valid.

Section 3.4

3.27 Using B's public key, A enciphers $c = \sum_{j=1}^{7} k_j m_j \equiv 110 \,(\mathrm{mod}\ 200)$ and sends c to B. Then B computes $d = r^{-1}c \equiv 163 \cdot 110 \equiv 130 \,(\mathrm{mod}\ 200)$. Solving the superincreasing subset sum problem, B gets $130 = 70 + 32 + 16 + 8 + 4$, so $x_j = 1$ for $3 \le j \le 7$ and $x_1 = x_2 = 0$. Thus, since $m_j = x_{\sigma(j)}$, then $m_j = 1$ for $1 \le j \le 5$ and $m_6 = m_7 = 0$, thereby recovering m.

3.29 A sets $m = 111$ as a bitstring of length three, then converts m to a 5-tuple as follows. $V_0 = 1$ since $m = 7 > \binom{4}{3}$. $V_1 = 1$ since $m = 3 = \binom{3}{3}$. $V_2 = 0$ since $m = 0 < \binom{2}{1} = 2$. $V_3 = 0$ since $m = 0 < \binom{1}{1} = 1$. $V_4 = 1$ since $m = 0 = \binom{0}{1}$. Thus, $V = (1, 1, 0, 0, 1)$, so A computes

$$c = \sum_{j=0}^{4} V_j c_j = c_0 + c_1 + c_4 = 52 + 59 + 79 \equiv 66 \,(\mathrm{mod}\ 124),$$

which gets sent to B, who computes

$$z = c - nd = 66 - 3 \cdot 29 \equiv 103 \,(\mathrm{mod}\ 124),$$

$$h(z) = (2x^2 + 2)^{103} \equiv 2x^2 + x + 4 \,(\mathrm{mod}\ x^3 + x + 1),$$

and

$$k(x) = h(x) + r(x) = x^3 + 2x^2 + 2x + 5,$$

which factors over $\mathbb{F}_5$ as,

$$k(x) = x(x + 3)(x + 4).$$

Thus, $r_1 = 0$, $r_2 = 4$, and $r_3 = 3$ from which B recovers m as

$$m = \sum_{j=1}^{n} V_{\sigma^{-1}(r_j)} \binom{p - 1 - \sigma^{-1}(r_j)}{n - \sum_{i=0}^{\sigma^{-1}(r_j)-1} V_i} =$$

$$V_1 \binom{3}{2} + V_0 \binom{4}{3} + V_4 \binom{0}{1} = 3 + 4 + 0 = 7.$$

3.31 We have

$$h(z) \equiv \ell(x)^z = \ell(x)^{c-nd} \equiv \ell(x)^{\sum_{j=0}^{p-1} V_j c_j - nd} \equiv$$

$$\ell(x)^{\sum_{j=0}^{p-1} V_j(a_{\sigma(j)}+d)-nd} \equiv \ell(x)^{\sum_{j=0}^{p-1} V_j a_{\sigma(j)}} \equiv \prod_{j=0}^{p-1} \ell(x)^{a_{\sigma(j)} V_j} \equiv$$

$$\prod_{j=0}^{p-1} (\ell(x)^{a_{\sigma(j)}})^{V_j} \equiv \prod_{j=0}^{p-1} (x + \sigma(j))^{V_j} \,(\mathrm{mod}\ r(x)),$$

where the last congruence comes from step 4 of the Chor-Rivest Key Generation Algorithm given on page 156. The latter has degree n modulo $r(x)$, as does $k(x)$. Thus,

$$k(x) \equiv h(x) + r(x) \equiv \prod_{j=0}^{p-1} (x + \sigma(j))^{V_j} \,(\mathrm{mod}\ r(x)).$$

Therefore, the roots of $k(x)$ are the $\sigma(j)$ for which $V_j = 1$, so applying σ^{-1} to them recovers all j such that $V_j = 1$.

3.33 We use induction on n. If $n = 2$, then $b_2 = 2$ and $b_1 = 1$ clearly provides the smallest. Assume that $b_j = 2^{j-1}$ provides the smallest for any $k < n$. Then the smallest $b_n > \sum_{j=1}^{n-1} 2^{j-1} = 2^{n-1} - 1$ is $b_n = 2^{n-1}$. Hence, any superincreasing sequence $\{a_1, a_2, \ldots, a_n\}$ must satisfy that their terms are at least as big, namely $a_j \geq 2^{j-1}$ for each $j = 1, 2, \ldots, n$.

3.35 They are $\{1, 2, 10\}$, $\{1, 3, 9\}$, $\{2, 11\}$, and $\{3, 10\}$. Note that the knapsack set $\mathcal{S}$ is not a superincreasing sequence since $b_3 = 3 \not> b_1 + b_2 = 3$.

Section 4.1

4.1 By Corollary 4.4, $\operatorname{ord}_n(m^e) = \operatorname{ord}_n(m)/\gcd(e, \operatorname{ord}_n(m))$. Therefore, $\operatorname{ord}_n(m^e) = \operatorname{ord}_n(m)$ if and only if $\gcd(e, \operatorname{ord}_n(m)) = 1$. In particular, if m is a primitive root modulo n, then $\operatorname{ord}_n(m^e) = \operatorname{ord}_n(m)$ if and only if $\gcd(e, \phi(n)) = 1$.

4.3 Let m be a primitive root modulo n. By Exercise 4.2, another primitive root must be of the form m^e with $1 \leq e \leq \phi(n)$. Thus, by Exercise 4.1, $\operatorname{ord}_n(m) = \operatorname{ord}_n(m^e)$ if and only if $\gcd(e, \phi(n)) = 1$, and there are precisely $\phi(\phi(n))$ such integers e.

4.5 If $g^{p-1} \not\equiv 1 \pmod{p^2}$, then g is a primitive root modulo p^2 since

$$p \,\Big|\, \operatorname{ord}_{p^2}(g^{p-1}) \,\Big|\, \phi(p^2) = (p-1)p.$$

Therefore, by induction on a, we may conclude that

$$p^{a-1} \,\Big|\, \operatorname{ord}_{p^a}(g^{p-1}) \,\Big|\, \phi(p^a),$$

for any $a \geq 2$. In other words, g is a primitive root modulo p^a. Now assume that $g^{p-1} \equiv 1 \pmod{p^2}$. Thus, by the Binomial Theorem,

$$(g+p)^{p-1} = \sum_{j=0}^{p-1} g^j p^{p-1-j} \binom{p-1}{j},$$

from which we get,

$$(g+p)^{p-1} \equiv g^{p-1} + p(p-1)g^{p-2} \equiv 1 - pg^{p-2} \pmod{p^2}.$$

Therefore, $(g+p)^{p-1} \not\equiv 1 \pmod{p^2}$, so $g+p$ is a primitive root modulo p^2, and as above, an induction on a shows that $g+p$ is a primitive root modulo p^a for any $a \geq 2$.

4.7 Let $\operatorname{ind}_a^p(c) = x$, $\operatorname{ind}_a^p(b) = y$, and $\operatorname{ind}_b^p(c) = z$. Since

$$b^{zy} \equiv (b^z)^y \equiv c^y \equiv (a^x)^y \equiv (a^y)^x \equiv b^x \pmod{p},$$

then by Fermat's Little Theorem, $zy \equiv x \pmod{p-1}$, as required.

4.9 Since $5x^5 \equiv 3 \pmod 7$, and 3 is a primitive root modulo 7, then

$$\operatorname{ind}_3^7(5) + 5\operatorname{ind}_3^7(x) \equiv \operatorname{ind}_3^7(3) \pmod 6.$$

Therefore,

$$5\operatorname{ind}_3^7(x) \equiv \operatorname{ind}_3^7(3) - \operatorname{ind}_3^7(5) = 1 - 5 \equiv 2 \pmod 6,$$

Thus,

$$\operatorname{ind}_3^7(x) \equiv 5^{-1} \cdot 2 \equiv 5 \cdot 2 \equiv 4 \pmod 6,$$

so by raising to appropriate powers we get $x \equiv 3^4 \equiv 4 \pmod 7$.

4.11 For any primitive root a modulo p^c, congruence (4.18) yields,

$$e \cdot \operatorname{ind}_a^{p^c}(x) \equiv \operatorname{ind}_a^{p^c}(b) \, (\operatorname{mod} \, \phi(p^c)).$$

The result now follows via Exercise 2.10 on page 73.

4.13 If $m^e \equiv -1 \, (\operatorname{mod} \, p)$, then $m^{2e} \equiv 1 \, (\operatorname{mod} \, p)$. Thus, $\operatorname{ord}_p(m) \mid 2e$. However, $2 \nmid \operatorname{ord}_p(m)$, so $\operatorname{ord}_p(m) \mid e$. Hence, $m^e \equiv 1 \, (\operatorname{mod} \, p)$, so $-1 \equiv 1 \, (\operatorname{mod} \, p)$ forcing $p = 2$, a contradiction.

4.15 If $m^{(p-1)/q} \equiv 1 \, (\operatorname{mod} \, p)$ for some prime $q \mid (p-1)$, then $\operatorname{ord}_p(m) \neq p-1 = \phi(p)$, so m is not a primitive root modulo p.

4.17 Let $M = \sum_{j=1}^{(p-1)/2} \lfloor jc/p \rfloor$ where $c \in \mathbb{Z}$ such that $p \nmid c$, and note that $\sum_{j=1}^{(p-1)/2} j = (p^2 - 1)/8$ by Theorem A.29 on page 283.

From (4.35) in the proof of Corollary 4.32,

$$(c - 1)(p^2 - 1)/8 = p(M - c) + 2 \sum_{j=1}^{c} r_j.$$

Therefore,

$$c \equiv M + \frac{(p^2 - 1)}{8}(c - 1) \, (\operatorname{mod} \, 2).$$

If $c = 2$, then $M = 0$, since $\lfloor 2j/p \rfloor = 0$ for all $j \in \mathbb{N}$ with $j < p/2$, so

$$c \equiv \frac{(p^2 - 1)}{8} \, (\operatorname{mod} \, 2).$$

This establishes the result via Gauss's Lemma 4.29.

4.19 By the Quadratic Reciprocity Law,

$$\left(\frac{3}{p} \right) \left(\frac{p}{3} \right) = (-1)^{\frac{3-1}{2} \frac{p-1}{2}} = (-1)^{\frac{p-1}{2}}.$$

If $\left(\frac{3}{p} \right) = 1$ and $p \equiv 1 \, (\operatorname{mod} \, 4)$, then $\left(\frac{p}{3} \right) = \left(\frac{3}{p} \right)$, by the Quadratic Reciprocity Law, so $\left(\frac{p}{3} \right) = (-1)^{\frac{p-1}{2}} = 1$. Thus, $p \equiv 1 \, (\operatorname{mod} \, 3)$, so we must have,

$$p \equiv 1 \, (\operatorname{mod} \, 12).$$

Similarly, if $p \equiv 3 \, (\operatorname{mod} \, 4)$, then $\left(\frac{p}{3} \right) = -\left(\frac{3}{p} \right)$, so $\left(\frac{p}{3} \right) = -1$. Therefore, $p \equiv 2 \, (\operatorname{mod} \, 3)$, so

$$p \equiv 11 \, (\operatorname{mod} \, 12).$$

In total, we have shown that $\left(\frac{3}{p} \right) = 1$ implies $p \equiv \pm 1 \, (\operatorname{mod} \, 12)$. Conversely, the Quadratic Reciprocity Law shows that $p \equiv \pm 1 \, (\operatorname{mod} \, 12)$ implies $\left(\frac{3}{p} \right) = 1$. Hence, $\left(\frac{3}{p} \right) = -1$ if and only if $p \equiv \pm 5 \, (\operatorname{mod} \, 12)$.

4.21 If $a \equiv x_p^2 \, (\operatorname{mod} \, p^t)$ whenever $p^t \| n$, then by the Chinese Remainder Theorem, there is a solution

$$y \equiv x^2 \equiv a \, (\operatorname{mod} \, n),$$

so a is a quadratic residue modulo n.

Section 4.2

4.23 Let $r|m$. It suffices to prove the result for the case where r is prime. Let $m = rt$ for some $t \in \mathbb{N}$. Thus, $rt = 0$ in R so r is a zero divisor in R (see Footnote A.2 on page 279). Let $I = \{x \in R : xt = 0\}$. Then I is an ideal in R and $r \in I$ (see Definition A.59 on page 295). Let M be a maximal ideal in R containing r (see Definition A.64), and set $F = R/M$, which is a field (see Theorem A.65). Since $\alpha^{s/p} - 1$ is a unit in R for any prime divisor p of s, then the order of α modulo M must be s. In other words, $\alpha^s - 1 = 0$ in M but $\alpha^j - 1 \neq 0$ in M for any nonnegative $j < s$. (Otherwise, there would be a unit in M forcing $M = R$, contradicting Definition A.64.) Since $f(x) \in \mathbb{Z}/m\mathbb{Z}[x]$ and $r = 0$ in F with $r|m$, then we may assume without loss of generality that $f(x) \in \mathbb{Z}/r\mathbb{Z}[x]$. Thus, $f(\alpha^r) = 0$, so $\alpha^r = \alpha^{m^j}$ for some nonnegative $j < k$. Since the order of α modulo M is s, then $r \equiv m^j \pmod{s}$.

4.25 By Exercise 2.24 on page 75, $\gcd(2^p - 1, 2^q - 1) = 2^{\gcd(p,q)} - 1 = 1$.

4.27 If n is prime, then by Exercise 4.26 we have the result with $m = m_q$ for all $q \mid (n-1)$. Conversely, if such m_q exist, then by the Chinese Remainder Theorem, we may find a solution $x = m$ to the system of congruences

$$x \equiv m_q \pmod{q^e}$$

for all primes q such that $q^e \| (n-1)$. Thus, the result now follows from Exercise 4.26.

4.29 Since $b = 3 \cdot 881 = 2643 > \lfloor\sqrt{n}\rfloor = 1856$, and $17^{n-1} \equiv 1 \pmod{3446473}$ but $17^{(n-1)/3} \equiv 1128877 \not\equiv 1 \pmod{n}$ and $17^{(n-1)/881} \equiv 1410633 \not\equiv 1 \pmod{n}$, then by Pocklington's Theorem n is prime.

4.31 Since $n = 26113 = 2^9 \cdot 51 + 1$, with $2^9 > 51$, then the hypothesis of Proth's Theorem is satisfied. Since 5 is a quadratic nonresidue modulo n, and $5^{(n-1)/2} = 5^{13056} \equiv -1 \pmod{n}$, then n is prime.

4.33 For n composite, $r(n) = 0$ since $n|(n-2)!$ in that case. If $n > 2$ is prime, then by Wilson's Theorem $n \mid [(n-1)! + 1]$. Thus, $n|(r(n) + 1)$ and $[(n-1)/2] \mid r(n)$. Since

$$r(n) < n(n-1)/2, \tag{S3}$$

then there exist integers $a \geq 2$ and $b \geq 0$ such that

$$r(n) = a(n-1)/2, \tag{S4}$$

and

$$r(n) + 1 = bn. \tag{S5}$$

Subtracting (S5) from (S4) and rearranging, we get

$$a - 2 = n(a - 2b),$$

so n divides $(a - 2)$. However, by (S3)–(S4), $a < n$, so this forces $a = 2$. Hence, by (S4), $r(n) = n - 1$.

4.35 $\mathfrak{F}_4$ is prime since $5^{(\mathfrak{F}_4-1)/2} = 5^{32768} \equiv -1 \pmod{65537}$.

4.37 Since $561 = 3 \cdot 11 \cdot 17$, $1729 = 7 \cdot 13 \cdot 19$, $10585 = 5 \cdot 29 \cdot 73$, and $294409 = 37 \cdot 73 \cdot 109$, then an easy check shows that for each of these n,

$$(p-1) \mid (n-1)$$

for each prime dividing n, so by Exercise 4.36, each n is a Carmichael Number.

Section 4.3

4.39 If $ab^{-1} \in H$ whenever $a, b \in H$, then $ab = a(b^{-1})^{-1} \in H$, $aa^{-1} = 1_G \in H$ and $1 \cdot a^{-1} = a^{-1} \in H$, so (a)–(c) of Exercise 2.39 are satisfied, which means that H is a subgroup of G. If H is a subgroup of G, then by the definition of a group, we must have that $ab^{-1} \in H$ for any $a, b \in H$.

Applying this result to $E(n)$, we get that if $a, b \in E(n)$, then since $\gcd(b, n) = 1$, there exists a b^{-1} modulo n. Thus,

$$\left(\frac{a \cdot b^{-1}}{n}\right) = \left(\frac{a}{n}\right)\left(\frac{b^{-1}}{n}\right) \equiv a^{(n-1)/2}(b^{-1})^{(n-1)/2} = (ab^{-1})^{(n-1)/2} \pmod{n}.$$

Hence, by the above result $E(n)$ is a multiplicative group.

4.41 It passes the test on any value of r since it is prime.

4.43 If $a, b \in F(n)$, then

$$(a \cdot b^{-1})^{n-1} \equiv a^{n-1}(b^{n-1})^{-1} \equiv 1 \pmod{n},$$

so $a \cdot b^{-1} \in F(n)$. Therefore, $F(n)$ is a multiplicative group by Exercise 4.39. That $\text{epsp}(a) \subseteq \text{psp}(a)$ is immediate from their respective definitions.

4.45 Let $p - 1 = 2^t m$ where $2 \nmid m$, $t \in \mathbb{N}$, p is prime and $d = \text{ord}_p(a)$ for a chosen $a \in \mathbb{N}$ with $2 \le a \le p - 2$. Therefore, by Proposition 4.3 , $d \mid (p-1)$. If $2 \nmid d$, then $d \mid m$. Thus, $a^m \equiv 1 \pmod{p}$. If $2 \mid d$, then $d = 2^j m_1 \mid 2^t m$ for some $j \in \mathbb{N}$, with $j \le t$ and $m_1 \mid m$. Also, since $a^{2^j m_1} \equiv 1 \pmod{p}$, then

$$a^{2^{j-1} m_1} \equiv \pm 1 \pmod{p}$$

by Exercise 2.8 on page 72. Yet, $a^{d/2} \equiv a^{2^{j-1} m_1} \not\equiv 1 \pmod{p}$, so

$$a^{2^{j-1} m_1} \equiv -1 \pmod{p}.$$

Since $m_1 \mid m$, then $m = m_1 m_2$ for some $m_2 \in \mathbb{N}$ and

$$a^{2^{j-1} m_1} \equiv (-1)^{m_2} \equiv -1 \pmod{p}.$$

Hence, $a \in S(n)$, and so the result follows by Proposition 4.68.

4.47 Let $a = 2$. Since $6600 = n - 1 = 2^3 \cdot 825$, then we compute

$$x = 2^{825} \equiv 2738 \pmod{6601}$$

in step (2). Since $x \not\equiv 1 \,(\mathrm{mod}\ n)$ we go to step (3) and compute for $j = 1$,

$$a^{2^j m} \equiv 2738^2 \equiv 4509 \,(\mathrm{mod}\ 6601),$$

then for $j = 2 = t - 1$,

$$a^{2^j m} \equiv 4509^2 \equiv 1 \,(\mathrm{mod}\ 6601).$$

Thus, we return the declaration that n is composite. In fact, as mentioned in the discussion on page 192, we get a nontrivial factor of n by looking at

$$\gcd(2^{t-2}m - 1, n) = \gcd(4509 - 1, 6601) = 161 = 7 \cdot 23.$$

Actually, $n = 6601 = 7 \cdot 23 \cdot 41$ is a Carmichael Number.

4.49 If $n = 65$, then $n \in \mathrm{epsp}(14)$ since

$$14^{(65-1)/2} = 14^{32} \equiv \left(\frac{14}{65}\right) = 1 \,(\mathrm{mod}\ 65),$$

but $14^m = 14 \not\equiv 1 \,(\mathrm{mod}\ 65)$, where $n - 1 = 2^t m = 2^6 \cdot 1$, and $14^{2^j} \not\equiv -1 \,(\mathrm{mod}\ 65)$, for $j = 0, 1, 2, 3, 4, 5$. Thus, $65 \notin \mathrm{spsp}(14)$.

4.51 Let $n \in \mathrm{psp}(2)$. Then n is an odd composite integer such that $2^{n-1} \equiv 1 \,(\mathrm{mod}\ n)$. Let $2^n - 2 = mn$ for some $m \in \mathbb{N}$. By Exercise 1.17 on page 29,

$$r = (2^n - 1) \,\big|\, (2^{mn} - 1).$$

Thus, $2^{mn} \equiv 1 \,(\mathrm{mod}\ r)$. Therefore, since $r - 1 = 2^n - 2 = mn$, then

$$2^{r-1} \equiv 1 \,(\mathrm{mod}\ r).$$

We have shown that $r \in \mathrm{psp}(2)$. Thus, for a given $n \in \mathrm{psp}(2)$, we can generate another, namely $2^n - 1 \in \mathrm{psp}(2)$. Thus, from one we can generate infinitely many. For instance since $n_0 = 341 \in \mathrm{psp}(2)$, then we generate infinitely many with $n_{j+1} = 2^{n_j} - 1 \in \mathrm{psp}(2)$ for $j \geq 0$.

4.53 Let $n = 1 + 2^t m$ where m is odd. If $a \in S(n)$, and $a^m \equiv 1 \,(\mathrm{mod}\ n)$, then $(-a)^m \equiv -1 \,(\mathrm{mod}\ n)$. If

$$a^{2^j m} \equiv -1 \,(\mathrm{mod}\ n)$$

for some $j = 0, 1, \ldots, t - 1$, then $(-a)^{2^j m} \equiv -1 \,(\mathrm{mod}\ n)$ if $j > 0$ and $(-a)^m \equiv 1 \,(\mathrm{mod}\ n)$ if $j = 0$. Conversely, if $n - a \in S(n)$, then the above argument can be reversed.

4.55 It suffices to prove the result when r is prime power by the multiplicativity property of the Euler Totient. If $a^{(n-1)/2} \equiv -1 \,(\mathrm{mod}\ n)$, then

$$a^{(n-1)/2} \equiv -1 \,(\mathrm{mod}\ r).$$

Thus,

$$|\mathrm{ord}_r(a)|_2 = |n - 1|_2. \tag{S6}$$

However, $\mathrm{ord}_r(a) \,\big|\, \phi(r)$ by Proposition 4.3. Hence,

$$|\mathrm{ord}_r(a)|_2 \leq |\phi(r)|_2. \tag{S7}$$

By (S6)–(S7),
$$|n - 1|_2 \le |\phi(r)|_2,$$
and $|n - 1|_2 = |\phi(r)|_2$ if and only if $|\text{ord}_r(a)/2|_2 = |\phi(r)/2|_2$ if and only if $a^{\phi(r)/2} \equiv -1 \pmod{r}$, which holds if and only if $\left(\frac{a}{r}\right) = -1$ by Euler's Criterion given in Example 4.27.

4.57 If n is prime, then by definition $\psi_D(n) = n - (D/n)$. Conversely, if $k = 1$ and $a_j > 1$, then $\psi_D(n)$ is a multiple of p, but $n - (D/n)$ is not. If $k > 1$, then

$$\psi_D(n) \le \frac{1}{2^{k-1}} \prod_{j=1}^{k} p_j^{a_j-1}(p_j + 1) = 2n \prod_{j=1}^{k} \frac{1}{2}\left(1 + \frac{1}{p_j}\right) \le$$

$$2n \cdot \frac{2}{3} \cdot \frac{3}{5} \cdots \le \frac{4n}{5} < n - 1,$$

since both $k > 1$ and $n > 5$.

Section 5.1

5.1 If we choose $B = 3$, then for $p = 2$, $m = 14$ and $a = 11216$. For $p = 3$, $m = 9$ and $a = 11835$. Here, $\gcd(a - 1, n) = 97$. Note that for $p = 97$, $p - 1 = 2^5 \cdot 3$ is B-smooth. The other prime factor of n is $q = 277$, which is not B-smooth since $q - 1 = 2^2 \cdot 3 \cdot 23$.

5.3 We use Footnote 5.7 to calculate that $k = 12$. Thus, the factor base consisting of those primes for which n is a quadratic residue are given by

$$\mathcal{F} = \{2, 7, 37, 43, 59, 79, 83, 101, 107, 127, 131, 149\}.$$

Also, $\lfloor\sqrt{39617}\rfloor = 199$. Using the quadratic sieve algorithm, we compute the data in the following table.

t	x_t	y_t	v_t
0	199	-2^4	$(1,0,0,0,0,0,0,0,0,0,0,0,0)$
-1	198	$-7 \cdot 59$	$(1,0,1,0,0,1,0,0,0,0,0,0,0)$
2	201	$2^4 \cdot 7^2$	$(0,0,0,0,0,0,0,0,0,0,0,0,0)$
-2	197	$-2^3 \cdot 101$	$(1,1,0,0,0,0,0,0,1,0,0,0,0)$
-6	193	$-2^6 \cdot 37$	$(1,0,0,1,0,0,0,0,0,0,0,0,0)$
6	205	$2^3 \cdot 7 \cdot 43$	$(0,1,1,0,1,0,0,0,0,0,0,0,0)$
-8	191	$-2^6 \cdot 7^2$	$(1,0,0,0,0,0,0,0,0,0,0,0,0)$
8	207	$2^5 \cdot 101$	$(0,1,0,0,0,0,0,0,1,0,0,0,0)$
10	209	$2^5 \cdot 127$	$(0,1,0,0,0,0,0,0,0,0,1,0,0)$
-12	187	$-2^3 \cdot 7 \cdot 83$	$(1,1,1,0,0,0,0,1,0,0,0,0,0)$
16	215	$2^4 \cdot 7 \cdot 59$	$(0,1,1,0,0,1,0,0,0,0,0,0,0)$
20	219	$2^3 \cdot 7 \cdot 149$	$(0,1,1,0,0,0,0,0,0,0,0,0,1)$
-22	177	$-2^5 \cdot 7 \cdot 37$	$(1,1,1,1,0,0,0,0,0,0,0,0,0)$
24	175	$2^7 \cdot 79$	$(0,1,1,0,0,0,1,0,0,0,0,0,0)$

A glance at the table tells us that for for $\mathcal{S} = \{2\}$ we are done since

$$y_2 \equiv 2^4 \cdot 7^2 \equiv 28^2 \pmod{n}$$

and
$$x_2^2 \equiv 201 \pmod{n},$$

so $\gcd(201 - 28, 39617) = 173$ and $\gcd(201 + 28, 39617) = 229$, so we have a complete factorization of n. We note that we can also get a set $\mathcal{S}$ quickly by observing the negative squares at $t = 0, -8$. Thus, for $\mathcal{S} = \{0, -8\}$, we get

$$\prod_{t \in \mathcal{S}} y_t \equiv 224^2 \pmod{n},$$

and

$$\prod_{t \in \mathcal{S}} x_t^2 \equiv 38009^2 \pmod{n}.$$

Thus, $\gcd(38009 - 224, 39617) = 229$, and $\gcd(38009 + 224, 39617) = 173$, so we again have a complete factorization $n = 173 \cdot 229$. Hence, we need not have calculated the full $k + 2 = 14$ rows in the table. Sometimes we can get lucky and find squares. The reader can peruse the above table to find as many sets $\mathcal{S}$ as possible that give a factorization.

5.5 We choose a factor base of primes for which n is a quadratic residue, $\mathcal{F} = \{2, 11, 23, 31, 43\}$. Since $\lfloor \sqrt{n} \rfloor = 80$, we compute the following table.

j	P_j	q_j	A_{j-1}	$(-1)^j Q_j$	v_j
0	0	80	1	1	$(0,0,0,0,0,0)$
2	77	39	81	4	$(0,0,0,0,0,0)$

We already have our set $\mathcal{S}$ for $j = 2$ since $Q_2 = 2^2 = x^2$. We compute $A_1 = 81 = y$, $\gcd(x - y, n) = \gcd(81 - 2, 6557) = 79$, and $\gcd(81 + 2, 6557) = 83$, thereby yielding a factorization, $n = 79 \cdot 83$.

5.7 The number of divisions in trial division is $O(\sqrt{n})$ and each division takes $O(\log_2^2 n)$ bit operations. Thus, the total is $O(\sqrt{n} \log_2^2 n)$.

Section 5.2

5.9 $\mathfrak{F}_6 = 274177 \cdot 67280421310721$.

5.11 In $\mathbb{Z}[\sqrt[3]{-7}]$,
$$7(7^{149} + 1) = 7^{150} + 7 = x^3 + 7,$$

where $x = 7^{50}$. Observe that for $F = \mathbb{Q}(\sqrt[3]{-7})$, $N_F(a + b\sqrt[3]{-7}) = a^3 - 7b^3$, so $N_F(x - \sqrt[3]{-7}) = x^3 + 7$, and we may apply the gcd method. An initial run shows that
$$\gcd(a^3 - 7b^3, 7^{150} + 7) = 10133$$

for $a = 47, b = 1804$, so 10133 is a factor of $7^{149} + 1$.

5.13 In $\mathbb{Z}[\sqrt[3]{-4}]$,
$$4(2^{457} + 1) = 2^{459} + 4 = x^3 + 4,$$

where $x = 2^{153}$. Observe that for $F = \mathbb{Q}(\sqrt[3]{-4})$, $N_F(a + b\sqrt[3]{-4}) = a^3 - 4b^3$, so $N_F(x - \sqrt[3]{-4}) = x^3 + 4$. An initial run shows that
$$\gcd(a^3 - 4b^3, 2^{459} + 4) = 3$$

for $a = b = 1$, so $3 | (2^{457} + 1)$.

5.15 In $\mathbb{Z}[\sqrt[3]{-2}]$,

$$4(2^{332} + 1) = 2^{333} + 2 = x^3 + 2,$$

where $x = 2^{111}$. As above, we look at

$$g = \gcd(a^3 - 2b^3, 2^{333} + 2).$$

An initial run reveals that at $a = 8$ and $b = 1$, $g = 34$, so $17|(2^{332} + 1)$.

5.17 Here, $n = 2^{488} + 1$, with

$$r = 2, \ s = -1, \ \text{and} \ t = 488,$$

and from (5.15), $d = 3$, so from (5.16),

$$k = 163, \ m = 2^{163}, \ \text{and} \ c = -2.$$

Hence,

$$f(x) = x^3 + 2, \ \text{with root} \ \alpha = \sqrt[3]{-2},$$

and $\mathbb{Z}[\alpha]$ is a UFD. Smoothness bounds need not be high since we find on an initial run that

$$\gcd(a + b \cdot 2^{163}, n) = 257 \ \text{for} \ a = 121, b = 17.$$

It turns out that

$$(2^{488} + 1)/257$$

is a 145-digit number which is a product of two primes, one having 49 digits and the other having 97 digits. See [28] for details.

Section 6.1

6.1 Suppose that $Y = A/B$, $X = C/D$ with nonzero $A, B, C, D \in \mathbb{Z}$ such that $Y^2 = X^3 - 4X$ holds. Set $x = AD^2$, $y = 2BCD$, and $z = 2C^3B^2D - A^2D^4$. Then

$$z^2 = 4B^4C^6D^2 - 4A^2B^2C^3D^5 + A^4D^8,$$

so

$$z^2 = 4B^2C^3D^2(B^2C^3 - A^2D^3) + A^4D^8. \tag{S8}$$

However, since $X^3 - Y^2 = 4X$, then $B^2C^3 - A^2D^3 = 4B^2CD^2$. Therefore, (S8) becomes

$$z^2 = (4B^2C^3D^2)(4B^2CD^2) + A^4D^8 = (2BCD)^4 + (AD^2)^4 = y^4 + x^4.$$

Conversely, if $x^4 + y^4 = z^2$ with nonzero $x, y, z \in \mathbb{Z}$, then by setting

$$X = \frac{2(x^2 + z)}{y^2}, \ \text{and} \ Y = \frac{4x(x^2 + z)}{y^3},$$

we get $Y^2 = X^3 - 4X$.

6.3 Let $c \in \mathbb{N}$ be the least value so that $a^4 + b^4 = c^2$ for some $a, b \in \mathbb{N}$. Since $\gcd(a, b) = 1$ necessarily, then we may assume without loss of generality that a is odd. If b is odd, then $c^2 = a^4 + b^4 \equiv 2 \pmod 4$, a contradiction. Therefore, b is even. By Exercise 6.2, there exist relatively prime integers u and v of opposite parity such that $a^2 = u^2 - v^2$, $b^2 = 2uv$ and $c = u^2 + v^2$. If u is even and v is odd, then $a^2 \equiv -1 \pmod 4$, a contradiction. Thus, u is odd and v is even, so let $v = 2d$. Hence, $b^2 = 2uv = 4ud$, so $(b/2)^2 = ud$ with $\gcd(u, d) = 1$. Therefore, we may set $u = w^2$ and $d = x^2$ for some $w, x \in \mathbb{N}$ with $\gcd(w, x) = 1$. Thus, $a^2 = u^2 - v^2 = w^4 - 4x^4$, or by rewriting

$$a^2 + (2x^2)^2 = (w^2)^2.$$

We may apply Exercise 6.2 again to this triple to get that there exist relatively prime $f, g, h \in \mathbb{N}$ such that

$$2x^2 = 2fg, \text{ and } w^2 = f^2 + g^2.$$

Therefore, $x^2 = fg$ with $\gcd(f, g) = 1$, so there are $k, \ell \in \mathbb{N}$ such that $f = k^2$ and $g = \ell^2$. Hence,

$$w^2 = k^4 + \ell^4, \text{ with } w \leq w^2 = u \leq u^2 < u^2 + v^2 = c,$$

contradicting the minimality of c.

(This is called *Fermat's Method of Infinite Descent*, or simply *Proof by Descent*. Fermat used this method to prove that any prime $p \equiv 1 \pmod 4$ is a sum of two squares in exactly one way. In a letter to Frenicle de Bessy (see Footnote 1.41 on page 24) dated June 14, 1641, Fermat stated that "if a prime $4n + 1$ is not a sum of two squares, there is a smaller prime of the same nature, then a third still smaller, etc., until the number 5 is reached, thus leading to a contradiction." Fermat also used this method to prove that the Fermat equation:

$$x^4 + y^4 = z^4$$

has no nontrivial integer solutions. This, of course, follows from this exercise. See also [144, pp. 273–274] for more information.)

6.5 If E denotes the curve given by $y^2 = x^3 - 3x + 2$, then

$$\Delta(E) = -16(4(-3)^3 + 27(2^2)) = 0,$$

so E is actually *not* an elliptic curve. The situation is depicted in the diagram below.

The term "curve", in this context, is used to mean a *plane algebraic curve*, which is the set of solutions in $\mathbb{C} \times \mathbb{C}$ of a polynomial equation

$$f(x, y) = 0 \text{ where } f(x, y) \in \mathbb{Q}[x, y].$$

Note that there is a *double* zero at $P = (1, 0)$, called a *node*. Any point that is a multiple zero is called a *singularity*. A point that is not singular is called *regular*. A curve having a singularity is called *singular*. If a curve has no singularities, then it is called *nonsingular* or *smooth*. The term "singularity" is justified by the fact that, at such a point, there is *not a unique tangent*. In order for a point to be regular on a curve, at least one of dy/dx or dx/dy has to be well defined. It can be shown that a cubic curve $y^2 = x^3 + ax + b$ is singular if and only if its discriminant (as given in Definition 6.1) is zero.

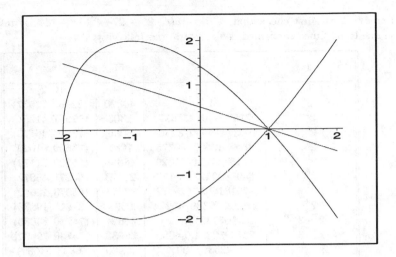

6.7 If $P, Q \in E(F)_t$, then there exist $n, m \in \mathbb{N}$ such that $nP = \mathfrak{o} = mQ$. Thus, if $\ell = \mathrm{lcm}(m, n)$, then $\ell(P - Q) = \mathfrak{o}$, so $P - Q \in E(F)_t$. Hence, by Exercise 4.39 on page 193, $E(F)_t$ is a subgroup of $E(F)$. (Although Exercise 4.39 is stated for the multiplicative case, the same naturally holds true for the additive case.)

6.9 By Nagell-Lutz's Theorem, since $4a^2 + 27b^2 = 27 \cdot 4$, then the only possible values for y when $P = (x, y)$ is a nontrivial torsion point are $y = \pm 1, \pm 2, \pm 3, \pm 6$. However, these yield $x^3 = 3$, $x^3 = 6$, $x^3 = 11$, and $x^3 = 38$, respectively, none of which are possible.

6.11 Assume that $n = ab/2$ for a Pythagorean triple (a, b, c), set $x = c^2/4$. Thus,

$$(a - b)^2/4 = x - n$$

and

$$(a + b)^2/4 = x + n,$$

so x and $x \pm n$ are squares of rational numbers. Conversely, if x and $x \pm n$ are all squares of rational numbers, we set

$$a = \sqrt{x + n} + \sqrt{x - n},$$

$$b = \sqrt{x + n} - \sqrt{x - n},$$

and

$$c = 2\sqrt{x}.$$

Then $a, b, c \in \mathbb{Q}$ and

$$a^2 + b^2 = c^2.$$

6.13 Suppose that (2) of Exercise 6.11 holds. Then

$$(x - n)x(x + n) = x^3 - n^2x = y^2$$

for some $y \in \mathbb{Q}$. However, $x \neq \pm n, 0$ since n is squarefree. This value of x yields the desired point. Conversely, if E has a point $P = (x_1, y_1)$ with $x_1, y_1 \in \mathbb{Q}$ and $P \neq (\pm n, 0), (0, 0)$ or $\mathfrak{o}$, then $y_1 \neq 0$ and $2P = (x_2, y_2) \neq \mathfrak{o}$. By Exercise 6.12, $x_2 \pm n$ and x are all squares of rational numbers, as required.

6.15 For $a = 2$ we first check that $\Delta = -16(4 \cdot 2^3 + 27 \cdot 4^2) = -7424$ is relatively prime to n. Once confirmed, we compute the following.

s	m	$\overline{m}$	$\overline{sP}$
1	$--$	$--$	$(0, 2)$
2	$1/2$	44320	$(22160, 77557)$
2^2	$736598401/77557$	1303	$(57987, 41254)$
2^3	$10087476509/82508$	6086	$(49598, 46775)$
2^4	$3689942407/46775$	7089	$(73690, 61029)$
2^5	$8145324151/61029$	48656	$(65822, 21177)$
2^6	$6498803527/21177$	27473	$(50278, 45072)$
2^7	$3791815927/45072$	77471	$(85873, 22612)$
2^8	$22122516389/45224$	9598	$(31215, 19270)$
$2^7 \cdot 3$	$1671/27329$	46624	$(74730, 84080)$
$2^8 \cdot 3$	$8376859351/84080$	20933	$(75730, 79002)$
$2^7 \cdot 3^2$	$-2539/500$	17900	$(6833, 31530)$
$2^8 \cdot 3^2$	$140069669/63060$	82917	$(19827, 40656)$
$2^7 \cdot 3^3$	$4563/6497$	24108	$(51720, 26225)$
$2^8 \cdot 3^3$	$4012437601/26225$	38390	$(65285, 56189)$
$2^7 \cdot 3^4$	$29964/13565$	32929	$(58427, 7360)$
$2^8 \cdot 3^4$	$10241142989/14720$	7507	$(41069, 88455)$
$2^7 \cdot 3^5$	$-81095/17358$	85167	$(77662, 31393)$
$2^8 \cdot 3^5$	$9047079367/31393$	64011	$(3661, 67097)$
$2^7 \cdot 3^6$	$-35704/74001$	44110	$(73366, 43624)$
$2^7 \cdot 3^5 \cdot 5$	$-23473/69705$	1401	$(24355, 14201)$
$2^8 \cdot 3^5 \cdot 5$	$1779498077/28402$	78224	$(18018, 21999)$
$2^7 \cdot 3^6 \cdot 5$	$-7798/6337$	80525	$(24485, 65590)$
$2^7 \cdot 3^5 \cdot 5^2$	$43591/6467$	65482	$(26835, 17653)$
$2^8 \cdot 3^5 \cdot 5^2$	$2160351677/35306$	3393	$(24348, 33)$
$2^7 \cdot 3^6 \cdot 5^2$	$17620/2487$	22746	$(32129, 25424)$
$2^7 \cdot 3^5 \cdot 5^3$	$25391/7781$	62578	$(55865, 35530)$
$2^8 \cdot 3^5 \cdot 5^3$	$9362694677/71060$	46857	$(55967, 60201)$
$2^7 \cdot 3^6 \cdot 5^3$	$24671/102$	11539	$(78189, 41407)$
$2^7 \cdot 3^5 \cdot 5^4$	$-9397/11111$	36720	$(25054, 42764)$
$2^7 \cdot 3^5 \cdot 5^3 \cdot 7$	$17437/30913$	70297	$(51577, 79909)$
$2^8 \cdot 3^5 \cdot 5^3 \cdot 7$	$7980560789/159818$	19071	$(2709, 19912)$
$2^7 \cdot 3^6 \cdot 5^3 \cdot 7$	$59997/48868$	44435	$(69853, 25188)$
$2^7 \cdot 3^5 \cdot 5^4 \cdot 7$	$1319/16786$	68779	$(80766, 73276)$
$2^7 \cdot 3^5 \cdot 5^3 \cdot 7^2$	$17788/26019$	81380	$(46679, 57918)$
$2^8 \cdot 3^5 \cdot 5^3 \cdot 7^2$	$6536787125/115836$	8885	$(49796, 80183)$

We have therefore exhausted all of the divisors of M with no factorization, since the additive group law did *not* fail.

6.17 Since $\lfloor \sqrt{n} \rfloor = 29 \geq p$, then by (6.28), we may choose $A = 29 + 1 + 2\lfloor \sqrt{29} \rfloor = 40$ and $B = 4$. Thus, $M = 2^5 \cdot 3^4$, where

$$5 = \lfloor \ln(40)/\ln(2) \rfloor, \text{ and } 3 = \lfloor \ln(40)/\ln(3) \rfloor.$$

The discriminant is easily checked to be relatively prime to n for $a = 1$, so using (6.8)–(6.10), we tabulate the following for the (E, P) pair $(y^2 = x^3 + x + 4, (0, 2))$:

s	m	$\overline{m}$	sP
1	$--$	$--$	$(0,2)$
2	$1/4$	225	$(281,602)$
2^2	$59221/301$	140	$(159,296)$
2^3	$18961/148$	37	$(152,862)$
2^4	$69313/1724$	752	$(628,786)$
2^5	$1183153/1572$	664	$(29,491)$
$2^4 \cdot 3$	$295/599$	14	$(438,76)$
$2^5 \cdot 3$	$575533/152$	149	$(648,99)$
$2^4 \cdot 3^2$	$23/210$	227	$(99,462)$
$2^5 \cdot 3^2$	$7351/231$	421	$(839,850)$
$2^4 \cdot 3^3$	$97/185$	399	$(39,104)$
$2^5 \cdot 3^3$	$1141/52$	385	$(711,88)$

Thus, we have reached $m = 2^5 \cdot 3^3$ without achieving a factorization. This forces us to abandon the elliptic curve $y^2 = x^3 + x + 4$ and seek a new one in the family. We now choose $a = 2$ and proceed with the (E, P) pair:

$$(y^2 = x^3 + 2x + 4, (0,2)).$$

However, $\gcd(\Delta(E), n) = \gcd(-7424, 899) = 29$. Thus, we have a factorization $899 = 29 \cdot 31$.

6.19 First we observe that $\gcd(n, 6) = \gcd(\Delta(E), n) = \gcd(47088, 1231) = 1$. Now we proceed to check n for primality. If n were prime, then Exercise 6.4 on page 249 tells us that $|\overline{E} \, (\mathrm{mod} \, n)| = 2 \cdot 619 = 2p$. Also,

$$1161 < n + 1 - 2\sqrt{n} < |\overline{E} \, (\mathrm{mod} \, n)| < 1302 < n + 1 + 2\sqrt{n}.$$

Therefore, conditions (a)–(b) of Theorem 6.30 are satisfied. Since

$$619 = 2^9 + 2^6 + 2^5 + 2^3 + 2^1 + 2^0,$$

we calculate up to 2^9 by doubling, then add distinct points until we achieve $\overline{619P}$.

s	m	$\overline{m}$	sP
1	$--$	$--$	$(0,1)$
2	$9/2$	620	$(328,985)$
2^2	$322761/1970$	1213	$(899,676)$
2^3	$606153/338$	1156	$(134,1037)$
2^4	$53877/2074$	226	$(337,1094)$
2^5	$85179/547$	302	$(667,188)$
2^6	$333669/94$	996	$(958,492)$
2^7	$917767/328$	1173	$(217,846)$
2^8	$11773/141$	1201	$(466,469)$
2^9	$651477/938$	457	$(1109,1120)$
$576 = 2^9 + 2^6$	$628/151$	852	$(723837,-615893400)$
$608 = 576 + 2^5$	$-307946794/361585$	557	$(-414255,231111366)$
$616 = 608 + 2^3$	$-231110329/414389$	954	$(1324237,-1263195299)$
$618 = 616 + 2$	$-421065428/441303$	3	$(-1324556,3973667)$
$619 = 618 + 1$	$-1986833/662278$	$--$	o

We were successful in this first run since $662278 = 2 \cdot 269 \cdot 1231$, which means that we cannot invert 662278 in $\mathbb{Z}/n\mathbb{Z}$, thereby yielding that $619P = \mathfrak{o}$, so 1231 is prime by Theorem 6.30.

6.21 Since $c = (\beta, \gamma) = ((11,1), (1,3))$, then

$$\mathfrak{D}_d(c) = (1,3) - 2(11,1) = (1,3) - (3,11) = (1,3) + (3,2) = (6,6) = m.$$

6.23 We first compute

$$ac_0 = 5(12,8) = (10,11) = (y_1, y_2).$$

Then we achieve the plaintext via:

$$(c_1 y_1^{-1}, c_2 y_2^{-1}) = (2 \cdot 10^{-1}, 8 \cdot 11^{-1}) = (2 \cdot 4, 8 \cdot 6) = (8,9) = m.$$

Section 6.2

6.25 Let A be a prover who knows the prime factors of a Blum Integer $n = pq$ and seeks to convince B, the verifier. This is accomplished as follows. B chooses a random integer $a \in \mathbb{N}$ and sends a^4 to A. Then A computes

$$b_1 \equiv (a^4)^{(p+1)/2} \pmod{p} \text{ and } b_2 \equiv (a^4)^{(q+1)/2} \pmod{q}.$$

Then A pieces them together via the Chinese Remainder Theorem to get

$$b = b_1 b_2 \equiv a^2 \pmod{n},$$

and sends a^2 to B. Since B can readily find a^2, then no information about the factors has been transmitted. Yet we know that extracting square roots is tantamount to factoring (see page 139). Thus, B is convinced (after possibly a number of iterations with various values of a), but has no information about the factors of n and cannot convey any information about them to a third entity.

6.27 A chooses $m = 3$ and sends $w = 9$ to B. B chooses $c = 1$, from which A computes $r = 3 \cdot 757 \equiv 2271 \pmod{n}$. B computes $r^2 \equiv 1321535 \pmod{n}$, which B verifies is the same as $wt_A^c \equiv 9 \cdot 573049 \pmod{n}$. So we set $a = 1$ and perform another round.

A chooses $m = 99$ and sends $w = 99^2 = 9801$ to B. B chooses $c = 0$, and A computes $r = 99 = m \cdot s_A^c \pmod{n}$, which gets sent to B. Then B computes $r^2 \equiv 99^2 = 9801 \equiv w \cdot t_A^c \pmod{n}$. Since this results in a being set to zero, then B accepts A's proof.

6.29 Let α be a generator of $(\mathbb{Z}/p\mathbb{Z})^*$ and let $\beta \in (\mathbb{Z}/p\mathbb{Z})^*$ be arbitrarily chosen by a TTP. Let $x \in (\mathbb{Z}/p\mathbb{Z})^*$ be random, and let b be the bit to which B will commit. Then B enciphers b via

$$f(b,x) \equiv \beta^b \alpha^x \pmod{p}.$$

The blob to which B commits is $y = f(b,x)$. This is unconditionally concealing. However, it is binding to B if and only if it is infeasible for B to compute $\ln_\alpha \beta$.

Bibliography

[1] L. Adleman, *Molecular computation of solutions to combinatorial problems*, Science **266** (1994), 1021–1024.

[2] L. M. Adleman, J. DeMarrais, and M.-D. A. Huang, *A subexponential algorithm for discrete logarithms over the rational subgroup of the Jacobians of large genus hyperelliptic curves over finite fields* in **Algorithmic Number Theory**: First Internat. Symp., ANTS-I, Springer-Verlag, Berlin, LNCS **877** (1994), 28–40.

[3] L. M. Adleman and M.-D. A. Huang, **Primality Testing and Abelian Varieties over Finite Fields**, Lecture Notes in Mathematics, Springer-Verlag, Berlin, **1512** (1992).

[4] L. Adleman and F. Leighton, *An $O(n^{1/10.89})$ primality testing algorithm*, Math. Comp. **36** (1981), 261–266.

[5] W. R. Alford, A. Granville, and C. Pomerance, *There are infinitely many Carmichael numbers*, Ann. Math. **140** (1994), 703–722.

[6] M. Anshel and D. Goldfeld, *Zeta functions, one-way functions, and pseudorandom number generators*, Duke Math. J. **88** (1997), 371–390.

[7] A. O. L. Atkin and R. G. Larson, *On a primality test of Solovay and Strassen*, SIAM J. Comput., **11** (1982), 789–791.

[8] D. Atkins, M. Graff, A. K. Lenstra, and P. C. Leyland, *The magic words are SQUEAMISH OSSIFRAGE* in **Advances in Cryptology — ASIACRYPT '94**, Springer-Verlag, Berlin, LNCS **917**, (1995), 263–277.

[9] L. Babai, *Trading group theory for randomness* in Proc. 17th Ann. ACM Symp. Theor. Comput. (1985), 421–429.

[10] L. Babai and S. Moran, *Arthur Merlin games: a randomized proof system and a hierarchy of complexity classes*, J. Comput. and Sys. Sci. **36** (1988), 254–276.

[11] E. Bach and J. Shallit, **Algorithmic Number Theory**, Vol. **1**, MIT Press, Cambridge, Massachusetts, (1997), (second printing).

[12] F. L. Bauer, **Decrypted Secrets**, Springer, Berlin (1997).

[13] A. H. Beiler, **Recreations in the Theory of Numbers**, Dover, New York, (1964).

[14] M. Bellare and O. Goldreich, *On defining proofs of knowledge* in **Advances in Cryptology** — CRYPTO '92, Springer-Verlag, Berlin, LNCS **740**, (1993), 390–420.

[15] C. H. Bennett and G. Brassard, *Quantum cryptography: public key distribution and coin tossing*, Proc. IEEE Internat. Conf. on Computers, Systems, and Signal Processing, Banjalore, India, (1984), 175–179.

[16] C. H. Bennett and G. Brassard, *An update on quantum cryptography* in **Advances in Cryptology** — CRYPTO '84, Springer-Verlag, Berlin, LNCS **196** (1985), 475–480.

[17] C. H. Bennett and G. Brassard, *Quantum public-key distribution system*, IBM Technical Disclosure Bull. **28** (1985), 3153–3163.

[18] C. H. Bennett and G. Brassard, *Quantum public key distribution reinvented*, SIGACT News **18** (1987), 51–53.

[19] C. H. Bennett and G. Brassard, *The dawn of a new era for quantum cryptography: the experimental prototype is working!*, SIGACT News **20** (1989), 78–82.

[20] C. H. Bennett, G. Brassard, S. Breidhart, and S. Weisner, *Quantum cryptography or unforgeable subway tokens* in **Advances in Cryptology** — CRYPTO '82 Proc., Plenum Press, New York (1983), 267–275.

[21] C. Bennett, G. Brassard, C. Crépeau, R. Jozsa, A. Peres, and W. Wootters, *Teleporting an unknown quantum state via dual classical and Einstein-Podolsky-Rosen channels*, Physical Review Letters **70** (1993), 1895–1899.

[22] C. H. Bennett, G. Brassard, C. Crépeau, and M.-H. Skubiszewska, *Practical quantum oblivious transfer* in **Advances in Cryptology** — CRYPTO '91, Springer-Verlag, Berlin, LNCS **576**, (1992), 351–366.

[23] C. H. Bennett, G. Brassard, and A. K. Ekert, *Quantum cryptography*, Scientific American **267** (1992), 50–57.

[24] C. H. Bennett, G. Brassard, and N. D. Mermin, *Quantum cryptography without Bell's theorem*, Physical Review Letters **68** (1992), 557–559.

[25] C. H. Bennett, G. Brassard, and J.-M. Robert, *How to reduce your enemy's information* in **Advances in Cryptography**— CRYPTO '85 Proc., Springer-Verlag, Berlin (1986), 468–476.

[26] C. H. Bennett, G. Brassard, and J.-M. Robert, *Privacy amplification by public discussion*, SIAM J. Comput., **17** (1988), 210–229.

[27] M. Ben-Or, S. Goldwasser, J. Kilian, and A. Wigderson, *Multi-prover interactive proofs: How to remove intractability*, Proc. 20th Ann. ACM Symp. Theor. Comput. (1988), 113–131.

[28] D. J. Bernstein and A. K. Lenstra, *A general number field sieve implementation*, in **The Development of the Number Field Sieve**, A. K. Lenstra and H. W. Lenstra Jr. (Eds.), Lecture Notes in Mathematics, Springer-Verlag, Berlin, **1554** (1993), 103–126.

[29] T. Beth, *Efficient zero-knowledge identification scheme for smart cards* in **Advances in Cryptology** — Eurocrypt '88, Springer-Verlag, Berlin, LNCS **330**, (1988), 77–84.

[30] E. Biham and A. Shamir, **Differential Cryptanalysis of the Data Encryption Standard**, Springer-Verlag, New York (1993).

[31] I. Blake, G. Seroussi, and N. Smart, **Elliptic Curves in Cryptography**, London Math. Soc. Lecture Note Series **265**, Cambridge University Press, Cambridge (1999).

[32] M. Blum, *Coin flipping by telephone: A protocol for solving impossible problems* in Proc. 24th IEEE Computer Conference (1982), 133–137.

[33] M. Blum, P. Feldman, and S. Micali, *Noninteractive zero-knowledge and its applications*, Proc. 20th Ann. ACM Symp. Theor. Comput. (1988), 103–112.

[34] D. Boneh and R. J. Lipton, *Quantum cryptanalysis of hidden linear functions* in **Advances in Cryptology** — CRYPTO '95, Springer-Verlag, Berlin, LNCS **963**, (1995), 424–437.

[35] D. Boneh and R. Venkatesan, *Breaking RSA may not be equivalent to factoring* in **Advances in Cryptology** — Eurocrypt '98, Springer-Verlag, Berlin, LNCS **1403**, (1998), 59–71.

[36] G. Brassard, *Quantum Information Processing: The Good, the Bad, and the Ugly* in **Advances in Cryptology** — CRYPTO '97, Springer-Verlag, Berlin, LNCS **1294**, (1997), 337–341.

[37] G. Brassard and C. Crépeau, *Quantum bit commitment and coin tossing protocols* in **Advances in Cryptology** — CRYPTO '90, Springer-Verlag, Berlin, LNCS **537**, (1991), 49–61.

[38] E. F. Brickell, *Solving low-density knapsacks* in **Advances in Cryptology** — CRYPTO '83 Proc., Plenum Press, New York (1984), 25–37.

[39] E. F. Brickell and K. S. McCurley, *An interactive scheme based on discrete logarithms and factoring*, J. Cryptology **5**, (1992), 29–39.

[40] J. Brillhart, D. H. Lehmer, J. L. Selfridge, B. Tuckerman, and S.S. Wagstaff Jr., **Factorizations of $b^n \pm 1$, $b = 2, 3, 5, 6, 7, 10, 11, 12$ up to High Powers**, Contemporary Math. **22**, Amer. Math. Soc., Providence, R.I., Second Edition (1988).

[41] J. Brillhart and J. Selfridge, *Some factorizations of $2^n \pm 1$ and related results*, Math. Comp. **21** (1967), 87–96.

[42] J. P. Buhler, H. W. Lenstra Jr., and C. Pomerance, *Factoring integers with the number field sieve*, in **The Development of the Number Field Sieve**, A. K. Lenstra and H. W. Lenstra Jr. (Eds.), Lecture Notes in Mathematics, Springer-Verlag, Berlin **1554** (1993), 50–94.

[43] K. W. Campbell and M. J. Weiner, *Proof that DES is not a group* in **Advances in Cryptology** — CRYPTO '92 Proc., Springer-Verlag, Berlin, LNCS **740** (1993), 518–526.

[44] R. D. Carmichael, *On the numerical factors of the arithmetic forms $\alpha^n \pm \beta^n$*, Ann. Math. **15** (1913–14), 30–70.

[45] A. Cauchy, *Mémoire sur diverses formules relatives à l'algèbre et à la théorie des nombres (suite)*, C.R. Acad. Sci. Paris **12** (1841), 813–846.

[46] C. Cavallar, B. Dodson, A. Lenstra, P. Leyland, W. Lioen, P. Montgomery, B. Murphy, and P. Zimmerman, *Factorization of RSA-140 using the number field sieve*, to appear.

[47] B. Chorr and R. L. Rivest, *A knapsack type public key cryptosystem based on arithmetic in finite fields* in **Advances in Cryptology** — CRYPTO '84, Springer-Verlag, Berlin, LNCS **196** (1985), 54–65.

[48] B. Chorr and R. L. Rivest, *A knapsack type public key cryptosystem based on arithmetic in finite fields* in IEEE Trans. Inform. Theory, **34** (1988), 901–909.

[49] J. Cirac and P. Zoller, *Quantum computations with cold trapped ions*, Physical Review Letters **74** (1995), 4091–4094.

[50] D. Coppersmith, *The Data Encryption Standard (DES) and its strength against attacks*, IBM J. R. and D. **38** (1994), 243–250.

[51] D. Coppersmith, H. Krawczyk, and Y. Mansour, *The shrinking generator* in **Advances in Cryptology** — CRYPTO '93, Springer-Verlag, Berlin, LNCS **773** (1994), 22–39.

[52] D. Coppersmith, A. Odlyzko, and R. Schroeppel, *Discrete logarithms in $GF(p)$*, Algorithmica **I** (1986), 1–15.

[53] D. Corey, A. Fahmy, and T. Havel, *Nuclear magnetic resonance spectroscopy: an experimentally accessible paradigm for quantum physics and computation*, PhysComp96, Boston University, New England Complex System Institute (1996), 87–91.

[54] D. A. Cox, **Primes of the Form** $x^2 + ny^2$, Wiley Interscience, New York, Toronto (1989).

[55] R. Crandall, K. Dilcher, and C. Pomerance, *A search for Wiefereich and Wilson primes.*, Math. Comp. **66** (1997), 433–449.

[56] A. J. C. Cunningham and H. J. Woodall, **Factorization of** $y^n \mp 1$**,** $y = 2, 3, 5, 6, 7, 10, 11, 12$ **Up to high powers** (n), Hodgson, London (1925).

[57] J. A. Davies, D. B. Holdridge, and G. L. Simmons, *Status report on factoring (at Sandia National Labs)* in **Advances in Cryptology** — Eurocrypt '84, Springer-Verlag, Berlin, LNCS **209**, (1985), 183–215.

[58] N. Demytko, *A new elliptic curve based analogue of RSA* in **Advances in Cryptology** — Eurocrypt '93, Springer-Verlag, Berlin, LNCS **765**, (1994), 40–49.

[59] D. Deuch, *Quantum communication thwarts eavesdroppers*, New Scientist, December 9, (1989), 25–26.

[60] L. E. Dickson, **History of the Theory of Numbers**, Vol. 1, Chelsea, New York, (1992).

[61] G. Di Crescenzo and R. Ostrovsky, *On concurrent zero-knowledge with pre-processing* in **Advances in Cryptology** — CRYPTO '99, Springer-Verlag, Berlin, LNCS **1666**, (1999), 485–502.

[62] W. Diffie and M. E. Hellman, *Multiuser cryptographic techniques*, Proc. AFIPS National Computer Conference (1976), 109–112.

[63] W. Diffie and M. E. Hellman, *New directions in cryptography*, IEEE Trans. Inform. Theory, **22** (1976), 644–654.

[64] W. Diffie and S. Landau, **Privacy on the Line** — **The Politics of Wiretapping and Encryption**, MIT Press, Cambridge (1998).

[65] J. D. Dixon, *Asymptotically fast factorization of integers*, Math. Comp. **36** (1981), 255–260.

[66] A. K. Ekert, *Quantum cryptography based on Bell's theorem*, Physical Review Letters **67** (1991), 661–663.

[67] A. K. Ekert, J. Rarity, P. Tapster, and G. Palma, *Practical quantum cryptography based on two-photon interferometry*, Physical Review Letters **69** (1992), 1293–1295.

[68] T. ElGamal, *A public key cryptosystem and signature scheme based on discrete logarithms* in **Advances in Cryptography**— CRYPTO '84, Springer-Verlag, Berlin, LNCS **196**, (1985), 10–18.

[69] T. ElGamal, *A public key cryptosystem and signature scheme based on discrete logarithms*, IEEE Trans. Inform. Theory, **31** (1985), 469–472.

[70] N. D. Elkies, *Elliptic and modular curves over finite fields and related computational issues* in *Computational perspectives on number theory.* Proc. Conf. in Honor of A. O. L. Atkin, Amer. Math. Soc. Internat. Press **7** (1988) 21–76.

[71] P. Erdös, *On the Converse of Fermat's theorem*, Amer. Math. Monthly **56** (1949), 623–624.

[72] G. Faltings, *The proof of Fermat's Last Theorem by R. Taylor and A. Wiles*, Notices Amer. Math. Soc. **42** (1995), 743–746.

[73] U. Feige, A. Fiat, and A. Shamir, *Zero-knowledge proofs of identity*, Proc. 19th Ann. ACM Symp. Theor. Comput. (1987), 210–217.

[74] U. Feige, A. Fiat, and A. Shamir, *Zero-knowledge proofs of identity*, J. Cryptology 1, (1988), 77–94.

[75] M. Fellows and N. Koblitz, *Self-witnessing polynomial-time complexity and prime factorization*, Designs, Codes and Cryptography **2** (1992), 231–235.

[76] R. P. Feynman *Simulating physics with computers*, Internat. J. Theoret. Phys. **21** (1982), 467–488.

[77] A. Fiat and A. Shamir, *How to prove yourself: Practical solution to identification and signature problems* in **Advances in Cryptology** — CRYPTO '86, Springer-Verlag, Berlin, LNCS **263**, (1987), 186–194.

[78] R. Flassenberg and S. Paulus, *Sieving in function fields*, to appear: Experimental Math.

[79] G. Frey and H.-G. Rück, *A remark concerning m-divisibility and the discrete logarithm problem in the divisor class group of curves*, Math. Comp. **62** (1994), 865–874.

[80] C. F. Gauss, **Disquisitiones Arithmeticae** (English edition), Springer-Verlag, Berlin (1985).

[81] A. Géradin *F. Proth*, Sphinx-Oedipe, **7** (1912), 50–51.

[82] N. Gershenfeld, I. Chuang, and S. Lloyd, *Bulk quantum computation*, Proc. 4th workshop on Physics and Computation, PhysComp96, Boston University, New England Complex System Institute (1996), 134.

[83] J. Gerver, *Factoring large numbers with a quadratic sieve*, Math. Comp. **41** (1983), 287–294.

[84] O. Goldreich, **Modern Cryptography, Probabilistic Proofs and Pseudo-randomness**, Springer-Verlag, Berlin (1999).

[85] O. Goldreich, S. Micali, and A. Wigderson, *Proofs that yield nothing but their validity and a methodology of cryptographic protocol design* in Proc. 27th IEEE Annual Symp. on the Foundations of Computer Science (1986), 174–187.

[86] O. Goldreich, S. Micali, and A. Wigderson, *How to prove all NP statements in zero-knowledge, and a methodology of cryptographic protocol design* in **Advances in Cryptology** — CRYPTO '86, Springer-Verlag, Berlin, LNCS **263**, (1987), 171–185.

[87] O. Goldreich, S. Micali, and A. Wigderson, *Proofs that yield nothing but their validity or all languages in NP have zero-knowledge proof systems*, J. Assoc. Comp. Mach. **38** (1991), 691–729.

[88] O. Goldreich, A. Sahai, and S. Vadhan, *Can statistical zero knowledge made non-interactive ? or On the relationship of SZK and NISK* in **Advances in Cryptology** — CRYPTO '99, Springer-Verlag, Berlin, LNCS **1666**, (1999), 467–484.

[89] S. Goldwasser and J. Kilian, *Almost all primes can be quickly certified*, Proc. 18th Ann. ACM Symp. Theor. Comput. (STOC), Berkeley (1986), 316–329.

[90] S. Goldwasser, S. Micali, and C. Rackoff, *The knowledge complexity of interactive proof systems*, Proc. 17th Ann. ACM Symp. Theor. Comput. (1985), 291–304.

[91] S. Goldwasser, S. Micali, and C. Rackoff, *The knowledge complexity of interactive proof systems*, SIAM J. Comput. **18** (1989), 186–208.

[92] S. W. Golomb, **Shift Register Sequences**, Holden-Day, San Francisco (1967). Reprinted by Aegean Park Press (1982).

[93] L. Grover, *A fast quantum mechanical algorithm for database search* in Proc. 28th Ann. ACM Symp. Theor. Comput. (1996), 212–219.

[94] R. K. Guy, **Unsolved Problems in Number Theory**, Vol. 1, Second Edition, Springer-Verlag, Berlin (1994).

[95] R. K. Guy, *How to factor a number*, in Proc. Fifth Manitoba Conf. on Numerical Math., Congressus Numerantium **16** (1976), 49–89.

[96] T. W. Hungerford, **Algebra**, Springer-Verlag, Berlin (1974).

[97] A. Hurwitz, *Question 801*, L'Intermédiaire Math. **3** (1896), 214.

[98] W. S. Jevons, **Principles of Science**, Macmillan and Co., London (1874).

[99] D. Johnson, *The NP-completeness column: an ongoing guide*, J. Algorithms **5** (1984), 433–447.

[100] D. Kahn, **The Codebreakers**, Macmillan, New York (1967).

[101] Lord Kelvin, *Nineteenth century clouds over the dynamical theory of heat and light*, Philosophical Magazine **2** (1901), 1–40.

[102] A. W. Knapp, **Elliptic Curves**, Math. Notes **40**, Princeton University Press, Princeton, New Jersey (1992).

[103] L. R. Knudsen, *Truncated and higher order differentials* in Fast Software Encryption, Second International Workshop, Springer-Verlag, Berlin, LNCS **1008** (1995), 196–211.

[104] D. E. Knuth, **The Art of Computer Programming**, Volume 2/ **Seminumerical Algorithms**, Third Edition, Addison-Wesley, Reading, Paris (1998).

[105] N. Koblitz, *Elliptic curve cryptosystems*, Math. Comp. **48** (1987), 203–209.

[106] N. Koblitz, *Hyperelliptic cryptosystems*, J. Cryptology **1** (1989), 139–150.

[107] S. Konyagin and C. Pomerance, *On primes recognizable in deterministic polynomial time*, The Mathematics of Paul Erdös, **1**, Springer-Verlag, Berlin (1996), 176–198.

[108] K. Koyama, U. Maurer, T. Okamato, and S. A. Vanstone, *New public-key schemes based on elliptic curves over the ring* $\mathbb{Z}_n$ in **Advances in Cryptology** — CRYPTO '91, Springer-Verlag, Berlin, LNCS **576**, (1992), 252–266.

[109] D. W. Kravitz, *Digital signature algorithm*, U.S. Patent no. 5,231,668, July 27 (1993).

[110] J. C. Lagarius and A. M. Odlyzko, *Solving low-density subset sum problems*, J. Assoc. Comp. Mach. **32** (1985), 229–246.

[111] X. Lai and J. L. Massey, *A proposal for a new block encryption standard* in **Advances in Cryptology** — Eurocrypt '90, Springer-Verlag, Berlin, LNCS **473**, (1991), 389–404.

[112] X. Lai, R. A. Rueppel, and J. Woollven, *A fast cryptographic checksum algorithm based on stream ciphers* in **Advances in Cryptology** — AUSCRYPT '92, Springer-Verlag, Berlin, LNCS **718**, (1993), 339–348.

[113] S. Landau, *Zero-knowledge and the department of defense*, Notices of the Amer. Math. Soc. **35** (1988), 5–12.

[114] S. K. Langford and M. E. Hellman, *Differential-linear cryptanalysis* in **Advances in Cryptology** — CRYPTO '94, Springer-Verlag, Berlin, LNCS **839**, (1994), 17–25.

[115] D. H. Lehmer, **Selected Papers of D. H. Lehmer**, Volumes I–III, D. McCarthy (Ed.), The Charles Babbage Research Centre, St. Pierre, Canada (1981).

[116] E. Lehmer, *On the infinitude of Fibonacci pseudo-primes*, Fibonacci Quart. **2** (1964), 229–230.

[117] D. H. Lehmer and R. E. Powers, *On factoring large numbers*, Bull. Amer. Math. Soc. **37** (1931), 770–776.

[118] A. K. Lenstra and H. W. Lenstra Jr., *Algorithms in number theory*, in **Handbook of Theoretical Computer Science**, J. van Leeuwen, ed., Elsevier Sci. Pub., (1990), 674–715.

[119] A. K. Lenstra, H. W. Lenstra Jr., M. S. Manasse, and J. M. Pollard, *The number field sieve*, in **The Development of the Number Field Sieve**, A. K. Lenstra and H. W. Lenstra Jr. (Eds.), Lecture Notes in Mathematics, Springer-Verlag, **1554**, (1993), 11–42.

[120] A. K. Lenstra, H. W. Lenstra, M. S. Manasse, and J. M. Pollard, *The factorization of the ninth Fermat number*, Math. Comp. **61** (1993), 319–349.

[121] A. K. Lenstra and M. S. Manasse, *Factoring by electronic mail* in **Advances in Cryptology** — EUROCRYPT '89, Springer-Verlag, Berlin, LNCS **434**, (1990), 355–371.

[122] H. W. Lenstra Jr., *Factoring integers with elliptic curves*, Ann. Math. **126** (1987), 649–673.

[123] H. W. Lenstra Jr., *On the Chor-Rivest knapsack cryptosystem*, Journal of Cryptology **3**, (1991), 149–155.

[124] H. W. Lenstra Jr., J. Pila, and C. Pomerance, *A hyperelliptic smoothness test. I*, Philos. Trans. Royal Soc. London **345** (1993), 397–408.

[125] D. E. G. Malm, *On Monte-Carlo primality tests*, Notices Amer. Math. Soc. **24** (1977), A-529, Abstract 77T-A222.

[126] C. Marand and P. Townsend, *Quantum key distribution over distances as long as 30 km*, Optic Letters **20** (1995), 1695–1697.

[127] M. Matsui, *Linear cryptanalysis method for the DES cipher* in **Advances in Cryptology** — EUROCRYPT '93, Springer-Verlag, Berlin, LNCS **765**, (1994), 386–397.

[128] M. Matsui, *The first experimental cryptanalysis of the Data Encryption Standard* in **Advances in Cryptology** — CRYPTO '94, Springer-Verlag, Berlin, LNCS **839** (1994), 1–11.

[129] U. M. Maurer, *Towards the equivalence of breaking the Diffie-Hellman protocol and computing discrete logarithms* in **Advances in Cryptology** — CRYPTO '94, Springer-Verlag, Berlin, LNCS **839** (1994), 271–281.

[130] U. M. Maurer and S. Wolf, *Diffie-Hellman oracles* in **Advances in Cryptology** — CRYPTO '96, Springer-Verlag, Berlin, LNCS **1109** (1996), 268–282.

[131] B. Mazur, *Rational points on modular curves, Modular Functions of One Variable V*, Lecture Notes in Mathematics, Springer Verlag, **601** Berlin (1977).

[132] K. S. McCurley, *A key distribution system equivalent to factoring*, J. Cryptology **1** (1988), 95–105.

[133] A. Menezes, T. Okamoto, and S. A. Vanstone, *Reducing elliptic curve logarithms to logarithms in a finite field*, IEEE Trans. Inform. Theory, **39** (1993), 1639–1646.

[134] A. J. Menezes, P. C. van Oorschot, and S. A. Vanstone, **Handbook of Applied Cryptography**, CRC Press, Boca Raton, New York, London, Tokyo (1997).

[135] A. J. Menezes and S. A. Vanstone, *Elliptic curve cryptosystems and their implementation*, J. Cryptology **6** (1993), 209–224.

[136] L. Merel, *Bornes pour la torsion des courbes elliptiques sur les corps de nombres*, Invent. Math. **124** (1996), 437–449.

[137] R. C. Merkle and M. E. Hellman, *Hiding information and signatures in trapdoor knapsacks*, IEEE Trans. Inform. Theory, **24** (1978), 525–530.

[138] C. H. Meyer and S. M. Matyas, **Cryptography: A New Dimension in Computer Data Security**, Wiley, New York, Toronto (1982).

[139] V. Miller, *Use of elliptic curves in cryptography* in **Advances in Cryptography** — CRYPTO '85 Proc., Springer-Verlag, Berlin, LNCS **218** (1986), 417–426.

[140] Z. Mo and J. P. Jones, *A new primality test using Lucas sequences*, unpublished, but contained within the Ph. D. thesis by Mo completed at the University of Calgary (1995).

[141] R. A. Mollin, (Editor), **Number Theory and Applications**, NATO ASI **C265**, Kluwer Academic Publishers, Dordrecht, the Netherlands (1989).

[142] R. A. Mollin, (Ed.), **Number Theory**, Proc. First Conf. of the Canadian Number Theory Association, Walter de Gruyter, Berlin, (1990).

[143] R.A. Mollin, **Quadratics**, CRC Press, Boca Raton, New York, London, Tokyo (1996).

[144] R. A. Mollin, **Fundamental Number Theory with Applications**, CRC Press, Boca Raton, New York, London, Tokyo (1998).

[145] R.A. Mollin, **Algebraic Number Theory**, Chapman and Hall/CRC Press, Boca Raton, New York, London, Tokyo (1999).

[146] M.A. Morrison and J. Brillhart, *A method of factoring and the factorization of F_7*, Math. Comp. **29** (1975), 183–205.

[147] A. M. Odlyzko, *Cryptanalytic attacks on the multiplicative knapsack cryptosystem and on Shamir's fast signature scheme*, IEEE Trans. Inform. Theory, **30** (1984), 594–601.

[148] A. Odlyzko, *Discrete logarithms: The past and the future*, AT&T Labs research preprint (1999).

[149] K. Ohta and T. Okamoto, *Practical extension of Fiat-Shamir scheme*, Electronics Letters **24** (1988), 955–956.

[150] K. Ohta and T. Okamoto, *A modification of the Fiat-Shamir Scheme* in **Advances in Cryptology** — CRYPTO '88, Springer-Verlag, Berlin, LNCS **403**, (1990), 232–243.

[151] S. J. D. Phoenix and P. D. Townsend, *Quantum cryptography:protecting our future networks with quantum mechanics* in **Cryptography and Coding**, 5th IMA Conf. Proc., Inst. Math. and its Apps. (IMA) (1995), 112–131.

[152] D. A. Plaisted, *Fast verification, testing, and generation of large primes*, Theoret. Comput. Sci. **9** (1979), 1–16. Errata in **14** (1981), 345.

[153] E. A. Poe, **Edgar Allan Poe Selected Works**, Random House, Toronto, New York (1990), 357–381.

[154] J. M. Pollard, *An algorithm for testing the primality of any integer*, Bull. London Math. Soc. **3** (1971), 337–340.

[155] J. M. Pollard, *Theorems on factorization and primality testing*, Proc. Cambridge Phil. Soc. **76** (1974), 521–528.

[156] J. M. Pollard, *Factoring with Cubic Integers* in **The Development of the Number Field Sieve**, A. K. Lenstra and H. W. Lenstra Jr. (Eds.), Lecture Notes in Mathematics, Springer-Verlag, Berlin, **1554** (1993), 4–10.

[157] C. Pomerance, *Analysis and comparison of some integer factoring algorithms*, in **Computational Methods in Number Theory, Part I**, H.W. Lenstra and R. Tijdeman, (Eds.), Mathematisch Centrum, (1982), 89–139.

[158] C. Pomerance, *The quadratic sieve factoring algorithm* in **Advances in Cryptology** — Eurocrypt '84, Springer-Verlag, Berlin, LNCS **209**, (1985), 169–182.

[159] C. Pomerance, *Very short primality proofs*, Math. Comp. **48** (1987), 315–322.

[160] C. Pomerance, *Factoring*, in **Cryptology and Computational Number Theory**, Proc. Symp. App. Math. **42**, Amer. Math. Soc. (1990), 27–47.

[161] C. Pomerance, *The number field sieve*, Proc. Symp. App. Math. **48**, (1994), 465–480.

[162] A. J. van der Poorten, **Notes on Fermat's Last Theorem**, Wiley, New York, Toronto (1996).

[163] V. R. Pratt, *Every prime has a succinct certificate*, SIAM J. Comput. **4** (1975), 214–220.

[164] F. Proth, *Théorèmes sur les nombres premiers*, Comptes Rendus Acad. des Sciences, Paris, **87** (1878), 926.

[165] J.-J. Quisquater and M. Guillou, *A practical zero-knowledge protocol fitted to security microprocessor minimizing both transmission and memory* in **Advances in Cryptology** — Eurocrypt '88, Springer-Verlag, Berlin, LNCS **330**, (1988), 123–128.

[166] J.-J. Quisquater, M. Guillou, and T. Berson, *How to explain zero-knowledge to your children* in **Advances in Cryptology** — CRYPTO '89, Springer-Verlag, Berlin, LNCS **435**, (1990), 628–631.

[167] M. O. Rabin, *Digital signatures* in **Foundations of Secure Computation**, Academic Press, New York, (1978), 155–168.

[168] M. O. Rabin, *Digitized signatures and public-key functions as intractable as factorization*, MIT/LCS/TR-212, MIT Lab. for Comp. Sci. (1979).

[169] J. Rarity, P. Ownes, and P. Tapster, *Quantum random number generation and key sharing*, J. Modern Optics **41** (1994), 2435–2444.

[170] M. Rivest, A. Shamir, and L. Adleman, *A method for obtaining digital signatures and public-key cryptosystems*, Comm. ACM, **21** (1978), 120–126.

[171] L. Rosenhead, *Henry Cabourn Pocklington*, Obituary Notices of the Royal Society, (1952), 555–565.

[172] A. Saloma, **Public-Key Cryptography**, Second Ed., Springer (1996).

[173] B. Schneier, **Applied Cryptography**, Wiley, New York, Toronto (1994).

[174] B. Schneier, J. Kelsey, D. Whiting, D. Wagner, C. Hall, and N. Ferguson, **The Twofish Encryption Algorithm**, Wiley, New York, Toronto (1999).

[175] C. P. Schnorr, *Efficient identification and signatures for smart cards* in **Advances in Cryptology** — CRYPTO '89, Springer-Verlag, Berlin, LNCS **435**, (1990), 239–252.

[176] R. Schoof, *Elliptic curves over finite fields and the computation of square roots mod p*, Math. Comp. **44** (1985), 483–494.

[177] R. Schoof, *Counting points on elliptic curves over finite fields*, J. Théorie des Nombres de Bordeaux **7** (1995), 219–254.

[178] A. Shamir, *Factoring numbers in $O(\log n)$ arithmetic steps*, Information Processing Letters, **8** (1979), 28–31.

[179] A. Shamir, *A fast signature scheme*, MIT/LCS/TM-107, MIT Lab. for Comp. Sci. (1978).

[180] A. Shamir, *A polynomial-time algorithm for breaking the basic Merkle-Hellman cryptosystem* in **Advances in Cryptology** — CRYPTO '82 Proc., Plenum Press, New York (1983), 279–288.

[181] A. Shamir, *A polynomial-time algorithm for breaking the basic Merkle-Hellman cryptosystem*, IEEE Trans. Inform. Theory, **30** (1984), 699–704.

[182] D. Shanks, **Number Theory**, Chelsea, New York (1985).

[183] C. E. Shannon, *Communication theory of secrecy systems*, Bell System Technical J., **28** (1949), 656–715.

[184] P. Shor, *Algorithms for quantum computation: discrete logarithms and factoring*, Proc. 35th IEEE Annual Symp. on Foundations of Computer Science (1994), 124–134.

[185] C. L. Siegel, *The integer solutions of the equation $y^2 = ax^n + bx^{n-1} + \cdots + k$*, J. London Math. Soc. **1** (1926), 66–68.

[186] T. Siegenthaler, *Decrypting a class of stream ciphers using ciphertext only*, IEEE Trans. Comput., **34** (1985), 81–85.

[187] J. H. Silverman, **The Arithmetic of Elliptic Curves**, *second printing of 1986 version*, Springer, New York, (1992).

[188] S. Singh, **The Code Book**, Doubleday, New York, London, Toronto (1999).

[189] J. H. Silverman and J. Tate, **Rational Points on Elliptic Curves**, Springer, New York, (1992).

[190] M. E. Smid and D. K. Branstad, *The data encryption standard: Past and future*, Proc. IEEE, **76** (1988), 550–559.

[191] J. L. Smith, *The design of Lucifer: A cryptographic device for data communications*, IBM Research Report RC 3326, IBM T.J. Watson Research Center, Yorktown Heights, N.Y., 10598, U. S. A., April 15, 1971.

[192] R. Solovay and V. Strassen, *A fast Monte-Carlo test for primality*, SIAM J. Comput., **6** (1977), 84–85. Erratum in **7** (1978), 118.

[193] D. R. Stinson, **Cryptography**, CRC Press, Boca Raton, London, Tokyo (1995).

[194] W. Teich, K. Obermayer, and G. Mahler, *Structural basis of multistationary quantum systems II. Effective few-particle dynamics*, Physical Review B **37** (1988), 8111–8120.

[195] P. Townsend, J. Rarity, and P. Tapster, *Single photon interference in 10 km-long optical fiber interferometer*, Electronics Letters **29** (1993), 634–635.

[196] P. Townsend, J. Rarity, and P. Tapster, *Enhanced single photon fringe visibility in a 10 km-long prototype quantum cryptography channel*, Electronics Letters **29** (1993), 1291–1293.

[197] C. Suetonius Tranquillus, **The Lives of the Twelve Caesars**, Corner House, Williamstown, Mass. (1978).

[198] S. A. Vanstone and R. J. Zuccherato, *Elliptic curve cryptosystems using curves of smooth order over the ring* $\mathbb{Z}_n$ in IEEE Trans. Inform. Theory, **43** (1997), 1231–1237.

[199] A. Weil, **Number Theory, an Approach Through History, from Hammurapi to Legendre**, Birkhäuser, Boston (1984).

[200] A. Weil, *L'arithmétique sur les courbes algébriques*, Acta Math. **52** (1928), 281–315.

[201] S. Weisner, *Conjugate coding*, SIGACT News **15** (1983), 78–88.

[202] E. W. Weisstein, **CRC Concise Encyclopedia of Mathematics**, CRC Press LLC, Boca Raton, London, Tokyo (1999).

[203] A. Wiles, *Modular elliptic curves and Fermat's Last Theorem*, Ann. Math. **142** (1995), 443–551.

[204] C. P. Williams and S. H. Clearwater, **Explorations in Quantum Computing**, Springer-Verlag, New York, (1998).

[205] H. C. Williams, *On numbers analogous to Carmichael numbers*, Canad. Math. Bull. **20** (1977), 133–143.

[206] H. C. Williams, *Primality testing on a computer*, Ars Combin. **5** (1978), 127–185.

[207] H. C. Williams, *A p + 1 method of factoring*, Math. Comp. **39** (1982), 225–234.

[208] H. C. Williams, **Édouard Lucas and Primality Testing**, Canadian Mathematical Society Series of Monographs and Advanced Texts, Vol. **22**, Wiley-Interscience, New York, Toronto, (1998).

[209] H.C. Williams and J.O. Shallit, *Factoring integers before computers*, Proc. Symp. App. Math. **48** (1994), 481–530.

[210] C. Zimmer, *Perfect gibberish*, Discover **13** (1992), 92–99.

Index

About the Author

Richard Anthony Mollin received both his B.A. and M.A. in mathematics from the University of Western Ontario in London, Ontario, Canada in 1971 and 1972, respectively. In 1975, he obtained his Ph.D. in mathematics from Queen's University, Kingston, Ontario, where he was born. Since then he has held several positions at Canadian universities, including being one of the first NSERC University Research Fellows, a position held at Queen's University in 1981–1982. Since 1982 he has been at University of Calgary where he is currently a full professor in the Mathematics Department. He has over 140 publications in algebra, number theory, and computational mathematics. This book is his fourth, with [143]–[145] as the other three texts, as well as two edited conference proceedings [141]–[142]. He resides in Calgary with his wife Bridget.